DNA–Protein Interactions

The Practical Approach Series

Related **Practical Approach** Series Titles

Bioinformatics: sequence, structure and databanks*
Functional Genomics*
Essential Molecular Biology VI 2/e*
RNA viruses
Differential Display
Protein Localization by Fluorescence Microscopy
Mouse Genetics and Transgenics
Apoptosis
Cell Growth, Differentiation and Senescence
DNA Viruses
Gene Targeting 2e
Crystallization of Nucleic Acids and Proteins
Steroid/Nuclear Receptor Superfamily
DNA Microarray Technology
Protein Expression
Transcription Factors 2/e
Chromosome Structural Analysis
Eukaryotic DNA Replication
In Situ Hybridization 2/e
RNA-Protein Interactions
Chromatin
Mutation Detection
PCR3:PCR In Situ Hybridization
Antisense Technology
Genome Mapping
DNA and Protein Sequence Analysis
Protein Structure Prediction
DNA Cloning 4: Mammalian Systems
DNA Cloning 3: Complex Genomes
Gene Probes 2
Gene Probes 1
Pulsed Field Gel Electrophoresis
PCR 2
DNA Cloning 2: Expression Systems
DNA Cloning 1: Core Techniques
The Cell Cycle
Gene Targeting
Human Genetic Disease Analysis

* indicates a forthcoming title

Please see the **Practical Approach** series website at

http://www.oup.co.uk/pas

for full contents lists of all Practical Approach titles.

Preface

The past decade has seen the parallel rapid development of both physical and chemical techniques for the study of biomolecules coupled with the realization that many biologically relevant reactions are mediated in the context of large macromolecular assemblies. This is especially relevant to the study of DNA–protein interactions where DNA recombination and replication, transcription and chromatin remodelling all require the participation of multicomponent systems. The study of the nature of the DNA transactions involved requires not only the characterization of these complexes as such, but also the analysis of the kinetics of the reaction. It is the characteristics of the latter, especially in the area of transcriptional regulation, that are of crucial important to the understanding of biological regulatory systems.

This book is intended to provide a guide to the use of recently developed techniques that are particularly suited to the study of the interactions of proteins with DNA. It is not intended to be exhaustive, since some relevant procedures, for example the determination of the structure of assemblies by NMR and crystallization are well described in other volumes in this series. Among the techniques covered here are those which include the use of phenomena such as surface plasmon resonance, fluorescence resonance energy transfer, fluorescence anisotropy as well as visualization of nucleoprotein complexes by scanning force microscopy and cryo-electron microscopy. In addition we have included articles on more established techniques, including microcalorimetry, various footprinting techniques and the quantitative estimation of affinity by electrophoretic mobility shift assays. Special emphasis has been placed on the study of reaction kinetics as applicable to protein–DNA interactions. In most chapters the authors have concentrated on specific examples but, in general, the techniques described have a much wider applicability.

It is with pleasure that we express our gratitude to all those who have contributed to and helped in other ways with this venture. We would particularly like to thank those colleagues who have shared our enthusiasm for this topic and have thought it worthwhile to help in spreading this rapidly moving field to a wider audience. Finally, our special thanks to our editors at Oxford University Press who throughout have maintained our momentum in the nicest possible way.

Paris and Cambridge M. B., A. T.

Contents

11 Solid-phase DNAse I footprinting

Raphael Sandaltzopoulos and Peter B. Becker

12 Hydroxyl radical footprinting

Annie Kolb, Tamara Belyaera and Nigel Savery

13 Radiolytic cleavage of DNA. Mapping of the protein interaction sites

Michel Charlier and Mélanie Spotheim-Maurizot

14 UV-laser photoreactivity of nucleoprotein complexes *in vitro*

Malcolm Buckle, Christophe Place and Iain K. Pemberton

18 Kinetic analysis of enzyme template interactions. Nucleotide incorporation by DNA dependent RNA and DNA polymerases

Bianca Sclavi and Pascal Roux

19 Kinetics of DNA interactions surface plasmon resonance Spectroscopy

Björn Persson, Malcolm Buckle and Peter G. Stockley

Protocol list

Abbreviations

A	absorption
ACF	autocorrelation function
AFM	atomic force microscope
ARB	alkali revealed breaks
BET	ethidium bromide
BSA	bovine serum albumin
cAMP	adenosine 3′.5′-cyclic monophosphate
CAP	catabolite gene activator protein
CCD	charge-coupled device
CRP	cAMP receptor protein (aka catabolite activator protein)
CV	coefficient of variation
DSC	differential scanning calorimetry
DMS	dimethyl subetimidate
DMSO	dimethyl sulfoxide
dNTPs	deoxynucleotide triphosphates
DSB/dsb	double-strand breaks
DTT	dithiothreitol
EDC	*N*-ethyl-*N*′-(diethylaminopropyl) carbodiimide
EDTA	ethylenediaminetetraacetic acid
EDTAcetate	ethylenediaminetetraacetate
EGTA	ethyleneglycol-bis(aminoethyl)-*N*,*N*,*N*′,*N*′-tetraacetic acid
EIA	enzyme immunoassay
ELISA	enzyme linked immunosorbent assay
EM	electron microscopy
EMSA	electrophoretic mobility shift assay
FCCS	fluorescence cross-correlation spectroscopy
FBS	fetal bovine serum
FIS	Factor for inversion stimulation
FCS	fluorescence correlation spectroscopy
FRET	Förster fluorescence resonance energy transfer
FSB	frank strand breaks
GST	glutathione S-transferase
HRC	heparin-resistant complexes

HEPES	*N*-2-hydroxyethylpiperazine-*N¢*-2-ethanesulfonic acid
HMG	high-mobility group protein
HPLC	high-pressure liquid chromatography
hUBF	human upstream binding factor
IHF	integration host factor
IPTG	isopropyl-β-D-thiogalactopyranoside
ITC	isothermal titration calorimeter
LB	Luria–Bertani (media)
LEF-1	lymphoid enhancer binding protein
MES	2-[*N*-Morpholino] ethanesulfonic acid
MOI	multiplicity of infection
NHS	*N*-hydroxy succinimide
NMR	nuclear magnetic resonance
NOE	nuclear Overhauser effect
NtrC	nitrogen regulatory protein C
PAGE	polyacrylamide gel electrophoresis
PBS	phosphate buffered saline
PCR	polymerase chain reaction
PEI	polyethyleneimine
PMSF	phenylmethylsulfonyl fluoride
poly[d(I–C)]	poly(deoxyinosinic-deoxycytidylic) acid
PNK	polynucleotide kinase
RU	resonance units
RNAP	RNA polymerase
SAM	S-adenosyl methionine
SFM	scanning force microscopy
SDS	sodium dodecyl sulfate
SDS-PAGE	sodium dodecyl sulfate–polyacrylamide gel electrophoresis
snRNP	small nuclear ribonucleoprotein
SRY	sex determining region *Y* protein (testis determining factor)
SSB/ssb	single-strand breaks
SPR	surface plasmon resonance
TAE	Tris-acetate-EDTA
TBP	TATA-element binding protein
TB	Tris-borate
TBE	Tris–borate–EDTA
TBS	Tris buffered saline
TCA	trichloroacetic acid
TE	Tris–EDTA
TRF1	telomere binding protein 1
TEMED	*N,N,N′,N′*-tetramethyl-ethylenediamine
5-TMRh	5-carboxytetramethylrhodamine
TMRh	see 5-TMRh
TEM	transmission electron microscope

Chapter 1
Expression systems

Reinhard Grisshammer, Christian Kambach and Christopher G. Tate
MRC Laboratory of Molecular Biology, Hills Road, Cambridge CB2 2QH, UK

1 Introduction

The over-expression of DNA-binding proteins or their domains is essential for their purification, characterization and three-dimensional structure determination. Unfortunately, isolation of functional protein in quantities sufficient for crystallization trials or for NMR experiments is still not routine. In this chapter, we discuss two expression systems that have been used successfully to produce large amounts of soluble protein, namely *Escherichia coli* and the baculovirus expression system. Many proteins can be expressed in *E. coli*, although it is often necessary to optimize expression parameters to achieve successful production of a desired protein. Methionine-auxotroph strains allow the expression of selenomethioninyl protein which can be used for multi-wavelength anomalous diffraction techniques during structure determination. However, some proteins, such as the eukaryotic telomere binding protein, TRF1, cannot be produced in *E. coli* for unknown reasons (T. de Lange and L. Fairall, personal communication). This protein was successfully expressed in insect cells (1), therefore, we will describe the baculovirus system in this chapter. Owing to space limitations, we will not consider yeast systems; the reader is referred to two excellent reviews containing detailed protocols (2, 3).

2 Expression in *Escherichia coli*

The bacterial host *E. coli* is the first choice of expression system for many researchers. It is simple to use and inexpensive to culture, it divides rapidly, and a wide selection of expression plasmids and strains are available. Basic procedures for expression in *E. coli* have been reviewed extensively, and the reader is referred to Ausubel *et al.*(4), Sambrook *et al.* (5), Coligan *et al.* (6) and to information provided by commercial suppliers. However, despite the extensive use of prokaryotic systems, the efficient expression of target genes or cDNAs in *E. coli* is far from routine. Strategies to obtain maximal yields of recombinant protein must consider many parameters, including protein stability and solubility, codon preference, or toxicity of the target protein to the bacterial cell. Most target proteins have been expressed in the cytoplasm. However, secretion into the periplasm and the medium have also been successful (7). In this section, we address selected problems and possible solutions associated with bacterial expression of foreign proteins. The discussion will not be limited to DNA-binding proteins in order to treat this topic in a more general manner. It should be emphasized that strategies which lead to the successful expression of one particular protein might not apply to a different protein, and that conditions must be optimized in each case.

2.1 Target genes and cDNAs

In contrast to eukaryotes, removal of non-coding pre-mRNA sequences does not occur in *E. coli*. Therefore, it is essential to use cDNA for the expression of eukaryotic proteins rather than their genes, because *E. coli* will treat an intron as part of the coding sequence. Furthermore, the 5′ and 3′ non-coding regions of the target cDNA have to be removed, which can be achieved easily using the PCR. The coding part of the cDNA has to be inserted into the bacterial vector in-frame with the start codon, which (in most cases) is provided by the expression plasmid.

The stability of recombinant proteins may be influenced by the nature of the amino-terminal amino acid (*N*-end rule) (8). Furthermore, removal of *N*-formyl methionine by methionyl aminopeptidase is dependent on the residue following (9), the degree of processing decreasing with increasing size of the second amino acid side-chain. One should therefore avoid certain residues such as leucine at the second position (with methionine being defined as the first position), because such residues may lead to rapid degradation of the recombinant protein by the ATP-dependent protease Clp after removal of *N*-formyl methionine (8).

2.2 Codon usage

The codon usage of genes differs considerably in eukaryotic and prokaryotic organisms (10, 11). Heterologous genes that contain a large number of codons that are rarely used in *E. coli* thus may be expressed inefficiently. The effect of biased codon usage is most severe when multiple rare codons occur near the 5′ end of the target cDNA. In addition, the occurrence of rare codons in *E. coli* correlates with a low level of their cognate tRNA species (10). Two strategies can be employed to minimize the effect of rare codon usage. First, rare tRNA species can be co-expressed with the target cDNA. This is of particular importance for DNA-binding proteins, which may contain a large number of positively charged amino acid residues. Co-expression of the *dna*Y gene encoding the minor arginine tRNA (AGA, AGG) (12) was needed to overproduce the DNA-binding domain of the yeast transcriptional activator GCN4 (13). Second, clusters of rare codons in the target cDNA can be altered without influencing the protein sequence. This approach was essential for obtaining full-length high-mobility group protein HMG-D whose acidic carboxyl terminus contains a series of glutamic acid and aspartic acid residues encoded by rare codons (14). If necessary, the entire target gene can be chemically synthesized from a series of deoxyoligonucleotides combined with PCR (15, 16).

2.3 Fusion proteins

Fusion proteins are often more stable and soluble compared with the native protein alone. In addition, fusion tags are particularly useful for purification (17). The most commonly used fusion partners for DNA-binding proteins are GST and histidine tags (*Table 2*). However, fusion tags may interfere with the functionality of the target protein or with structure determination (18, 19). In this case, affinity peptides can be removed proteolytically after purification if a cleavage site for a specific protease (e.g. tobacco etch virus protease; see ref. 20) has been included between the target protein and the tag. Alternatively, the foreign cDNA can be expressed directly, usually under the control of a strong, inducible promoter and a sequence for efficient translation initiation (see for example pET vectors, *Table 1*).

2.4 Prokaryotic expression vectors

A large variety of expression plasmids are commercially available. It is beyond the scope of this section to discuss their advantages and disadvantages in full. The reader is referred to Williams *et al.* (21) and to information from commercial suppliers for details about a particular vector system. In general, inducible expression systems are preferable to constitutive expression systems because high levels of heterologous proteins are often lethal to the bacterial host. *Table 1* shows the features of some commonly used expression plasmids.

A well-designed prokaryotic expression vector consists of optimally configured elements such as an inducible promoter, a transcription terminator for stabilization of the mRNA and vector, a ribosomal binding site, a translational enhancer (optional), a stop codon, an antibiotic resistance gene for phenotypic selection of the vector and an origin of replication which determines the plasmid copy number (for a review see ref. 7). Most, or all, of these features are optimized in commercially available expression plasmids. One of several factors affecting plasmid stability and expression levels is the type of promoter used for transcription of the target cDNA. Therefore, we will discuss briefly characteristics of some commonly used promoters.

Transcription of the target gene can be driven by two basic systems. Promoters such as the *tac* promoter are recognized by *E. coli* RNA polymerase. In contrast, the bacteriophage T7 ϕ10 promoter is recognized by T7 RNA polymerase but not by *E. coli* RNA polymerase. Strong, inducible *E. coli* RNA polymerase-based promoters cannot be completely repressed, even in the absence of inducer. Low-level expression of a potentially toxic target protein in non-induced cells may lead to plasmid instability. The T7ϕ10 promoter system is a potential alternative in such cases. The target gene is cloned into a pET vector, and a *recA*$^-$ cloning host that does not contain the T7 RNA polymerase gene (e.g. DH5α) is transformed with this construct. Once established, plasmids are transferred into

Table 1 Features of some prokaryotic vector systems used for expression of nucleic acid binding proteins[a]

Vector	Expression host	Promoter	Origin of replication	Antibiotic resistance	Fusion tag	Supplier	Reference
pQE series	Any host with pREP4	T5N25/O[b]	pMB1	Ampicillin	None, optional with His tags, DHFR	Qiagen	39
pREP4	Repressor plasmid	n.a.	p15A	Kanamycin	n.a.	Qiagen	40
pGEX series	Any host	*tac*[c]	pMB1	Ampicillin	GST, protease cleavage site	Pharmacia	41
pET series	DE3 lysogen	T7ϕ10[d]	pMB1	Ampicillin or kanamycin	None, optional with His tags, protease cleavage site, T7 gene 10 tag, S-tag, signal peptide	Novagen	22
pLys, pLysE	Coding for T7 lysozyme	n.a	p15A	Chloramphenicol	n.a	Novagen	22

[a] Abbreviations: n.a. not applicable; DHFR, mouse dihydrofolate reductase.

[b] Strong coliphage T5 promoter fused with two *lac* operator sequences, recognized by the *E. coil* RNA polymerase. Transcription is controlled by the Lac repressor protein which is overproduced by the compatible repressor plasmid pREP4.

[c] Hybrid promoter consisting of –35 region of the *trp* promoter followed by –10 region of the *lac* promoter, recognized by the *E. coil* RNA polymerase. Transcription is controlled by the Lac repressor protein which is encoded by the *lacI*^{q} gene located on pGEX vectors.

[d] Bacteriophage T7 promoter ϕ10 is recognized by the T7 RNA polymerase but not by the *E. coil* RNA polymerase. λDE3 lysogen strains contain a chromosomally integrated copy of the T7 RNA polymerase gene under the control of the IPTG-inducible lacUV5 promoter.

Table 2 Examples of some DNA-binding proteins or their domains expressed in *Escherichia coil*

Protein[a]	Expression host	Vector system	Fusion tag	Problems/comments	Solution	Reference
Chicken histone GH5	BL21(DE3)	pET-3c	None	Inefficient expression	Include first 3 codons of H5 coding region	31
Full-length chicken histone *H5*	BL21(DE3)	pET-3c	None	No expression	Not reported	32
Rat HMG-1	BL21(DE3) pLysE	pT7-7	None	Proteolysis	Not reported	34
Drosophila HMG-D	BL21(DE3	pET-24a	None	Incomplete HMG-D, rare GAG and GAT codons	Change codon usage, express at 25°C	14
DNA-binding domain of yeast GCN4	HMS174(DE3) pLysS	pET-3a	None	Rare AGA and AGG codons	Coexpress dnaY	13
b/HLH/Z domain of murine Max	BL21(DE3) pLysS	Not indicated	None			42
Human RAR-LBD	BL21(DE3)	pET-15b	N : His6	Inclusion body formation	Coexpress RXR-LBD, soluble heterodimer	38
Drosophila ecdysone receptor LBD	DH5α	pGEX-KT	N : GST	Inclusion body formation	Coexpress USP-LBD, soluble heterodimer	38
Human RXR-LBD	DH5α	pGEX-KT	N : GST	Soluble		38
Human RXR-LBD	BL21(DE3)	pET-15b	N : His6	Soluble		38
Full-length RAR	BL21(DE3)	pET-15b	N : His6	Inclusion body formation, severe proteolysis	Coexpress full-length RXR, less degradation, RAR/RXR complex extremely toxic by binding to host DNA (?)	J.W.R. Schwabe, pers. comm.
Zinc finger peptide of murine Zif268	BL21(DE3) pLysE	pET-3a	None	Inclusion body formation	Refolding	26

[a] Abbreviations: b/HLH/Z, basic/helix–loop–helix/leucine zipper domain; GH5, globular domain of chicken histone H5; HMG, high-mobility group protein; LBD, ligand-binding domain; N, amino-terminus; RAR, retinoic acid receptor; RXR, retinoid X receptor; USP, Drosophila ultraspiracle (homologous to human RXR)

expression hosts containing a chromosomal copy of the T7 RNA polymerase gene under the control of the *lac*UV5 promoter (λDE3 lysogens). Unfortunately, the *lac*UV5 promoter allows basal expression of T7 RNA polymerase in the absence of inducer and, consequently, the target gene under control of the T7ϕ10 promoter is also expressed. Co-expression of T7 lysozyme, a natural inhibitor of T7 RNA polymerase, may be necessary to counteract deleterious effects of toxic target proteins in non-induced cells. This can be achieved by co-transformation with plasmids pLysS or pLysE, expressing low or high levels of T7 lysozyme, respectively (22).

One feature of successful over-expression is the delicate balance between the amount of transcript and translational machinery available in the bacterial cell. The presence of too little mRNA can easily become the rate-limiting step for protein production, while too much mRNA may cause uncoupling of transcription from translation. The latter problem can arise particularly when using the T7 system because the T7 RNA polymerase transcribes DNA about seven times faster than the *E. coli* RNA polymerase, leading to naked mRNA stretches that are easily degraded by ribonucleases. A selection procedure described by Miroux and Walker (23) might be employed to optimize the expression of a particular target protein. Alternatively, mRNA levels can be optimized by choosing a different expression plasmid containing a stronger or weaker promoter, respectively (see also Section 2.8.3).

2.5 *E. coli* hosts

A *recA*$^-$ strain is recommended for cloning purposes to avoid insert loss by recombination. For expression, most vector systems can be used independently of the host strain (see *Table 1*). T7 promoter-based systems require λDE3 lysogen strains or infection with bacteriophage CE6 carrying the T7 RNA polymerase gene (22). In several cases, the mutant strain C41(DE3) has been found to be superior to its parent BL21(DE3) as a host for protein expression (23). Strains deficient in one or several proteases may prove advantageous for the production of a recombinant full-length protein. The *E. coli* B strain BL21 lacks the Lon protease and the OmpT outer membrane protease, which are present in many K12 strains.

The combination of a particular strain and plasmid often determines the expression level and solubility of the desired protein (23–25). However, since these interactions are not fully understood, optimization of expression has to be determined empirically.

2.6 Growth conditions

The type of culture media, incubation temperature and level of aeration affect the growth rate and the metabolic state of host cells and, consequently, may influence the expression level and properties of recombinant proteins. In particular, these parameters influence whether foreign proteins are expressed in soluble form or as insoluble aggregates (inclusion bodies). A combination of several changes should be considered to obtain optimal results (see Section 2.8.3).

If a protein is aggregated, it must be refolded to study its functional properties. For example, the structure of the zinc finger domain of Zif268, in complex with DNA, has been determined at high resolution starting from inclusion bodies followed by refolding under anaerobic conditions (26). However, refolding is not always successful and it is best to establish *in vivo* conditions which give rise to soluble target protein rather than aggregates.

For initial experiments, rich media such as LB or 2× TY (5) should be used and expression monitored at several intervals after induction. LB medium contains less tryptone and yeast extract than 2× TY, and its use may reduce the rate of protein synthesis, thus increasing the amount of correctly folded target protein. Changing to LB medium was one of the parameters required to express the snRNP subcore complex D_3B in a soluble form (C. Kambach, unpublished data, see Section 2.8.3).

Some proteins form inclusion bodies when expressed at 37 °C, but become soluble by changing the growth temperature to 30 °C or below (27). This is not only due to the increased physical stability of many proteins at lower temperature, but also because the rate of protein synthesis and protein folding are different at lower temperature. Temperatures as low as 20 °C or 15 °C might improve the folding of some proteins further.

Aeration can also influence the state in which a recombinant protein is obtained. Carp haemoglobin becomes soluble when cell growth is carried out under poor aeration (N. Komiyama, personal communication). In contrast, a protein component of the small nuclear ribonucleoprotein complex is soluble only under good aeration (J. Avis, personal communication). For good aeration, the volume of the medium should not exceed 20% of the total volume of the flask.

2.7 Co-expression of proteins

Many proteins form complexes with other proteins *in vivo* to exert their function. Expression of each individual protein component alone can lead to inclusion body formation or to low levels of expression. In contrast, co-expression of all components can give protein complexes which are both soluble and functional. Examples are the expression of tetrameric human haemoglobin (28) and chicken skeletal muscle troponin complex (29).

Co-expression of several target cDNAs can be achieved by constructing a polycistronic expression cassette. The *order* of the cDNAs in a polycistronic operon can greatly influence expression levels of individual protein components, as shown for the δε-subcomplex of bovine F_1-ATPase (30) and for the snRNP subcore complex D_1D_2 (Christian Kambach, unpublished results). In the latter case, expression of a D_1D_2 operon results in 10–100-fold higher yields than expression of a D_2D_1 operon. Therefore, cloning strategies should permit insertion of the cDNAs in an arbitrary order to allow all possible combinations. A general cloning strategy is shown in *Figure 1*. Each of the individual target cDNAs (cDNA-1 or cDNA-2) are inserted into a cloning vector, such as pUC18, which contains a set of restriction enzyme sites A-C and a ribosomal binding site.

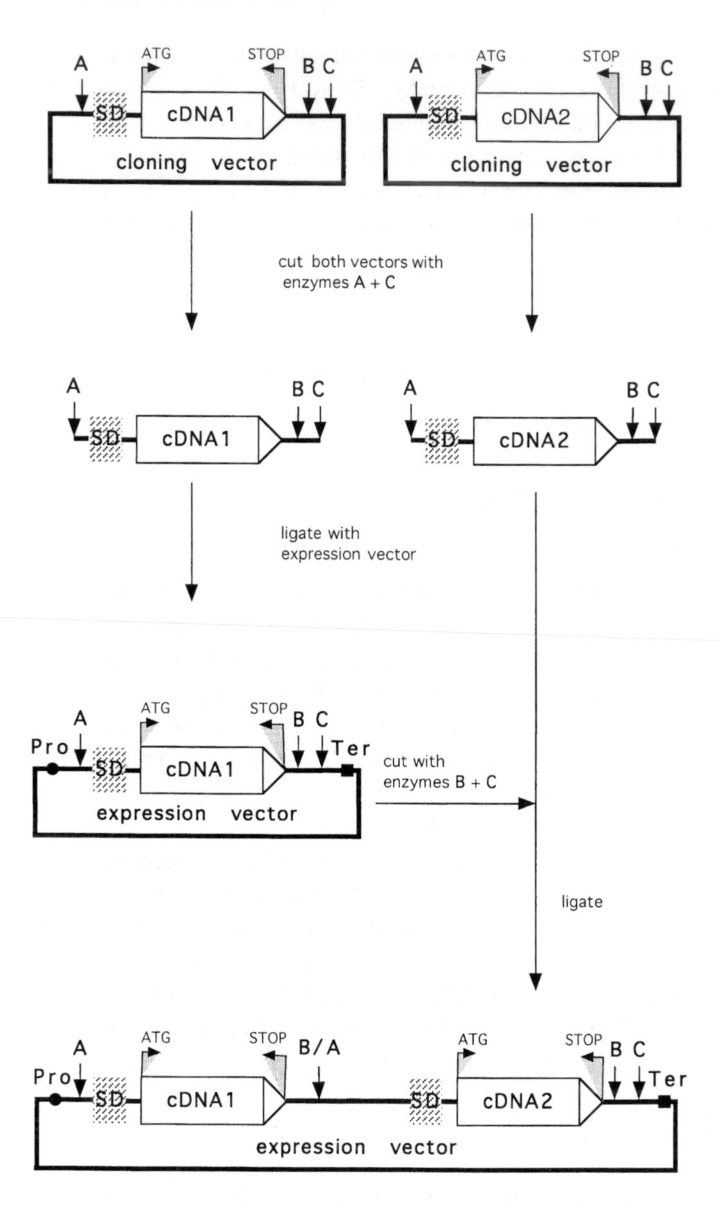

Figure 1 Scheme for the construction of a polycistronic operon. Details of the general cloning strategy are discussed in Section 2.7. Each of the individual target cDNAs are inserted into a cloning vector which contains a set of restriction enzyme sites A-C. cDNA-1 is then cloned into the expression vector using sites A and C. The cDNA-2 fragment is obtained by digest with enzymes A and C, and inserted into the expression vector containing cDNA-1 cut with enzymes B and C. A possible polylinker region of the cloning vector is shown below (see ref. 22). Restriction enzyme sites A (*Bam*HI, GGATCC), B (*Bgl*II, AGATCT), C (*Hin*dIII, AAGCTT) and *Nde*I (CATATG) are underlined. Note that sites A and B have compatible overhangs. The Shine–Dalgarno sequence (SD), start codon (ATG) and stop codons (STOP) are shown in bold type. The promoter and transcriptional terminator are abbreviated as Pro and Ter, respectively.

5′-GGATCC AATAATTTTGTTTAACTTTAAG**AAGGAG**ATATA CAT**ATG** ... **TAATGA**GATCTCGCC AAGCTT-3′

It is essential that sites A and B have compatible overhangs. cDNA-1 is then cloned into the expression vector using sites A and C. The cDNA-2 fragment is obtained by digest with enzymes A and C, and inserted into the expression vector containing cDNA-1 cut with enzymes B and C. *Bam*HI and *Bgl*II are, for example, a good choice for A and B, respectively. This strategy allows the construction of operons with two or more different cDNAs and facilitates changing their order.

2.8 Troubleshooting

There are three general types of problems that are commonly encountered when expressing foreign proteins in *E. coli*:

- absence of the target protein upon induction
- presence of truncated forms of a desired protein
- formation of insoluble aggregates

We will discuss these topics assuming that a chosen expression plasmid containing the target cDNA can be stably maintained in non-induced cells. In cases where expression with a given strain is not reproducible, a plasmid stability test is recommended (6) to find out whether plasmid loss occurs due to 'leaky' expression of a potentially toxic target protein. Rearrangements within the expression plasmid can be investigated by restriction enzyme and DNA sequence analyses.

2.8.1 Absence of target protein after induction

The lack of expression of a recombinant protein might be caused by a variety of reasons, such as mRNA instability, poor translation or proteolytic degradation. As a first measure, another promoter system or different host strain can be tested. Switching from the original cDNA to a synthetic gene with optimized codon usage can sometimes overcome problems associated with the mRNA sequence, as successfully shown for the snRNP D_1 protein (Jo Avis, personal communication). However, deciphering the exact cause for lack of protein expression can be very complex – as experienced in the expression of chicken histones (31, 32).

Poor translational efficiency may be overcome by employing a two-cistron expression system (33) or by maximizing the AT content of the 5′ coding region just downstream of the start site. However, it is not entirely clear whether optimization of the AT content was responsible for good expression of the globular domains of chicken histones H1 and H5, since many optimal codons for *E. coli* are also AT rich (32).

The presence of clusters of 'rare' codons and subsequent depletion of the corresponding tRNA pool may also cause failure to express a full-length target protein. Co-expression of the corresponding tRNA gene, site-directed mutagenesis or chemical synthesis of the target gene might be considered (see Section 2.2). These approaches may often prove successful. However, it is notable that in certain cases, such as the chicken histone H5, substitution of rare codons with those optimal for *E. coli* did not noticeably improve expression (32).

Proteolysis occurring during expression may be diminished by changing to a protease-deficient host strain (see Section 2.5). Unfortunately, hosts deficient in multiple proteases show decreased viability and thus impaired growth properties.

2.8.2 Presence of truncated forms of a desired protein

Truncated proteins may arise because of codon usage (14) (see Section 2.2) or because of limited proteolysis occurring within the host cell (34). In the latter case, changing to host strains deficient in one or several proteases can sometimes alleviate the problem. Alternatively, protease-sensitive sites within the target protein can be identified by *N*-terminal sequence analysis of fragments and by mass spectrometry, and then substituted or eliminated by mutagenesis of the cDNA. However, great care, has to be taken to ensure functionality of the mutant protein. If the primary sequence of the desired protein is to be retained, then tagging the amino and/or carboxyl termini may stabilize the full-length protein (for review see ref. 35). In persistent truncation, the longest proteolytically stable fragment may be determined by *in vitro* proteolysis, combined with mass spectrometry. With luck, this protein fragment may retain its biological function. Such a protein fragment can then be expressed in recombinant form for biochemical and/or structural studies (36).

2.8.3 Inclusion body formation

Attempts to increase the solubility of proteins formed as inclusion bodies have included co-expression of chaperones, manipulation of growth conditions and/or a fusion protein strategy (for review see ref. 37). Sometimes, several changes must be combined to obtain soluble protein, as shown for example for the snRNP subcore complex D_3B. The following measures were required to produce the protein complex up to 70% in soluble form:

1. Use of the pQE vector system, which has a lower plasmid copy number and weaker T5 promoter than the T7-based expression plasmid used initially (see Section 2.4).
2. Low temperature during expression (see Section 2.6).
3. Change from 2× TY to LB medium (see Section 2.6).

These measures probably reduce the rate of protein production and allow the slow formation of the complex in soluble form (C. Kambach, unpublished observation).

Some nuclear hormone receptors exert their biological activity by heterodimer formation. Expression of the ligand-binding domain of the retinoic acid receptor alone led to formation of insoluble aggregates. However, co-expression with its partner, the retinoid X receptor, resulted in a dramatic increase in the production of soluble and stable heterodimer (38).

Protocol 1

Basic procedure for expression of recombinant proteins in *E. coli*

Equipment and reagents

- Orbital refrigerated shaker with temperature control, plate incubator
- Ampicillin stock solution (100 mg/ml in H_2O) (selection for plasmids pGEX, pET or pQE, for example)
- Chloramphenicol stock solution (34 mg/ml in ethanol) (selection for plasmids pLysS or pLysE in the T7 expression system)
- LB medium (10 g/l tryptone, 5 g/l yeast extract, 10 g/l NaCl, pH 7.5)
- Kanamycin stock solution (10 mg/ml in H_2O) (selection for plasmid pREP4 in the pQE vector system)
- LB plates containing 100 μg/ml ampicillin (and optionally 34 μg/ml chloramphenicol or 25 μg/ml kanamycin)
- IPTG stock solution (1 M in H_2O) (store at −20 °C)

Method

1 Transform competent host cells with desired expression plasmid(s) and plate on selective LB plates. Incubate plates overnight at 37 °C. Pick single colony, prepare plasmid at small scale (miniprep) and check plasmid integrity by restriction enzyme analysis. Re-streak confirmed clone.

2 Inoculate 50 ml of LB medium (room temperature or prewarmed, 500 ml conical flask) containing the appropriate antibiotics with a single colony.[a]

3 Incubate flask with aeration at 37°C until culture reaches 0.5–0.7 at OD_{600}.

4 Remove 1 ml of the non-induced cell suspension for analysis.

5 Induce the remaining culture by adding IPTG to a final concentration of 0.4 mM. Continue incubation at 37 °C.

6 Remove 1 ml samples of induced culture after 1, 2, 3 and 5 h for analysis.

7 Analyse samples by SDS–PAGE (and Western blot, if necessary).

[a] Alternatively, transform competent *E. coli* cells with the desired recombinant plasmid before each expression experiment, grow transformants on a selective plate to a small size only, then inoculate 50 ml of selective LB medium with a small colony. Do not re-streak and do not allow cells to reach stationary phase by incubation overnight. This procedure is preferred to counteract plasmid loss.

3 The baculovirus expression system

3.1 Introduction

The baculovirus expression system has become one of the most popular expression systems for eukaryotic proteins because of its ease of use and the large amounts of protein often expressed. Commercialization has resulted in a multitude of vectors with convenient restriction sites for the expression of both native and fusion proteins. In addition, there have been a number of improvements in the way that recombinant viruses are constructed, and deletion of nonessential genes from the baculovirus has further improved its performance. All the methods described here were originally developed by Summers and coworkers (43), and they are discussed in depth by O'Reilly *et al.* (44) and King and Possee (45).

Autographa californica multicapsid nuclear polyhedrosis virus (AcMNPV) is the baculovirus that is most frequently used. It infects only a few species of Lepidoptera, and cannot replicate in mammalian cells. The life cycle of the virus can be divided into two distinct phases. In the first phase, the baculovirus infects an insect cell and replicates itself using virus-encoded genes under the control of immediate-early, early and late promoters. The new viruses bud from the cell about 12 h after the onset of infection. The second phase starts from about 15 h post-infection, when the very-late genes, encoding polyhedrin and protein p10, are transcribed. Polyhedrin forms a matrix around virus particles, providing a protective coat for the viruses when they are released from the dead insect and are exposed to the environment. In the laboratory, this stage is not essential for virus propagation, so both the polyhedrin and p10 genes can be replaced by heterologous genes. In the most commonly used systems, which will be discussed here, the polyhedrin gene is replaced by a cDNA under the control of the polyhedrin promoter. When cultured insect cells are infected with the recombinant virus, the protein of interest will be expressed maximally 2–4 days post-infection.

Using the baculovirus expression system is not as easy as expressing proteins in bacteria, because tissue culture facilities are required to grow insect cells. In addition, it takes about 3–4 weeks to produce large titred stocks of a new recombinant baculovirus, starting from a cDNA in a bacterial plasmid. It is also considerably more expensive compared with bacterial expression systems. Therefore, most researchers only use the baculovirus expression system as a last resort, when bacterial expression systems have failed or because specific post-translational modifications unique to eukaryotes are required on the protein to be studied.

3.2 Choice of transfer vector and baculovirus DNA

There is a wide variety of commercially available vectors (*Table 3*). Vectors usually contain very-late promoters (polyhedrin or p10) because these are extremely strong and lead to very high protein expression. However, under conditions where protein folding is rate limiting for the expression of a functional protein (usually where a post-translational modification is essential for folding), it may be advantageous to use a weaker promoter expressed earlier in the baculovirus life cycle when the cell is most healthy. The basic protein promoter (late promoter) and the immediate-early promoter *ie1* are both weaker than the very-late promoters and could be used to obtain smaller amounts of functional protein if required. Regardless of the promoter used, all the transfer vectors are designed

Table 3 Commercially available baculovirus transfer plasmids

Plasmid	Promoter[a]	Fusion protein[b]	Supplier	Comments
pBlueBac4.5	pol	None	Invitrogen	Blue recombinant plaques
pBacPAK8/9	pol	None	Clontech	Expression occurs after 20 h p.i.
pAcSG2	pol	None	Pharmingen	
pAcMP2/3	bp	None	Pharmingen	Expression occurs 12–24 h p.i.
pAcP(-)IE1–6	ie1	None	Novagen	Expression occurs immediately on virus infection
pAcUW51	1 pol + 1 p10	None	Pharmingen	2 proteins expressed simultaneously
pBac4x-1	2 pol + 2 p10	His6 (C) off one of the promoters	Novagen	4 proteins expressed simultaneously
pACHLT	pol	His6 (N)	Pharmingen	Thrombin cleaves off tag
pBacPAK-His	pol	His6 (N)	Clontech	
pBlueBacHis2	pol	His6 (N)	Invitrogen	Blue recombinant plaques
pBac-1	pol	His6 and/or S-tag (N), optional His6 (C)	Novagen	pBacGus is identical, but with a β-glucuronidase reporter gene
pAcGHLT	pol	GST and His6 (N)	Pharmingen	Thrombin cleavage removes GST
pBAC-7, 8, 9	pol	Cellulose-binding domain (N or C)	Novagen	Purification of fusion protein on cellulose columns
BioColors	pol	Green, blue or yellow fluorescent protein (N)	Clontech	Thrombin cleavage removes fluorescent protein
pMelBac	pol	Signal sequence (N)	Invitrogen	Secretes recombinant protein
pBACsurf-1	pol	Signal sequence (N) and optional gp64 (C)	Novagen	gp64 fusion displays the recombinant protein on the virus surface
pAcGP67	pol	Signal sequence (N)	Pharmingen	Secretes recombinant protein
pPbac/pMbac	pol	Signal sequence (N)	Stratagene	Secretes recombinant protein
pBSV-8His	pol	Signal sequence (N) and His8 (C)	Boehringer	Secretes recombinant protein

[a] pol, polyhedrin promoter; bp, basic protein promoter; ie1, immmediate–early promoter 1.

[b] N, fusion is made at the *N*-terminus of the desired protein; C, fusion is made at the *C*-terminus of the desired protein

to insert the cDNA to be expressed in the position of the polyhedrin gene in the baculovirus genome. However, it is important to ensure that the transfer vector is compatible with the linearized baculovirus DNA used in the co-transfection step, i.e. there are identical DNA sequences in the plasmid and baculovirus DNA that will allow recombination *in vivo*. Systems that allow colour selection of recombinant virus plaques (Invitrogen) contain the β-galactosidase gene in the viral DNA and the transfer vector and must be used in conjunction with each other for successful recombination; conversely, they are not compatible with products that do not allow colour selection.

If the protein to be expressed needs to be purified, then it may be worth considering attaching an affinity tag to the protein using a variety of baculovirus transfer vectors (*Table 3*). Fusion partners include poly-histidines, glutathione S-transferase and a chitin-binding domain. If the cDNA is not fused downstream of an affinity tag, it is important to consider the effect of any 5′ untranslated region on expression. It is generally considered that the shorter this region is the less likely it will affect protein expression. In addition, the sequence around the initiator methionine in baculovirus genes (ATA/CAAA/CATGAA) (46) is different from the consensus Kozak sequence (CCACCATGG) (47), and may need to be optimized for high-level expression.

3.3 Choice of host cells and growth conditions

Four cell lines from two different insects are commercially available for baculovirus expression. The most commonly used cell line, Sf9, was derived from the Sf21 cell line that was originally isolated from the fall army worm (*Spodoptera frugiperda*). Sf9 cells are smaller, and grow to higher cell densities, than Sf21 cells, and both cell lines are easy to grow in suspension. When grown in flasks, Sf9 and Sf21 cells adhere lightly to the plastic surface and can be removed by banging the flask sharply on a hard surface. In contrast, the cell lines derived from *Trichoplusia ni* (Hi5 and MG1) are more adherent and require a brief EDTA treatment (Hi5) or trypsin treatment (MG1) to remove them from the flask for passaging. They are also more difficult to grow in suspension, although Hi5 cells can be adapted to growth in suspension in the absence of microcarrier beads. The advantage of these cell lines is that they often produce far more protein than Sf9 or Sf21 cells, but it is often the case that a large percentage of this protein may be misfolded and inactive. Sf21 cells are considered the cell line of choice for expressing *functional* protein.

Insect cells are usually grown at 27 °C under normal atmospheric concentrations of O_2 and CO_2, so a thermostatically-controlled incubator is the only requirement for growing cells (*Protocol 2*). Sf9 and Sf21 cells can be grown in suspension in two ways, either in magnetically-stirred glass bottles (60–70 r.p.m.) or in shaker flasks placed in an orbital shaker (125–150 r.p.m.). There is no advantage for cell growth or recombinant protein expression with either system, but shaker flasks are far cheaper to purchase than stirred bottles. There are many types of insect cell culture media available, either as powders or in a ready-to-use liquid form.

When making medium from a powdered stock, always use high-quality distilled water e.g. from a MilliQ system (Millipore). Include antibiotics (200 μg/ml penicillin/streptomycin) and a fungicide (e.g. 2.5 μg/ml Fungizone), and a surfactant (0.1% Pluronic F-68) to protect the cells from shear forces when they are grown in suspension. If fetal bovine serum is used (e.g. when using TNMFH medium), then it is essential to purchase a high-quality serum, or one that has been specifically tested for insect cell growth. If serum-free medium is to be used, then it is often easier to obtain preadapted cells for that specific medium from the manufacturer rather than trying to do it yourself. Note that cells grown in serum-free medium are more adherent than those grown in the presence of serum and, therefore, should not be left standing unstirred if they are grown in suspension. Cells grown in a proprietary serum-free medium may sometimes express more protein than cells grown in medium containing 5–10% serum. However, regardless of the medium used, it is *essential* that the cells are healthy (doubling every 20–24 h), otherwise expression levels will be low.

Protocol 2

Growing insect cells

Equipment and reagents

- 27 °C incubator with magnetic stirrers or an orbital shaker
- Stirrer bottles (Techne) for magnetic stirrers or conical flasks for orbital shaker
- Laminar flow hoods (category 1–2 containment, depending on the protein overexpressed)
- Bench-top centrifuge
- Tissue culture plastic ware (pipettes, 25 cm^2 and 75 cm^2 flasks) (Corning)
- Sterile medium e.g. TNMFH (Sigma), Sf900II (Gibco BRL)
- Additives to medium e.g. FBS, penicillin/streptomycin, Fungizone, Pluronic F-68, additional lipids (all from Gibco BRL)

Method

1. Thaw cells from liquid nitrogen rapidly in a 37 °C water bath until only a small lump of ice remains.
2. Wash the vial with 70% ethanol and pipette the contents (1×10^7 cells) into 15 ml of medium at 4 °C in a 75 cm^2 flask.
3. Place the flask in a 27 °C incubator for 2 h to allow the healthy cells to adhere. Remove the supernatant and replace with 15 ml of fresh medium prewarmed to 27 °C.
4. Grow the cells for 3–4 days until the cells are confluent.
5. Remove the cells by banging the flask sharply on a hard surface. Dilute cells into pre-warmed medium (27 °C) to give a cell density of 0.4×10^6 cells/ml. The medium should contain 10% FBS and 0.1% pluronic F-68.
6. Grow cells in either a stirrer bottle or a shake flask at 27 °C until they reach a density of 2×10^6 cells/ml. Some cells may not reach this density immediately after thawing, so split the cells after 3–4 days when they are still healthy.

Protocol 2 continued

7 Decant the cells into a 50 ml polypropylene tube and centrifuge at 146 g (21 °C) for 5 min. Decant the supernatant carefully as the cell pellet is relatively loose.

8 Resuspend the cell pellet *gently* in 5 ml of pre-warmed medium by gently pipetting the cells up and down using a 10 ml pipette.

9 Dilute the cells to a final cell density of 0.5×10^6 cells/ml, and continue to grow them.

10 Split cells every 2 days (i.e. when they reach 2×10^6 cells/ml); cells to be left over the weekend can be diluted to 0.4×10^6 cells/ml. Cells benefit from being spun down and resuspended in fresh medium every 1–2 weeks.

11 After 2–3 months of passaging cultures of Sf9 cells, the doubling time gradually becomes longer and the cells do not grow to high cell densities, resulting in a decrease of recombinant protein expression. Discard the cells and thaw out a new batch.

12 Freezing cells for storage in liquid N_2: when the new batch of cells is growing well, centrifuge 200 ml of cells at 2×10^6 cells/ml (146 g, 5 min, 21 °C) and resuspend the pellet in medium containing 10% FBS and 10% DMSO at 10^7 cells/ml. Place 1-ml aliquots in cryovials and freeze slowly by placing the tubes in a foam rack and transferring the rack to a −20 °C freezer (1 h), then to a −80 °C freezer (overnight) and finally into liquid nitrogen for long-term storage (years).

Insect cells are far less robust than either bacteria or yeast and must be handled very gently. Always prewarm the medium before adding it to the cells; excessive cold (10–15 °C) will shock the cells and reduce their growth rate for a week or two. Excessive heat (> 30 °C) is also detrimental. Healthy cells will double in density every 20–24 h and when stained with trypan blue (900 μl of cells plus 100 μl of 0.1% trypan blue in PBS) greater than 97% of cells should be colourless (dead cells are stained blue). Cells can be grown in suspension for 2–3 months, until their doubling time starts to increase. At this point, discard the cells and thaw another batch. After obtaining cells for the first time from either a manufacturer or a cell culture collection, it is advisable to freeze down a quantity of cells as soon as possible.

3.4 Constructing a recombinant baculovirus

There are two steps to make a recombinant baculovirus

(1) Construction of a transfer plasmid in *E. coli* that contains the cDNA encoding the protein of interest downstream from the polyhedrin promoter. The polyhedrin promoter may be replaced by other baculovirus promoters (*Table 3*) but the flanking DNA is the same.

(2) Co-infection of insect cells with the transfer plasmid and linearized baculovirus DNA (see *Table 4*).

Table 4 Commercially available linearized baculovirus genomic DNA

Linearized baculovirus DNA	Supplier
Bac Vector 3000[a]	Novagen
Baculogold	Pharmingen
BacPAK6	Clontech
Bac-N-Blue[b]	Invitrogen

[a] Viral genes deleted for generating better production of recombinant proteins.

[b] Used for generating blue recombinant plaques with Invitrogen vectors only.

The recombinant baculovirus is created by homologous recombination between the transfer plasmid and the linearized baculovirus DNA inside the insect cell (*Protocol 3*). The efficacy of this step has been assured by deleting an essential gene from the baculovirus during the linearization step, and by placing this gene on the bacterial transfer plasmid. This ensures that greater than 90% of the viable viruses produced from the recombination step contain the cDNA to be expressed.

Protocol 3

Co-transfection of linearized baculovirus DNA and the transfer plasmid

Equipment and reagents

- Standard tissue culture equipment and medium (*Protocol 2*)
- Linearized baculovirus DNA and transfection reagents (Clontech, Invitrogen, Novagen or Pharmingen)
- Miniprep of the transfer vector containing the cDNA to be expressed

Method

The method for co-transfection of baculovirus DNA and the transfer plasmid is supplied by the manufacturer with the linearized baculovirus DNA, and it should be followed exactly. The transfer plasmid should be purified on a small-scale using a kit (e.g. Qiagen) or on a large-scale using a CsCl gradient. The co-transfection step will take 4–5 days and results in 3–5 ml of medium (transfection supernatant) containing a mixture of recombinant viruses, a large percentage of which will contain the cDNA to be expressed

The co-transfection step described above produces about 5 ml of a mixture of recombinant and non-recombinant viruses of low titre. The next step is to isolate a homogeneous population of recombinant viruses and to amplify them to provide sufficient quantities for expression experiments. A homogeneous viral population is obtained by performing a plaque purification (*Protocol 4*). A single viral plaque well separated from other plaques is picked and then amplified (*Protocol 5*). Normally, if five independent plaques are picked, amplified and tested for whether they express the cDNA insert, at least four will express the protein of interest (*Protocol 6*). For large-scale expression experiments at least 100 ml of virus is made (*Protocol 7*) and titred (*Protocol 8*) to determine the number of viral particles per ml.

Protocol 4

Plaque purification

Equipment and reagents

- Standard tissue culture equipment and medium (*Protocol 2*)
- 60 × 15 mm Falcon Easy Grip tissue culture dish (Becton Dickinson)
- Sterile 6 ml Falcon round-bottomed tubes with caps (Becton Dickinson)
- Agarose (e.g. SeaPlaque agarose from FMC Bioproducts or BacPlaque agarose from Novagen)
- 2 × TNMFH medium (Sigma)
- Fetal bovine serum (Gibco BRL)

Method

1. Dilute exponentially growing insect cells to 0.4×10^6 cells/ml.
2. Place 3 ml of diluted cells in a 60 mm × 15 mm sterile dish and leave to adhere for 15 min.
3. Make a serial dilution of the transfection supernatant (10^{-3}–10^{-6}) in polypropylene tubes (6 ml). Each dilution is made by adding 100 μl of the previous dilution to 900 μl of fresh medium.
4. Remove the medium from the adhered cells in the culture dishes by tilting the dish and aspirating off the liquid.
5. Add the diluted virus (900 μl) to each plate and leave for 1 h, with an occasional tilting of the plates to ensure the cells remain wet.
6. Melt sterile 2% agarose in H_2O in a microwave and transfer 25 ml to a 50 ml tube. Place tube in a 37 °C water bath. Place 20 ml of twice normal-strength medium and 5 ml of fetal bovine serum in another 50 ml tube, and also put in a 37 °C water bath. Both tubes should be at 37 °C for at least 30 min to ensure temperature equilibration.
7. Immediately before use, pour the 2× medium into the 2% agarose and mix the solution well by inversion.
8. After 1 h of incubation, the virus is aspirated from the cells, ensuring that all the liquid is removed by tilting the dishes.
9. Pipette 4 ml of freshly mixed, molten agarose/medium gently onto the inside bottom edge of a tilted culture dish containing the cells; allow the molten agarose/medium to flow slowly over the cells until all the 4 ml is in the dish.
10. Replace the top on the dish and leave the agarose to set for at least 30 min without being moved.
11. Incubate culture dishes at 27 °C, upright (cells at the bottom), in a plastic box containing some damp tissue paper to provide a humid environment. Plaques should be visible after 4–5 days.

Protocol 4 continued

12 If plaques are present[a], they are easily seen by the naked eye by looking at the bottom of the plate while illuminated from the top; large plaques on plates containing the lower dilutions are white opaque rings around a small (fraction of a millimetre) clearing.

13 Using a low-powered inverted microscope, plaques can be found on the plates containing higher dilutions of virus. Ideally, pick plaques from plates containing 1–10 plaques/plate. Using the microscope, label five well-spaced plaques with a marker pen.

14 Pick plaques using a small sterile pipette tip on an automatic pipettor, and transfer the small plug of agarose to 1 ml of medium.

[a] If no plaques are visible on any of the dishes, and the cells look healthy and form a confluent layer, then the most likely explanation (assuming virus was added) is that too many cells were added to the dish. This will result in all plaques being overgrown. To optimize the number of cells per dish, add increasing volumes of cells (0.4×10^6 cells/ml, range 1.5–3 ml) to each dish, and make up the difference in volume to 3 ml with fresh medium. After the cells have adhered, overlay with agarose and incubate the dishes at 27 °C. Check the dishes daily. Ideally, the cells in one of the dishes should become confluent on day 4–5; in this case, the correct number of cells was added.

Protocol 5

Virus amplification from plaques

Equipment and reagents

- Standard tissue culture equipment and medium (*Protocol* 2)
- 25 cm^2 flasks (Corning)

Method

1 The 1 ml of virus produced from the plaque purification needs to be amplified before it can be used for expression experiments. This is achieved by adding the 1 ml of virus to 1×10^6 cells in 4 ml of medium in a 25 cm^2 flask.

2 Incubate the flask for 3–4 days at 27 °C. Cells should be visibly infected; cells will increase in diameter by about 50%, they will have an enlarged nucleus, be grey in colour as opposed to yellow-green and they will lift from the flask when it is shaken gently.

3 Decant the medium from the flask into a 15 ml tube and pellet the cells (228 g, 21 °C, 5 min). The supernatant is the first-passage virus stock. This virus stock should be titred to determine the number of virus particles/ml (see below).

4 Freeze 1 ml aliquots in cryovials at −80 °C for long-term storage. Short-term storage (about 1 year) is at 4 °C in the dark.

Protocol 6

Testing protein expression from first-passage virus

Equipment and reagents

- Standard tissue culture equipment and medium (*Protocol 2*)
- Equipment for SDS-PAGE and Western blotting
- 10 mM Tris pH 7.5 buffer containing appropriate protease inhibitors e.g. PMSF, leupeptin, pepstatin (Sigma)

Method

1. Infect 10^6 cells in 5 ml of medium in a 25 cm^2 flask by adding 100 μl of first-passage virus.
2. Incubate at 27 °C for 2–3 days until the cells look infected.
3. Harvest the cells from the medium by centrifugation (228 g, 21 °C, 5 min) and resuspend in 500 μl of 10 mM Tris pH 7.5 buffer containing appropriate protease inhibitors.
4. Analyse 10 μl samples by SDS-PAGE (Coomassie blue staining) and/or by Western blotting. Include a control of cells infected with a baculovirus with no cDNA insert.

Protocol 7

Production of second-passage virus stocks

Equipment and reagents

- Standard tissue culture equipment and medium (*Protocol* 2)

Method

1. Spin down 50 ml of 2×10^6 cells/ml of exponentially growing Sf9 cells (146 g, 21 °C, 5 min) and decant the supernatant. Resuspend pellet in 10 ml of fresh medium.
2. Add first-passage virus to the 1×10^8 cells to give an MOI of 0.3–0.5 (i.e. for every virus particle there is between two and three insect cells).
3. Leave cells for 1 h at room temperature and then dilute to 1×10^6 cells/ml with fresh medium (100 ml final volume).
4. Incubate the cells for 2–3 days until the cells are clearly infected, but have not lysed.
5. Centrifuge the cells (228 g, 21 °C, 5 min) and decant the supernatant into a fresh tube.
6. Viral supernatants can be filtered through a 0.22 μm filter if required, although this may decrease the virus titre slightly (the titre should be $2–5 \times 10^8$ pfu/ml; see *Protocol 8*)

Protocol 7 continued

7 Make third-passage virus if litres of virus are needed to infect cells grown in large-scale fermentors by scaling up the volumes in steps 1–3. Avoid multiple passages of a virus as this can decrease the subsequent level of protein expression.

Protocol 8

Determining the titre of a virus stock by the end-point dilution method

Equipment and reagents

- Standard tissue culture equipment and medium (*Protocol 2*)
- 60 × 10 μl well microtitre plates (Gibco BRL)
- Sterile 6 ml polypropylene round-bottomed tubes with caps (Falcon, Becton Dickinson)

Method

1 Make a serial dilution of the virus stock (range 10^{-4}–10^{-8}) by consecutively transferring 20 μl of virus to 180 μl of fresh medium in a 6 ml tube.

2 Add an equal volume of cells (180 μl of 0.25 × 10^6 cells/ml) to each tube and mix thoroughly by gently shaking the tube.

3 Fill 10 wells of a 60 × 10 μl well microtitre plate with each dilution (10 μl per well).

4 Incubate the plates at 27 °C in a sealed box to prevent evaporation from the plates.

5 Determine after 6 days the number of infected wells for each dilution and calculate the titre of the virus as shown in the example below.

Virus	10^{-4}	10^{-5}	10^{-6}	10^{-7}	10^{-8}
Number of infected wells	10	10	9	3	0
Number of uninfected well	0	0	1	7	10
Cumulative infected wells		0+3+9+10=22	0+3+9=12	0+3=3	0
Cumulative uninfected wells		0	0+1=1	0+1+7=8	0+1+7+10=18
% infected wells		(22/22+0)*100=100	(12/12+1)*100=92.3	(3/3+8)*100=27.3	(0/0+18)*100=0

To find the dilution of virus that would infect 50% of the wells (which in this example is somewhere between the 10^{-6} and 10^{-7} dilution), the following formula is used:

PD = –[(Percentage of wells infected at the dilution above 50%) – 50%]/[Percentage of wells infected at the dilution abot 50% – percentage of wells infected at the diltution below 50%].

PD is the proportionate distance and always has a negative value. In the example above, PD = –(92.3–50)/(92.3–27.3) = –0.651, which means that if the virus was diluted to $10^{-6.651}$ then 50% of the wells would be infected. To convert this into a virus titre, the following calculation is made: Titre = $[1/(0.005*10^{-6.651})]*0.69 = 6.2 \times 10^8$ plaque forming units (pfu) per ml where 0.005 is the volume of diluted virus added to each well and 0.69 is a conversion factor.

3.5 Optimization of expression

Once a recombinant virus has been constructed and 100 ml of titred second-passage virus is available, then expression can be optimized (*Protocol 9*). The major factor influencing the level of protein expression is the MOI, or the number of virus particles added per cell. Infection of cells should be conducted initially with an MOI of 2, 5 and 10 on a small scale, and the amount of protein expressed should be measured, preferably with a functional assay. If the amount of functional protein expressed at an MOI of 10 is higher than at an MOI of 5, then try infections at higher MOIs, until no further increase in protein expression is observed. The other parameter that needs to be defined is the length of time between infection and harvesting. This will depend on the characteristics of the particular protein expressed. Some proteins may still be expressed 4 days after infection, whereas other proteins will experience proteolysis if the cells are left to day 3. Other proteins that require post-translational modifications for activity may be expressed maximally at day 2 if *functional* protein is measured. Leaving the cells longer results in the accumulation of misfolded, inactive protein. One other factor, that can be altered to improve expression, is the type of cell used. As mentioned above, Hi5 cells could be used for large-scale expression in shaker flasks and they sometimes express more protein than Sf9 cells. It is worth noting here that Sf9 cells differ in their characteristics depending on their source, which can result in different expression levels and time courses of expression. When comparing expression levels between various cell lines, ensure that all the cultures are healthy and doubling in cell density every 24 h, because expression levels will invariably decrease if the cells are growing poorly.

Protocol 9

Expression of protein from the polyhedrin promoter

Equipment and reagents

- Standard tissue culture equipment and medium (*Protocol 2*)

Method

Small-scale infections:

1. Add 2×10^6 cells in 5 ml of medium to a 25 cm^2 flask or 1×10^7 cells in a 75 cm^2 flask.
2. Add the appropriate amount of virus and incubate at 27 °C for 2–4 days.
3. Harvest by banging the flask on a hard surface to remove any adherent cells, decant into a 15 ml tube and centrifuge to harvest the cells (228 g, 21 °C, 5 min).

Large-scale infections:

1. Spin down 1×10^8 cells (146 g, 21 °C, 5 min), decant the supernatant and resuspend the cells gently in 10 ml of fresh, prewarmed medium.

Protocol 9 continued

2 Add the appropriate amount of virus and stand the tube at room temperature for 1 h.

3 Add 90 ml of fresh medium to give a final cell concentration of 1×10^6 cells/ml and place in a 250 ml glass spinner bottle or a 500 ml shake flask. It is very important that flasks are not overfilled, because the infected cells have a large oxygen demand and so they need a relatively large surface to volume ratio to allow good gaseous exchange.

4 Incubate for 2–4 days and harvest the cells by centrifugation (228 g, 21 °C, 5 min).

Acknowledgements

The authors thank Jacqueline Milne, Kiyoshi Nagai, John Schwabe, Helena Taylor and Louise Tierney for critical comments on the manuscript.

References

1. Bianchi, A., Smith, S., Chong, L., Elias, P., and de Lange, T. (1997). *EMBO J.* **16**, 1785.
2. Guthrie, C. and Fink, G. R. (ed.) (1991). *Methods in enzymology*, Vol. 194. Academic Press, London.
3. Romanos, M. A., Scorer, C. A., and Clare, J. J. (1995). In *DNA cloning 2: a practical approach, expression systems* (ed. D. M. Glover and B. D. Hames), p. 123. IRL Press, Oxford.
4. Ausubel, F. M., Brent, R., Kingston, R. E., Moore, D. D., Seidman, J. G., Smith, J. A., and Struhl, K. (ed.) (1994). *Current protocols in molecular biology*. John Wiley and Sons, New York.
5. Sambrook, J., Fritsch, E. F., and Maniatis, T. (ed.) (1989). *Molecular cloning: a laboratory manual*. Cold Spring Harbor Laboratory Press, Cold Spring Harbor.
6. Coligan, J. E., Dunn, B. M., Ploegh, H. L., Speicher, D. W., and Wingfield, P. T. (ed.) (1997). *Current protocols in protein science*. John Wiley & Sons, New York.
7. Hannig, G. and Makrides, S. C. (1998). *Trends in Biotechnology*, **16**, 54.
8. Tobias, J. W., Shrader, T. E., Rocap, G., and Varshavsky, A. (1991). *Science* **254**, 1374.
9. Hirel, P.-H., Schmitter, J.-M., Dessen, P., Fayat, G. and Blanquet, S. (1989). *Proc. Natl. Acad. Sci. USA* **86**, 8247.
10. Ikemura, T. (1982). *J. Mol. Biol.*, **158**, 573.
11. Nakamura, Y., Gojobori, T., and Ikemura, T. (1998). *Nucl. Acids Res.* **26**, 334.
12. Brinkmann, U., Mattes, R. E., and Buckel, P. (1989). *Gene* **85**, 109.
13. König, P. and Richmond, T. J. (1993). *J. Mol. Biol.* **233**, 139.
14. Payet, D. and Travers, A. (1997). *J. Mol. Biol.* **266**, 66.
15. Stemmer, W. P. C., Crameri, A., Ha, K. D., Brennan, T. M., and Heyneker, H. L. (1995). *Gene* **164**, 49.
16. Hale, R. S. and Thompson, G. (1998). *Prot. Expr. Purif.*, **12**, 185
17. Nilsson, J., Ståhl, S., Lundeberg, J., Uhlén, M., and Nygren, P.-Å. (1997). *Prot. Expr. Purif.* **11**, 1.
18. Büning, H., Gärtner, U., von Schack, D., Baeuerle, P. A., and Zorbas, H. (1996). *Anal. Biochem.* **234**, 227.

19. Lindner, P., Guth, B., Wülfing, C., Krebber, C., Steipe, B., Müller, F., and Plückthun, A. (1992). *Methods: Companion Methods Enzymol.* **4**, 41.
20. Parks, T. D., Leuther, K. K., Howard, E. D., Johnston, S. A., and Dougherty, W. G. (1994). *Anal. Biochem.* **216**, 413.
21. Williams, J. A., Langeland, J. A., Thalley, B. S., Skeath, J. B., and Carroll, S. B. (1995). In *DNA cloning 2: a practical approach, expression systems* (ed. D. M. Glover and B. D. Hames), p. 15. IRL Press, Oxford.
22. Studier, F. W., Rosenberg, A. H., Dunn, J. J., and Dubendorff, J. W. (1990). In *Methods in enzymology* (ed. D. V. Goeddel). Vol. 185, p. 60. Academic Press, London.
23. Miroux, B. and Walker, J. E. (1996). *J. Mol. Biol.* **260**, 289.
24. Ejdebäck, M., Young, S., Samuelsson, A. and Karlsson, B. G. (1997). *Prot. Express. Purif.* **11**, 17.
25. Doherty, A. J., Ashford, S. R., Brannigan, J. A., and Wigley, D. B. (1995). *Nucl. Acids Res.* **23**, 2074.
26. Pavletich, N. P. and Pabo, C. O. (1991). *Science* **252**, 809.
27. Lin, K., Kurland, I., Xu, L. Z., Lange, A. J., Pilkis, J., El-Maghrabi, M. R., and Pilkis, S. J. (1990). *Prot. Express. Purif.* **1**, 169.
28. Hoffman, S. J., Looker, D. L., Roehrich, J. M., Cozart, P. E., Durfee, S. L., Tedesco, J. L., and Stetler, G. L. (1990). *Proc. Natl. Acad. Sci. USA* **87**, 8521.
29. Malnic, B. and Reinach, F. C. (1994). *Eur. J. Biochem.* **222**, 49.
30. Orriss, G. L., Runswick, M. J., Collinson, I. R., Miroux, B., Fearnley, I. M., Skehel, J. M., and Walker, J. E. (1996). *Biochem. J.* **314**, 695.
31. Graziano, V., Gerchman, S. E., Wonacott, A. J., Sweet, R. M., Wells, J. R. E., White, S. W., and Ramakrishnan, V. (1990). *J. Mol. Biol.* **212**, 253.
32. Gerchman, S. E., Graziano, V., and Ramakrishnan, V. (1994). *Prot. Express. Purif.* **5**, 242.
33. Schoner, B. E., Belagaje, R. M. and Schoner, R. G. (1990). In *Methods in enzymology* (ed. D. V. Goeddel), Vol. 185, p. 94. Academic Press, London.
34. Bianchi, M. E. (1991). *Gene* **104**, 271.
35. Murby, M., Uhlén, M., and Ståhl, S. (1996). *Prot. Express. Purif.* **7**, 129.
36. Cohen, S. L., Ferré-D'Amaré, A. R., Burley, S. K. and Chait, B. T. (1995). *Prot. Sci.* **4**, 1088.
37. Hockney, R. C. (1994). *Trends Biotechnol.* **12**, 456.
38. Li, C., Schwabe, J. W. R., Banayo, E., and Evans, R. M. (1997). *Proc. Natl. Acad. Sci. USA* **94**, 2278.
39. Bujard, H., Gentz, R., Lanzer, M., Stueber, D., Mueller, M., Ibrahimi, I., Haeuptle, M.-T., and Dobberstein, B. (1987). In *Methods in enzymology* (ed. R. Wu), Vol. 155, p. 416. Academic Press, London.
40. Certa, U., Bannwarth, W., Stüber, D., Gentz, R., Lanzer, M., Le Grice, S., Guillot, F., Wendler, I., Hunsmann, G., Bujard, H., and Mous, J. (1986). *EMBO J.* **5**, 3051.
41. Smith, D. B. and Johnson, K. S. (1988). *Gene* **67**, 31.
42. Ferré-D'Amaré, A. R., Prendergast, G. C., Ziff, E. B., and Burley, S. K. (1993). *Nature* **363**, 38.
43. Summers, M. D. and Smith, G. E. (1987). A manual of methods for baculovirus vectors and insect cell culture procedures. Texas A & M University, College Station, Texas.
44. O'Reilly, D. R., Miller, L. K., and Luckow, V. A. (1992). *Baculovirus expression vectors: a laboratory manual.* W. H. Freeman & Co, New York.
45. King, L. A. and Possee, R. D. (1992). *The baculovirus expression system: a laboratory guide.* Chapman & Hall, London.
46. Ayres, M. D., Howard, S. C., Kuzio, J., Lopez-Ferber, M. and Possee, R. D. (1994) *Virology* **202**, 586.
47. Kozak, M. (1991) *J. Biol. Chem.* **266**, 19867.

Chapter 2
Gel electrophoresis and bending in cisplatin-modified linear DNA

Jean-Marc Malinge, Annie Schwartz and Marc Leng
Centre de Biophysique Moleculaire, CNRS, rue Charles Sadron, F–45071 Orléans Cedex 2, France

1 Introduction

Recent studies have shown that some natural DNA fragments are bent and that the bend can be deduced from the DNA electrophoretic mobility in polyacrylamide gels (1–4). Structural distortions are induced in DNA by the covalent binding of compounds such as mutagens, carcinogens or antitumour drugs (5). The purpose of this chapter is to show the relations between electrophoretic mobility and some aspects of the structural distortions induced in linear DNA by the binding of the antitumour drug cisplatin (*cis*-diamminedichloroplatinum(II)). This chapter comprises of four parts which deal, respectively, with: (i) gel electrophoresis in polyacrylamide gels; (ii) platination of oligonucleotides; (iii) ligation of the platinated oligonucleotides into multimers and their separation by gel electrophoresis; (iv) quantitative analysis of the results.

2 Gel electrophoresis

2.1 Background on gel electrophoresis

Most DNA molecules are isotropically flexible which means that the axis of the double helix has no preferred direction of deviation from linearity. However, it is now well-proved that some DNA molecules are bent, due to either a sequence-directed effect or a ligand-induced effect (1–4, 6, 7). The discovery of the anomalous electrophoretic mobility of a restriction fragment of *Leishmania tarantolae* kinetoplast DNA in polyacrylamide gel under non-denaturing conditions (this restriction fragment migrates almost normally in agarose gel) has revealed the great utility of polyacrylamide gels in the study of bent molecules (8). Although a quantitative theory explaining gel migration anomaly is still lacking, qualitative agreement with experiments has been found by several authors. Lumpkin and Zimm (9) derived a relationship between the rate of migration and the end-to-end distance of the molecule

$$\mu = QE < h_x/L > 1/\zeta$$

where Q is the effective charge on the DNA molecule, E a constant electric field applied to DNA in the gel, μ and ζ are respectively the mean velocity and the frictional coefficient for translation of DNA along a 'tube' in the gel (the tube is determined by constraints that the gel fibres impose on the longitudinal motion of DNA), L is the DNA contour length and h_x is the projection of the end-to-end vector of the DNA molecule on the field direction. The angled brackets represent the average over an ensemble of conformations. This formula predicts that a non-linear DNA molecule migrates slower than a linear molecule because of its smaller end-to-end distance. Moreover a bend located in the middle of the DNA fragment slows down the electrophoretic migration more than a bend near the extremity. This has been exploited to identify the locus and nature of bends in natural DNA fragments (circular permutation test).

In a DNA molecule containing two or more identical bends, the end-to-end distance of the molecule depends on the position of the bends relative to the helix repeat of DNA. When the bends are repeated in phase with the helix repeat, the DNA axis is always curved in the same direction with a planar curvature of the helix. A DNA molecule with bends repeated in phase has an end-to-end distance smaller than that of the same molecule with bends not in phase (the overall curvature being no longer planar) and thus has a slower electrophoretic mobility.

There are several approaches to relate electrophoretic mobility and structural distortions induced in DNA by adducts (2). A convenient one is to study DNA molecules containing adducts regularly spaced as a function of the size of the spacers (10). Such constructions (multimers) are obtained by ligation of oligonucleotides containing a single adduct. The comparison of the electrophoretic migrations of the adducted and unadducted multimers allows one to determine the adducted-DNA helical repeat. In addition, the bend magnitude is deduced from comparison with the migration of well-defined bent DNA (in general multimers of 10-mer oligonucleotides containing several A residues in a row). In order to know the bend direction, other constructions are needed such as multimers of oligonucleotides containing a known bend and the bend under study.

2.2 Experimental procedure

Vertical electrophoresis system consists of tanks, glass plates (20 × 40 cm), spacers (0.4 mm thick) and a commercially available power pack. A 1× TBE 8% (w/v) acrylamide solution (29:1 acrylamide/*N*,*N*′-methylenebisacrylamide, 89 mM Tris-borate, 2 mM EDTA) plus ammonium persulfate and *N*,*N*,*N*′,*N*′-tetramethylenediamine is poured between the two glass plates and then a comb is inserted to create wells for samples loading, as described in several publications (1–3). During the electrophoresis it is advisable to thermostatically control the gel. However, in many cases, to running the gels in a room kept at constant temperature is adequate, provided that the voltage is low.

3 Chemistry of cisplatin and transplatin

3.1 Reactivity of cisplatin and transplatin with DNA

Cisplatin is among the most widely used human anticancer chemotherapeutic agents. It is generally accepted that the cytotoxic effects of cisplatin result from the formation of DNA bifunctional adducts in which the platinum residues are covalently bound to the base residues. Cisplatin forms several types of lesions in DNA including mainly intrastrand and interstrand crosslinks (11–13). *In vivo* and *in vitro* the major adducts are 1,2-intrastrand cross-links at d(GpG) and d(ApG) sites, *cis*-{$Pt(NH_3)_2$[d(GpG)-*N7*(1),*N7*(2)]} and *cis*-{$Pt(NH_3)_2$[d(ApG)-*N7*(1),*N7*(2)]} respectively. The interstrand cross-links are minor adducts preferentially formed at d(GpC).d(GpC) sites between the two guanine residues on opposite strands (*cis*-{$Pt(NH_3)_2$[d(GpC).d(GpC)-*N7*(G),*N7*(G)]}.

In several studies, properties of cisplatin are compared with those of transplatin (*trans*-diamminedichloroplatinum(II)), the stereoisomer of cisplatin. Transplatin reacts with DNA but is clinically inactive. Numerous results (11–13) support that formation of the adducts in the reaction between DNA and cisplatin or transplatin proceeds in a two-solvent assisted reaction as summarized in *Figure 1*. The first attack on DNA by the drugs occurs preferentially on the N7 of guanine residues and monofunctional adducts are formed. The monofunctional adducts can further react with the flanking base residues. The closure of the transplatin monofunctional adducts into bifunctional adducts is slow ($t_{1/2}$, the half-life of the adducts, is larger than 20 h). These bifunctional adducts are mainly interstrand crosslinks between the platinated guanine residues and their

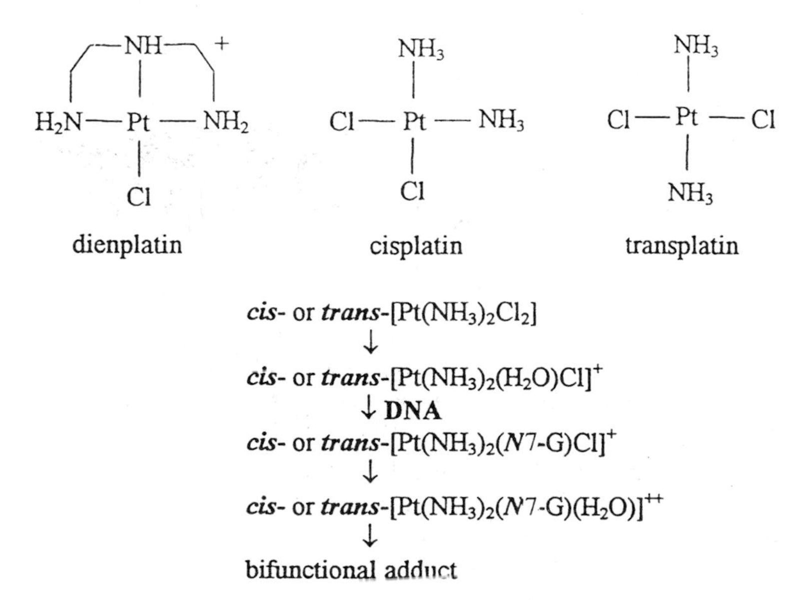

Figure 1. The formulae of cisplatin, transplatin (*cis*- or *trans*-diamminedichloroplatinum(II)) and dienplatin (diethylenetriammineplatinum(II)) are shown at the top. Below the formulae are the main steps of the reaction between DNA and cisplatin or transplatin.

complementary cytosine residues. In the same conditions, the closure of the cisplatin monofunctional adducts is fast ($t_{1/2}$ in the range of a few hours) and several kinds of crosslinks are formed. Comparative studies are also quite often done with dienplatin (diethylenetriammineplatinum(II)) which forms only monofunctional adducts (11–13).

3.2 Platination of the oligonucleotides

The strategy for preparing a double-stranded oligonucleotide containing a single adduct at a given position is to first react the platinum(II) complex with the single-stranded oligonucleotide and then to hybridize the platinated strand with the complementary strand. A difficulty in the reaction between platinum(II) complexes and single-stranded oligonucleotides is that the complexes react with the four base residues. The order of their affinity for ribonucleotides is GMP > AMP > CMP >> UMP (14). The reaction with C-residues, and also to some extent with A residues, can be prevented by working at low pH (below the p*K*s of C and A residues, 4.2 and 3.7 respectively). At a pH 3–3.6, the platination occurs almost exclusively on the N7 of G residues within single-stranded oligonucleotides containing A, C, T and G residues. There is no simple way to platinate A and C residues without platination of G residues (one can replace the G residues by 7-deazaG residues but not by I residues, which are as reactive as G residues). Thus the sequences of the single-stranded oligonucleotides are designed as a way of containing a single site to be platinated flanked by pyrimidine-rich residues.

Platination of purified oligonucleotides (a method of purification is given in *Protocol 1*, step 4) is usually done in $NaClO_4$ solution. NaCl (at least high concentrations of NaCl) should not be used since it has been proved that in the reaction with DNA, hydrolysis of chloride ligands of platinum(II) complexes is the rate-limiting step. The adducts decrease the thermal stability of the duplexes and thus it is preferable to work with 19–24-mers. This also avoids adducts interfering with the action of the ligase during the synthesis of the multimers.

Protocol 1

Formation of intrastrand crosslinks

Equipment and reagents

- Platinum (II) complex (Johnson Matthey)
- Dimethylformamide
- TE buffer (10 mM Tris-HCl pH 7.5, 1 mM EDTA)
- 1.5 ml Eppendorf tubes
- Mono Q 5 × 50 mm column (Pharmacia)
- C18 cartridge (Millipore)

Method

1 Prepare a 3×10^{-2} mM solution (at least 10 nmol) of the oligodeoxynucleotide containing the target site in 10 mM $NaClO_4$ and 5 mM acetate buffer pH 3.5.

Protocol 1 continued

2 Weigh about 0.2 mg of platinum (II) complex and dissolve the powder in dimethylformamide at a final concentration of 1 mM in an 1.5 ml Eppendorf tube.

3 Add to the oligonucleotide solution one equivalent of the platinum(II) derivative (the platinum solution can be diluted in water if necessary) and incubate the mixture for 15 h, at 37 °C, in the dark.

4 Purify the major product which should correspond to the platinated oligonucleotide by strong anion exchange HPLC (Mono Q 5 mm × 50 mm column, Pharmacia) on a HPLC system with a 30-min linear gradient of 0.2–0.8 M NaCl, 10 mM NaOH pH 12 and a flow rate of 1 ml/min. The platinated products are eluted a few minutes earlier than the starting unplatinated oligonucleotides.

5 Neutralize the platinated oligonucleotide solution by adding glacial acetic acid and Tris-HCl pH 7.5 to a final concentration of 10 mM and 100 mM, respectively (check that the pH is about 8) and desalt the platinated oligonucleotide solution by passage through a C18 cartridge (Millipore).

6 After lyophilization, resuspend the pellet in TE buffer and determine the oligonucleotide strand concentration by UV absorption from A_{260} values and theoretically estimated extinction coefficient.

Protocol 2

Formation of interstrand crosslinks

Equipment and reagents

- Platinum(II) derivatives (3 mM)
- Dimethylformamide
- 5 mM acetate buffer pH 3.5
- Mono Q 5 × 50 mm column (Pharmacia)

Methods

1 React the platinum(II) derivatives (3 mM) with one equivalent of $AgNO_3$ in dimethylformamide during 15 h at room temperature in the dark to form the reactive *cis* or *trans*-$[Pt(NH_3)_2(H_2O)Cl]^+$. Then remove AgCl precipitate by 10 min centrifugation at 8000 g with a bench-top centrifuge.

2 Mix the single-stranded oligonucleotide (at least 30 nmoles; final concentration 1 mM) with the reactive species *cis*- or *trans*-$[Pt(NH_3)_2(H_2O)Cl]^+$ at a molar input platinum/oligonucleotide ratio of 5 in 10 mM $NaClO_4$, 5 mM acetate buffer pH 3.5 and incubate for 20 min at 37 °C in the dark. Stop the reaction by adding NaCl to a final concentration of 0.2 M and put the sample on ice.

3 Purify the major product, which should correspond to the monofunctionally cisplatin or transplatin modified oligonucleotide by strong anion exchange HPLC (Mono Q 5 mm × 50 mm column, Pharmacia) on an HPLC system with a 30-min linear gradient of 0.2 M–0.8 M NaCl, Tris-HCl pH 7.5 and a flow rate of 1 ml/min.

Protocol 2 continued

4 Anneal the monofunctionally platinated oligonucleotide with the complementary oligonucleotide (minimal concentration in duplex 10^{-2} mM) in 0.2 M NaCl, 10 mM Tris-HCl, pH 7.5 at 4 °C for 15 h in the dark.

5 Dialyse the monofunctionally platinated duplex against 0.1 M $NaClO_4$, 10 mM Tris-HCl, pH 7.5 at 4 °C in the dark.

6 Incubate the platinated duplex at 37 °C for 20 h in the dark.

7 Purify the crosslinked duplex by strong anion exchange HPLC (Mono Q 5 mm × 50 mm column, Pharmacia) on an HPLC system with a 30-min linear gradient of 0.2–0.8 M NaCl, 10 mM NaOH pH 12 and a flow rate of 1 ml/min. The cross-linked duplex is eluted a few minutes later than the starting platinated and unplatinated single-stranded oligonucleotides. Desalt and determine the interstrand cross-linked duplex concentration as above (see *Protocol 1*, steps 5 and 6).

3.3 Analysis of the adducts

After purification of the platinated oligonucleotides it is necessary to check the nature of the adducts. First, atomic absorption spectroscopy allows one to verify that platinated oligonucleotides have one platinum atom per strand. Second, the nature of the adducts is confirmed by Maxam–Gilbert specific reactions (15) and by endonuclease P1 (16) and alkaline phosphatase digestions followed by C18 reverse-phase HPLC with coinjection of standard platinated deoxyribonucleosides (17).

Once formed, the adducts can be considered as stable even in the presence of NaCl under physiological conditions. There are a few exceptions, but they might be of importance (18–20). One concerns the transplatin 1,3-intrastrand cross-links (19). These adducts, which are easily formed in the reaction between transplatin and single-stranded oligonucleotides containing the triplets d(GNG) (N being a nucleotide residue) are stable as long as the platinated oligonucleotides are single-stranded. They rearrange into interstrand crosslinks when the platinated oligonucleotides are paired with their complementary strands. The bound platinum within single- and double-stranded DNAs can be removed by some reagents that have a high affinity for the 'soft' platinum(II) centre, such as cyanide ions and sulfur-containing nucleophiles.

3.4 Synthesis of the multimers

The sequences of the oligonucleotides are chosen to provide double-stranded oligonucleotides with non-self-complementary ends (one or two residues) which leads to ligation with a unique polarity (20–26). An example, discussed in Section 4, is the duplex with a unique d(GC).d(GC) site for formation of a cisplatin interstrand crosslink (26)

CTTCTCCTT**GC**TCTCCTTCTCTC
AAGAGGAA**CG**AGAGGAAGAGAGG

Protocol 3
Ligation of the oligonucleotides

Equipment and reagents

- [γ-^{32}P]ATP (New England Biolabs)
- T4 DNA ligase (New England Biolabs)
- T4 polynucleotide kinase (New England Biolabs)

Methods

1. Label 0.3–0.6 nmol of the single-stranded oligonucleotides (or the interstrand cross-linked duplexes) by incubating the oligonucleotides with 1 μl of [γ-^{32}P]ATP (specific activity 110 TBq/mmol) and 1 μl of T4 polynucleotide kinase (10 000 U/ml, New England Biolabs) in 70 mM Tris-HCl pH 7.6, 10 mM $MgCl_2$, 5 mM DTT within a final volume of 10 μl at 37 °C, for 30 min in the dark.
2. Add to the mixture 1 μl of ATP 20 mM, 1 μl of kinase and incubate further at 37 °C for 30 min.
3. Mix equimolar amounts of 5′-phosphorylated complementary strands, heat the sample mixture at 65 °C for 2 min and cool slowly to 15 °C in a water bath over a period of 4 h for annealing.
4. Add to the solution T4 DNA ligase (400 U, New England Biolabs) in 10 mM $MgCl_2$, 10 mM DTT, 1 mM ATP, 50 mM Tris-HCl pH 7.5 and incubate at 15 °C for 2 h within a total volume of 20 μl.
5. Prerun the gel at 15 V/cm until the temperature in the gel and the electrical current are constant.
6. Mix the DNA samples with gel loading buffer (bromophenol blue, xylene cyanol and 7% (w/v) sucrose in water) and load them (2–3 μl) onto a 8% (w/v) polyacrylamide gel with a microcapillary flat tip. Load also ^{32}P end-labelled marker DNA fragments (pBR322 DNA cleaved by endonuclease *Hpa*II).
7. Run the gel at 15 V/cm until the bromophenol blue dye marker reaches the bottom of the gel.
8. Determine the migration distances of the multimers and of the DNA molecular weight marker on the gel either from autoradiography of the gel exposed to X-Ray film (R type, 3 M) with the use of an intensifying screen at −80 °C or by phosphor-imager analysis.

4 Analysis of the results

As an example, some results (26) relative to the multimers of oligonucleotides (21–24 bp) containing a single cisplatin interstrand crosslink are shown in *Figure 2*. The autoradiogram of a 8% polyacrylamide gel of the ligation products of the platinated and unplatinated oligonucleotides (21–24 bp) is shown (*Figure 2A*). It has been verified that the unplatinated multimers have normal electrophoretic

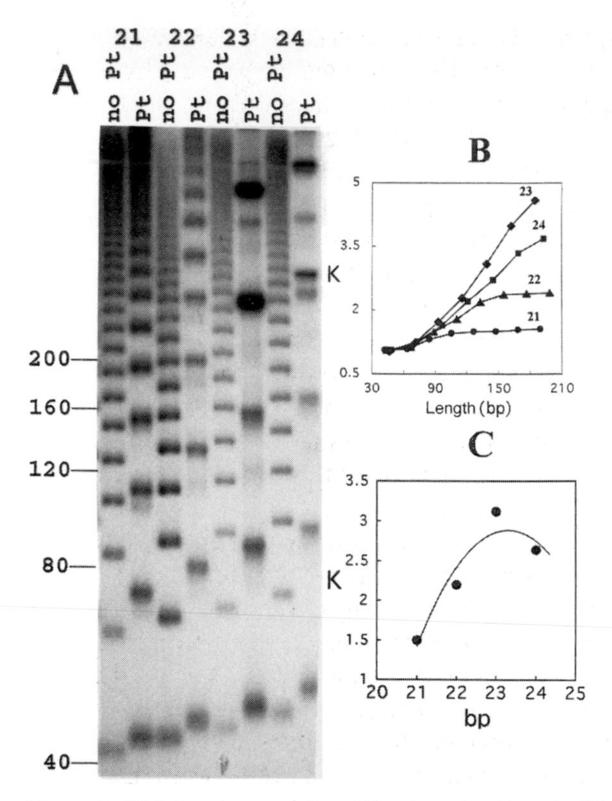

Figure 2. (A) Autoradiogram of an 8% polyacrylamide gel of the multimers of the 21–24 bp duplexes containing or not containing a single cisplatin interstrand crosslink. Lanes: no Pt, unplatinated multimers; Pt, platinated multimers. (B) Variation of the K factor versus sequence length; the curves for the multimers of the 21–24 bp duplexes are denoted 21, 22, 23 and 24 respectively. (C) Variation of the factor K versus interplatinum distance (bp) for the platinated multimers with a total length of 140 bp (from (26) with permission).

mobility compared with DNA weight markers (not shown). The platinated multimers migrate more slowly than the corresponding unplatinated multimers. The results are usually analysed by plotting the K factor (apparent chain length to sequence length; the apparent length of each multimer is defined as the length of the DNA marker having the same electrophoretic mobility) versus sequence length expressed in bp (*Figure 2B*). The value of the bend angle is calculated from the empirical relation given by the following equation

$$K-1 = [(9.6 \times 10^{-5}L^2) - 0.47](RC)^2 \qquad 2$$

where L represents the length of a particular multimer with the relative mobility K and RC the curvature relative to a DNA bending induced at the A tract of six A residues. Application of equation (2) to the 115 bp and 138 bp multimers of the ligated duplexes (23 bp) leads to a curvature of 1.25. The average bend angle per helix repeat calculated by multiplying the relative curvature by the absolute value of the A_6 tract bend (18° ± 3°; 24) is about 45°.

It should be noted that this equation is valid for molecules in size range 120–170 bp which migrate in 8% polyacrylamide gels, 29:1 acrylamide/*N,N'*-methylenebisacrylamide. Also it applies to the multimers in which the bends are in phase or almost in phase with the helical repeat (27, 28).

The intense bands observed close to the top of the gel (*Figure 2A*) with the ligation products of the duplexes (23 bp) correspond to DNA circles. These circular products are formed as a consequence of the bending that allows the ends of platinated DNA fragments of definite length to close covalently by ligation. These results and the maximum retardation observed for the linear multimers of the duplexes (23 bp) suggest that the natural 10.5 bp repeat of B-DNA and that of DNA fragments modified by the interstrand crosslinks are different as a consequence of DNA unwinding. The exact helical repeat of the platinated DNA determined by fitting the experimental values of the K-factor versus the distance between the interstrand crosslinks with the curve of the form $ax^2 + bx + c$ (23, 27, 28) is 23.3 ± 0.04 bp (*Figure 2C*). The difference between the helical repeat of B-DNA and the DNA containing the interstrand cross-links is $[(23.3 \pm 0.04) - 2(10.5 \pm 0.05)] = 2.3 \pm 0.09$ bp. Given that there are 360° per 10.5 bp, the DNA unwinding due to one interstrand crosslink is $79 \pm 4°$.

The direction of the bend has been determined by constructing a series of multimers of oligonucleotides containing a single interstrand crosslink and oligonucleotides containing a A_5 tract, these tracts being located either in phase or out of phase with the interstrand crosslinks (24). The multimers in which the A_5 tracts are in phase with the crosslinks migrate slower than those in which the A_5 tracts and the interstrand cross-links are out of phase. The conclusion is that the interstrand crosslinks as the A tracts bend the axis of the DNA double helix towards the minor groove (24).

Similar experiments have been done on multimers containing other platinum(II) adducts and the results are given in *Table 1*. Interestingly, in the case of the major intrastrand crosslink and of the interstrand cross-link of cisplatin, the bending and the unwinding determined by gel electrophoresis are in good agreement with the respective values deduced by nuclear magnetic resonance (18, 29, 30) and X-ray crystallographic analysis (31, 32).

Table 1 Extent of DNA bending and unwinding induced by platinum(II) adducts

	Sites	Bending	Unwinding	References
Cisplatin	d(G*pG*/CpC)	32–34°	13°	21-23
	d(A*pG*/CpT)	34–36°	13°	22, 23
	d(G*TG*/CAC)	30–35°	23°	22, 23
	d(G*pC/G*pC)	45°	79°	24, 26
Transplatin	d(G*TG*/CAC)	26°	45°	25, 33
	d(G*/C*)	26°	12°	34
Dienplatin	d(G*/C)	0°	6°	35, 36

References

1. Diekmann, S. (1992). In *Methods Enzymol.* **212**, 30.
2. Crothers, D. M. and Drak, J. (1992). In *Methods Enzymol.* **212**, 46.
3. Hagerman, P. J. (1990). *Annu. Rev. Biochem.* **59**, 755.
4. Trifonov, E. N. and Ulanovsky, L. E. (1988). In *Unusual DNA Structures* (ed. R. D. Wells ans S. C. Harvey), pp. 173–187. Springer-Verlag, New-York.
5. Leng, M. (1990). *Biophys. Chem.* **35**, 155.
6. Travers, A. A. (1989). *Annu. Rev. Biochem.* **58**, 427.
7. Travers, A. A. (1995). In *DNA–protein: structural interactions* (ed. D. M. J. Lilley), pp. 49–75. IRL Press, Oxford.
8. Marini, J. C., Levene, S. D., Crothers, D. M., and Englund, P. T. (1982). *Proc. Natl. Acad. Sci. USA*, **79**, 7664.
9. Lumpkin, O. J. and Zimm, B. H. (1982). *Biopolymers*, **21**, 2315.
10. Koo, H.-S., Wu, H.-M., and Crothers, D. M. (1986). *Nature* **320**, 501.
11. Lepre, C. A. and Lippard, S. J. (1990). In *Nucleic Acids and Molecular Biology* (ed. F. Eckstein and D. M. J. Lilley), Vol. 4, pp. 9–38. Springer-Verlag, Berlin.
12. Reedijk, J. (1992). *Inorg. Chem. Acta*, **198**, 873.
13. Sip, M. and Leng, M. (1993). In *Nucleic acids and molecular biology* (ed. F. Eckstein and D. M. J. Lilley), Vol. 7, pp. 1–14. Springer-Verlag, Berlin.
14. Mansy, S., Chu, G. Y. H., Duncan, R. E., and Tobias, R. S. (1978). *J. Am. Chem. Soc.* **100**, 607.
15. Maxam, A.M. and Gilbert, W. (1977). *Proc. Natl. Acad. Sci. USA* **74**, 560.
16. Fichtinger-Schepman, A. M. J., van der Veer, J.L., Lohman, P. H., and Reedijk, J. (1985) *Biochemistry* **24**, 707.
17. Eastman, A. (1986) *Biochemistry* **25**, 3912.
18. Yang, D., van Boom, S. S. G. E., Reedijk, J., van Boom, J. H., and Wang, A. H.-J. (1995). *Biochemistry* **34**, 12912.
19. Dalbiès, R., Payet, D., and Leng, M. (1994). *Proc. Natl. Acad. Sci USA*, **91**, 8147.
20. Pérez, C., Leng, M., and Malinge, J.-M. (1997). *Nucl. Acids Res.* **25**, 896.
21. Rice, J. A., Crothers, D. M., Pinto, A. L., and Lippard, S. J. (1988). *Proc. Natl. Acad. Sci. USA* **85**, 4158.
22. Bellon, S. F. and Lippard, S. J. (1990). *Biophys. Chem.* **35**, 179.
23. Bellon, S. F., Coleman, J. H., and Lippard, S. J. (1991). *Biochemistry* **20**, 8026.
24. Huang, H., Zhu, L., Reid, B. R., Drobny, G. P., and Hopkins, P. B. (1995). *Science* **270**, 1842.
25. Anin, M.-F. and Leng, M. (1990). *Nucl. Acids Res.* **18**, 4395.
26. Malinge, J.-M., Pérez, C., and Leng, M. (1994). *Nucl. Acids Res.* **22**, 3834.
27. Koo, H.-S., Drak, J., Rice, J. A., and Crothers, D. M. (1990). *Biochemistry* **29**, 4227.
28. Drak, J. and Crothers, D. M. (1991). *Proc. Natl. Acad. Sci USA* **88**, 3074.
29. Herman, F., Kozelka, J., Stoven, V., Guittet, E., Girault, J.-P., Huynh-Dinh, T., Igolen, J., Lallemand, J.-Y., and Chottard, J.-C. (1990). *Eur. J. Biochem.* **194**, 119.
30. Paquet, F., Pérez, C., Leng, M., Lancelot, G., and Malinge, J.-M. (1996). *J. Biomolec. Struct. Dynam.* **14**, 670.
31. Takahara, P. M., Rosenzweig, A. C., Frederick, C. A., and Lippard, S. J. (1995). *Nature*, **377**, 649.
32. Coste, F., Malinge, J.-M., Serre, L., Shepard, W., Roth, M., Leng, M., and Zelwer, C. (1999). *Nucl. Acids Res.* **27**, 1837.
33. Boudvillain, M., Dalbiès, R., Aussourd, C., and Leng, M. (1995). *Nucl. Acids Res.* **23**, 2381.
34. Brabec, V., Sip, M., and Leng, M. (1993). *Biochemistry* **32**, 11676.

35. Marrot, L. and Leng, M. (1989). *Biochemistry* **28**, 1454.
36. Keck, M. V. and Lippard, S. J. (1992). *J. Am. Chem. Soc.* **114**, 3386.

Chapter 3
Use of DNA microcircles in protein–DNA binding studies

Dominique Payet
Centre d'Immunologie de Marseille Luminy, Parc Scientifique et Technologique de Luminy, Case 906, F–13288 Marseille Cedex 9, France and MRC Laboratory of Molecular Biology, Hills Road, Cambridge CB2 2QH, UK

1 Introduction

In many protein-mediated DNA transactions such as site-specific recombination and transcription initiation, the DNA is tightly bent or otherwise distorted in order to bring DNA sequences that are separated by many double-helical turns into close spatial proximity. For example, in the core nucleosome 145 bp of DNA are wrapped around the histone octamer so that the entering and exiting duplexes are close to each other (1). Similarly DNA looping can be induced by proteins bound at separate distant sites. One example is the tetrameric *lac* repressor that interacts simultaneously with two sites and can form a tightly bent repression loop of 90 bp (2).

Below the DNA persistence length of around 150 bp, naked DNA is not sufficiently flexible to circularize. However certain relatively non-specific DNA-binding proteins such as the HMG-domain proteins, exemplified by the vertebrate HMG-1 and the *Drosophila* HMG-D proteins and the bacterial IHF protein bend DNA by $\approx 90°$ and $\approx 160°$, respectively (3, 4). One suggested function of these architectural proteins is to facilitate the formation of the nucleoprotein complexes that are required for the enzymatic manipulation of DNA. This chapter describes methods to obtain DNA microcircles, which can be regarded as the equivalent of the natural highly bent DNA microloops. These DNA microcircles can be further used to study the interaction of tightly bent DNA with proteins.

2 DNA circularization

The ligase-mediated circularization assay is usually used to demonstrate the DNA-bending properties of proteins. In this assay, bending is determined by the circularization of a DNA fragment shorter than the persistence length (around 150 bp), which cannot circularize in absence of DNA-bending protein. In the basic protocol, a DNA fragment is incubated with the DNA-bending protein and then with T4 DNA ligase. This approach is based on the principle that a bend inducer brings the ends of the DNA fragment closer together and thereby facilitates cyclization by DNA ligase. The rate of DNA circularization depends on the concentration and the bending properties of the DNA-bending protein. It is also important to distinguish between circular and linear DNA products of this reaction. This is achieved by incubation with exonuclease III which digests linear DNA but not closed circular DNA. Therefore, the final treatment with exonuclease III differentiates between closed circular and linear DNA and is a crucial step in the purification of DNA circles.

We developed an alternative method that uses short oligonucleotides (> 10 bp). This approach can quickly assess the DNA-bending properties of a molecule and also produces small circles of different sizes. This method was used to determine the bending angle induced by bifunctional adducts formed by the antitumoral drug *cis*-dichlorodiammineplatinum (II) (5) and can be applied to study the DNA bending properties of non-covalently DNA-bound molecules such as drugs (6) or proteins (7).

2.1 Oligonucleotide purification

Protocol 1 describes the gel purification of oligonucleotides. This protocol can be slightly modified depending on the DNA used. The percentage of the polyacrylamide gel decreases with the length of the DNA fragment ranging from 20% for short DNA oligonucleotides (< 25 bases) to 8% for around 100 bases long DNA molecules. If the DNA fragment is double-stranded the gel should be run in native condition and the power should be lowered to 8 W. *Protocol 1* is suitable for the purification of at least 0. 1 mg of oligonucleotides. For smaller quantities, for example, DNA fragment excised from plasmid, it is more convenient to radiolabel the DNA fragment and then to gel-purify the DNA as described in *Protocol 2* steps 8–11.

Protocol 1

Purification of oligonucleotides

Equipment and reagents

- 38% acrylamide/2% bis-acrylamide stock solution
- Glass plates (20 × 20 cm), spacer thickness 1.5 mm
- Vertical electrophoresis system and a power pack
- UV spectrophotometer

Methods

1. Prepare a 20% polyacrylamide gel and run for 30 min with the power set at 20 W.
2. Mix the oligonucleotide (1–2 mg/gel) with 1 volume of denaturing loading buffer. Heat the solution at 80 °C for 1 min and chill in ice.
3. Electrophorese the oligonucleotides at 20 W (in these conditions the bromophenol blue migrates at approximately the same position as a decamer).

Protocol 1 continued

4. At the end of the electrophoresis, remove both glass plates and wrap the gel in Clingfilm. Visualize the oligonucleotides by fluorescence: place the gel on a screen such as Dupont Cronex lightning-plus BF and illuminate at 254 nm. Excise the band corresponding to the full-length oligonucleotides.
5. Incubate the gel slices overnight at 37 °C in 0. 3 M sodium acetate.
6. Pump the solution with a syringe and filter using a 0.45 μm syringe filter in order to remove the remaining pieces of polyacrylamide gel.
7. Desalt the oligonucleotide either by ethanol precipitation or by using C18 Sep-Pack cartridges (Waters).
8. Resupend the oligonucleotide in water and determine the DNA concentration by measuring the absorbance at 260 nm.

Protocol 2

Synthesis of circles with short oligonucleotides

Equipment and reagents

- The enzymes T4 polynucleotide kinase (10 units/μl), T4 DNA ligase (400 units/μl), exonuclease III (100 units/μl) and their respective buffers are from New England Biolabs
- [γ-^{32}P]ATP solution at 10 mCi/ml, specific activity 3000 Ci/mmol (Amersham)
- 40% acrylamide/bis-acrylamide stock solution
- loading buffer (50% glycerol, 1 mM EDTA, 0.1% bromophenol blue and 0.1% xylene cyanol)
- Glass plates (20 × 38 cm), spacer thickness 0. 4 mm
- Vertical electrophoresis system and a power pack
- TE (1 mM EDTA, 10 mM Tris-HCl pH 8.0)

Method

1. Phosphorylate each strand of the duplex separately in 10 μl of T4 DNA kinase buffer (70 mM Tris-HCl pH 7.6, 10 mM $MgCl_2$, 5 mM dithiothreitol) supplemented with 1 μl of [γ-^{32}P]ATP and 1 μl of T4 DNA kinase. Incubate at 37 °C for 30 min.
2. Anneal the two complementary strands by mixing the solutions for 20 min at room temperature.
3. Incubate the DNA solution with the bending molecule in 50 μl of T4 DNA ligase buffer (50 mM Tris-HCl pH 7.5, 10 mM $MgCl_2$, 10 mM dithiothreitol, 1 mM ATP, 25 μg/ml BSA) at room temperature for 15 min. Then add 1 μl of T4 DNA ligase and incubate at room temperature for at least 1 h.
4. To stop the reaction add SDS (1% w/v final concentration) and sodium acetate (c_{final} = 0.3 M).

Protocol 2 continued

5 Deproteinize the solutions by adding an equal volume of phenol–chloroform (1:1) equilibrated in TE. Vortex and centrifuge for 3 min. in a bench-top centrifuge. Keep the aqueous phase (upper phase). To remove traces of phenol, add two volumes of ether saturated with H_2O, vortex, centrifuge for 2 min., discard the upper (ether) phase.

6 To precipitate DNA, add two volumes of ethanol, chill at −70 °C for 15 min, centrifuge at 12 000 g for 15 min, remove the supernatant. To the pellet add 70% ethanol, centrifuge at 12 000 g for 5 min, remove the supernatant, and dry the pellet, for example, in a DNA Speed Vac (Savant Instruments).

7 Resuspend the pellet in 10 μl of exonuclease III buffer (66 mM Tris-HCl pH 8.0, 0.66 mM $MgCl_2$) add 1 μl of exonuclease III, incubate at 37 °C for 1 h. Remove the enzyme by phenol extraction as in step 5 and precipitate DNA as in step 6, resuspend the pellet in 4 μl of TE, add 1 μl of loading buffer.

8 Load the sample onto a native 8% polyacrylamide gel and carry out the electrophoresis for ≈ 4 h at 15 V/cm in 90 mM Tris/borate, 1 mM EDTA pH 8.4 (1× TBE), in these conditions the bromophenol blue migrates as a 45 bp fragment.

9 Autoradiograph the wet gel and excise from the gel the bands of interest.

10 Elute the DNA circles by incubating the gel pieces in 0.3 M sodium acetate overnight at 37 °C.

11 Precipitate the DNA as in step 6. Resuspend the circles in 50 μl of TE.

2.2 Oligonucleotide design

In order to facilitate ligation, oligonucleotides have cohesive ends consisting of one or two overhanging bases. The sequence of the oligonucleotides depends of the nature of bend-inducing protein, although for the less-specific DNA-bending proteins such as the bacterial histone-like protein HU and the chromosomal proteins of the HMG1 class this is not essential. The binding site is generally located in the centre of the duplex and different length of the duplex can be tested in order to establish the optimum DNA length for circularization. Typically, the helical repeat of the DNA double helix is 10.5 bp per turn and, therefore, by adjusting the length of the duplex the binding sites can be orientated in or out of phase. The in-phase orientation favours the planar curvature of DNA and therefore its circularization.

The ligase has two functions: in a first step, it mediates the polymerization of the short duplex and when the DNA polymer is sufficiently bent, it circularizes the DNA. To minimize the synthesis of long linear ligation products, which can result in a shift of the circular products towards larger sizes, the DNA concentration should not be too high. High DNA concentration is correlated with high concentration of DNA ends and therefore the intermolecular ligation is favoured with respect to intramolecular ligation. In this experiment, DNA concentration is kept in (or below) the micromolar range.

Protocol 3

Synthesis of DNA circles using a DNA fragment

Equipment and reagents

- See *Protocol 2*

Method

1. Dephosphorylate the DNA fragment. Incubate the DNA in 50 μl of dephosphorylation buffer with 1 unit of alkaline phosphatase (Boehringer Mannheim) for 30 min at 37 °C.
2. Stop the reaction by adding 50 μl of 1:1 phenol–chloroform equilibrated in TE. Vortex, centrifuge for 3 min. Keep the aqueous phase (upper phase) and repeat twice the phenol extraction in order to remove the phosphatase.
3. Add two volumes of ether saturated with H_2O, vortex, centrifuge for 2 min, discard the ether (upper) phase.
4. Precipitate DNA as in *Protocol 2*, step 6.
5. Resuspend the pellet in 10 μl of T4 polynucleotide kinase buffer. Add 1 μl of T4 polynucleotide kinase and 1 μl of [γ-^{32}P]ATP. Incubate at 37 °C for 30 min.
6. Follow *Protocol 2* from steps 3 to 11.

3 Examples

3.1 DNA circularization by HMG-D

The HMG domain proteins form a very large class of proteins which contains the chromosomal high mobility group proteins such as HMG1 and HMG2, but also some transcription factors such as hUBF, SRY, LEF-1(8–14). These HMG domain proteins are typical architectural proteins. They contain one or more HMG domain which is a DNA-binding motif of around 75 residues (15). This motif binds DNA via the minor groove, prefers to bind pre-bent DNA and bends DNA (16–20).

The biological function of the HMG1/2 proteins family is not completely elucidated: recent results suggest that it might be a co-activator in different biological processes such as transcription or recombination (21–23). They could be considered as DNA chaperones which distort the DNA in a conformation facilitating the formation of a nucleoprotein complex (24, 25).

HMG-D is one of the *Drosophila* counterparts of the mammalian HMG1/2 proteins and contains just one HMG domain (26, 27). We have assessed the DNA-bending capabilities of HMG-D by ligase-mediated circularization assays as described in the *Protocol 2*. In the absence of HMG-D the ligation leads to the formation of a ladder of bands corresponding to the different size of the duplex multimers. In presence of HMG-D, the ladder is perturbed by the presence of intense bands which correspond to DNA circles as confirmed by the treatment

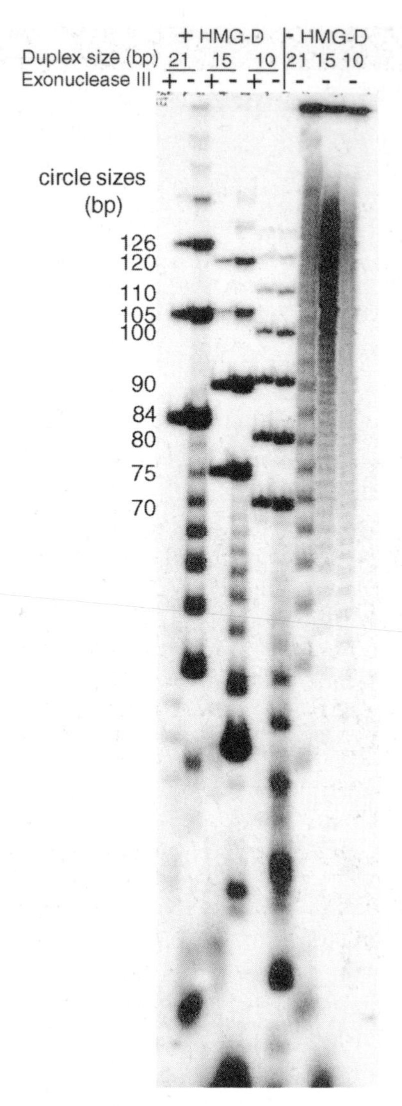

Figure 1 Multimerization–circularization assay. Three duplexes of 10, 15 and 21 bp – d(TAATATTGAA).d(TATTCAATAT), d(GCAAATATTGAAAAC).d(GCGTTTTCAATATTT) and d(GCGCAAAATATTGAAAACGCC).d(GCGGCGTTTTCAATATTTTGC), respectively – were ligated in presence or in absence of the bending molecule, HMG-D, as described in *Protocol 2*. The ligation product was then digested with exonuclease III (+); – indicates no exonuclease III. The size of the circles formed is indicated.

by Exonuclease III (*Figure 1*). The circles have been purified and used for the HMG-D–circle interaction.

3.2 HMG-D–circle interaction

The HMG-1 like proteins bind to distorted DNA structures such as four-way junctions, *cis*-platinated DNA, DNA bulges and negatively supercoiled DNA (7,

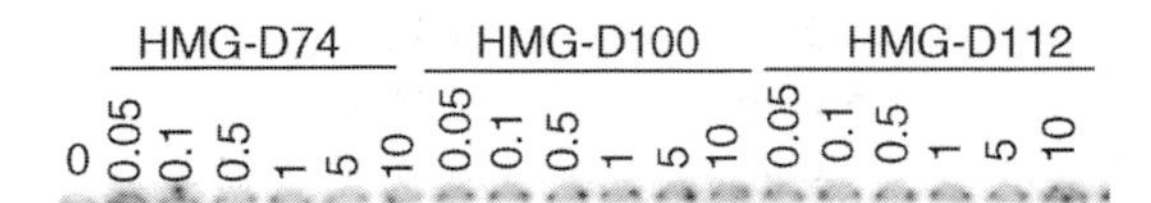

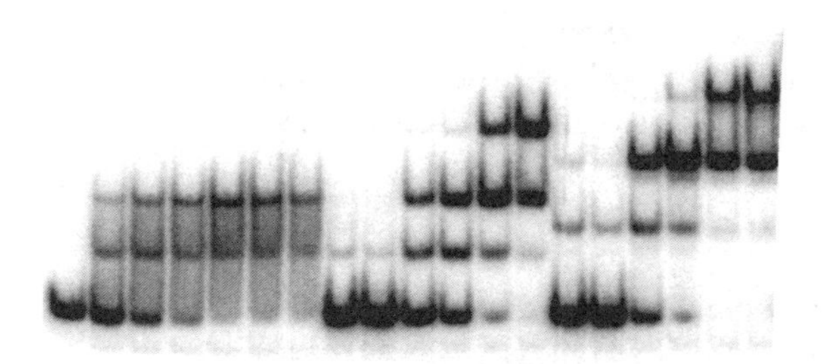

Figure 2 Electrophoretic mobility shift assay of HMG-D with a 75-bp circle. Radiolabelled 75-bp circles ($\approx 55.10^{-12}$ M) were incubated with increasing concentrations of HMG-D74, HMG-D100 and HMG-D112. The HMG-D concentration (in nM) is indicated at the top of each lane.

28–30). The HMG domain of HMG-D has low affinity for DNA, binds cooperatively to long DNA fragment and, in contrast with most of the HMG domain of other proteins, fails to bind to sharp DNA bends.

Protocol 4

Electrophoretic mobility shift assays (EMSA)

Equipment and reagents

- Binding buffer: 20 mM HEPES, pH7.9, 50 mM KCl, 2 mM $MgCl_2$, 1 mM DTT, 100 μg/ml BSA. The composition of the binding buffer may be adjusted according to the nature of the protein studied
- Gel dryer
- 30% acrylamide/bis-acrylamide stock solution
- Glass plates (20 × 20 cm), spacer thickness 0.7 mm

Method

1. Incubate for 15 min, in 10 μl of binding buffer, around 5000 c. p. m. of radiolabelled circles (synthesized and purified as described in *Protocol 2*) with different concentrations of proteins.
2. Add 1 μl of loading buffer and load on a native 8% polyacrylamide gel and electrophorese at 125 V in 0.5× TBE for 1–2 h, depending on the circle size.
3. Dry the gel under vacuum and autoradiograph.

A 75 bp circle is expected to have an average curvature of 50° per double helical turn and be untwisted. We have investigated by EMSA the binding of three recombinant forms of HMG-D, the full-length protein (HMG-D112), the protein lacking the acidic tail (HMG-D100) and the HMG domain of HMG-D (HMG-D74). The results shown in *Figure 2* indicate that HMG-D has a very high affinity for 75 bp circles ($K_D < 1$ nM). The HMG domain of HMG-D binds with at least three orders of magnitude tighter to the 75 bp circles than to other DNA structures. Moreover HMG-D75 and HMG-D100 bind the circle with approximately the same affinity, indicating that one function of the basic region is to facilitate DNA binding. This could be accomplished by the interaction of the basic residues with the phosphate groups on the inside of the bend (31, 32).

References

1. Luger, K., Mader, A. W., Richmond, R. K., Sargent, D. F., and Richmond, T. J. (1997). *Nature* **389,** 251.
2. Lewis, M. *et al.* (1996). *Science* **271,** 1247.
3. Lorenz, M., Hillisch, A., Payet, D., Buttinelli, M., Travers, A. A., and Diekmann, S. (1999). *Biochemistry*, **28**, 12150.
4. Rice, P. A., Yang, S., Mizuuchi, K., and Nash, H. A. (1996) *Cell* **87,** 1295.
5. Rice, J. A., Crothers, D. M., Pinto, A. L., and Lippard, S. J. (1988). *Proc. Natl. Acad. Sci. USA* **85,** 4158.
6. Salzberg, A. A. and Dedon, P. C. (1997). *J. Bio. Struct. Dynam.* **15,** 277.
7. Payet, D. and Travers, A. (1997). *J. Mol. Biol.* **266,** 66.
8. Ner, S. S. (1992). *Curr. Biol.* **2,** 208.
9. Jantzen, H.-M., Admon,A., Bell, S. P., and Tjian, R. (1990). *Nature* **344,** 830.
10. Gubbay, J., Collignon, J., Koopman, P., Capel, B., Economou, A., Munsterberg, A., Vivian, N., Goodfellow, P., and Lovell, B. R. (1990). *Nature* **346,** 245.
11. Sinclair, A. H., Berta, P., Palmer, M. S., Hawkins, J. R., Griffiths, B. L., Smith, M. J., Foster, J. W., Frischauf, A. M., Lovell, B. R., and Goodfellow, P. N. (1990). *Nature* **346,** 240.
12. Travis, A., Amsterdam, A., Belanger, C., and Grosschedl, R. (1991). *Genes Dev.* **5,** 880.
13. Grosschedl, R., Giese, K., and Pagel, J. (1994). *Trends Genet.* **10,** 94.
14. Bustin, M., Lehn, D. A. and Landsman, D. (1990). *Biochim. Biophys. Acta* **1049,** 231.
15. Bianchi, M. E. (1995). In *DNA–protein: structural interactions* (ed. D. M. J. Lilley), p. 177. IRL Press, Oxford.
16. Love, J. J., Li, X. A., Case, D. A., Giese, K., Grosschedl, R., and Wright, P. E. (1995). *Nature* **376,** 791.
17. Werner, M. H., Huth, J. R., Gronenborn, A. M., and Clore G. M. (1995). *Cell* **81,** 705.
18. Giese, K., Cox, J., and Grosschedl, R. (1992). *Cell* **69,** 185.
19. Giese, K., Amsterdam, A., and Grosschedl, R. (1991). *Genes Dev.* **5,** 2567.
20. Paull, T. T., Haykinson, M. J., and Johnson, R. C. (1993). *Genes Dev.* **7,** 1521.
21. vanGent, D. C., Hiom, K., Paull, T. T., and Gellert, M. (1997). *EMBO J.* **16,** 2665.
22. Zappavigna, V., Falciola, L., Citterich, M. H., Mavilio, F., and Bianchi, M. E. (1996). *EMBO J.* **15,** 4981.
23. Onate, S. A., Prendergast, P., Wagner, J. P., Nissen, M., Reeves, R., Pettijohn, D. E., and Edwards, D. P. (1994). *Mol. Cell. Biol.* **14,** 3376.
24. Crothers, D. E. (1993). *Curr. Biol.* **3,** 675.
25. Travers, A. A., Ner, S. S., and Churchill, M. E. A. (1994). *Cell* **77,** 167.

26. Ner, S. S. and Travers, A. A. (1994). *EMBO J.,* **13,** 1817.
27. Wagner, C. R., Hamana, K., and Elgin, S. C. R. (1992). *Mol. Cell. Biol.* **12,** 1915.
28. Bianchi, M. E., Beltrame, M., and Paonessa, G. (1989). *Science* **243,** 1056.
29. Pil, P. M. and Lippard, S. J. (1992). *Science* **256,** 234.
30. Sheflin, L. G., and Spaulding, S. W. (1989). *Biochemistry* **28,** 5658.
31. Love, J. J., Xiang, L. Case, D. A, Giese, K., Grosschedl, R., and Wright, P. E. (1995). *Nature* **376,** 791.
32. Travers, A. A. (1995). *Nature Struct. Biol.* **2,** 615.

Chapter 4
Use of topology to measure protein-induced DNA bend and unwinding angles

Leonard C. Lutter, Christopher E. Drabik, and Herbert R. Halvorson
Molecular Biology Research Program, Henry Ford Hospital, Detroit, MI 48202-3450, USA

1 Introduction

This chapter describes details of the rotational variant method for measuring protein-induced changes in bend and unwinding angles. To summarize the approach, a series of plasmids is constructed such that each contain tandem repeats of the binding site for a protein that bends its site upon binding. The members of the series differ by the progressive addition of single base-pair increments to the repeat length, meaning that the rotational setting between adjacent binding sites differs by 34° when a construct with an n base pair repeat is compared with one with an $n + 1$ base-pair repeat. Binding of the protein to this series of plasmids induces a family of superhelices with pitches based on the repeat length, and relaxation by topoisomerase causes a change in the DNA linking number that reflects the specific form of superhelix generated in each plasmid. The set of differences between the linking numbers of a plasmid with and without bound protein can then be used to calculate the bend induced, as well as any change in duplex winding upon protein binding. The advantage of this method is that the measurement derives from first principles and requires no comparison standards. Details of the conceptual basis of the approach as well as a short description of the methodology are published in Lutter *et al.* (1). This chapter expands on the methods needed to perform such a measurement. In addition, improvements developed since that publication will be described, as well as ongoing efforts aimed at further facilitating the method.

The approach involves several stages, each of which employs distinct methodologies. Stage 1 entails construction of a series of plasmids containing tandemly repeated protein binding sites, with the length of the repeat unit increased in single base-pair increments in successive plasmids. Stage 2 involves binding of the protein, relaxation by topoisomerase, and separation of the extracted DNA as a series of topoisomers using electrophoresis in an agarose gel. Stage 3 involves measurement of the linking number differences from the gel images. Stage 4 involves determining the bend and duplex winding angle changes from that set of linking number differences.

2 Plasmid construction

Plasmid construction is the most labour-intensive aspect of the method and, in its original version (1), comprises three steps. In the first step, the DNA-binding site of the protein of interest is cloned into a standard vector. This then serves as a template in the second step in which a set of PCR primers that flank the cloning site are used to generate a series of monomer fragments that differ in length by single base-pair increments and contain asymmetric *Ava*I termini. In the third step each monomer fragment is cloned as tandem repeats into a vector with the same asymmetric *Ava*I termini. The asymmetric *Ava*I sites ensure head to tail oligomerization of the monomers (2). This section focuses on aspects of plasmid construction specific to the rotational variant method. The reader is directed to other references for various routine cloning procedures (3–5).

2.1 Designing the plasmid

There are several design considerations for the rotational variant plasmid set. First, the binding sites need to be separated by sufficient length of DNA (usually several protein diameters) to avoid direct steric interference between proteins binding to adjacent sites (e.g. see *Figure 5* in ref. 1). Second, the number of repeats needs to be adequate to generate a significant linking number change (ΔL). A change of one to two in the linking number is readily detected by the gel method. Thus, for a protein that bends its site by, for example 90°, plasmids should be constructed with at least four to eight repeats. Third, the range of repeat length variation should ideally span greater than a duplex turn, in 1–2 bp increments. This means that at least six to seven plasmids should be constructed in order to span the full range of superhelix structures and facilitate data fitting.

2.2 Initial cloning of the protein-binding site

In the original version of the procedure (1) a restriction fragment containing the binding site for the catabolite gene activator protein (CAP) was blunt-cloned into the *Sma*I site of the vector pBluescript II KS+ (Stratagene). Alternatively, the binding site can be synthesized by oligonucleotide synthesis, an approach used in a subsequent project that measured the A-tract bend angle (unpublished observations), and then cloned. Finally, a strategy that completely eliminates this initial cloning is described at the end of this section.

2.3 Generating of the monomer fragment

The construct containing the blunt-cloned binding site is then used as a template for a set of PCR primers to generate a series of monomer fragments differing by 1-bp increments and terminated by asymmetric *Ava*I sites. *Figure 1*

```
5'
G
C
T
G
C ┐
T |
C | Ava I
G |
G |
G ┘
 -GGTGGCGGCCGCTCTAG 3'               SKA12
  -GTGGCGGCCGCTCTAGA                 SKA11
   -TGGCGGCCGCTCTAGAA                SKA10
    -GGCGGCCGCTCTAGAAC               SKA9
     -GCGGCCGCTCTAGAACT              SKA8
      -CGGCCGCTCTAGAACTA             SKA7
       -GGCCGCTCTAGAACTAG            SKA6
        -GCCGCTCTAGAACTAGT           SKA5
         -CCGCTCTAGAACTAGTG          SKA4
          -CGCTCTAGAACTAGTGG         SKA3
           -GCTCTAGAACTAGTGGA        SKA2
            -CTCTAGAACTAGTGGAT       SKA1
             -TCTAGAACTAGTGGATC      SKA0
              -CTAGAACTAGTGGATCC     SKAm1
               -TAGAACTAGTGGATCCC    SKAm2
                -AGAACTAGTGGATCCCC   SKAm3
                 -GAACTAGTGGATCCCCC  SKAm4
                  -AACTAGTGGATCCCCC  SKAm5
                   -ACTAGTGGATCCCCC  SKAm6
                    -CTAGTGGATCCCCC  SKAm7
                     -TAGTGGATCCCCC  SKAm8
5'GGTGGCGGCCGCTCTAGAACTAGTGGATCCCCCGGGCTGCAGGAATTCGATATCAAGCTTATCGATAC 3'
3'CCACCGCCGGCGAGATCTTGATCACCTAGGGGGCCCGACGTCCTTAAGCTATAGTTCGAATAGCTATG 5'
       NotI   XbaI  SpeI  BamHI SmaI  PstI  EcoRI EcoRV HindIII ClaI
                   KSAm6            CCCGACGTCCTTAA-
                   KSAm3            CCCGACGTCCTTAAGCT-
                   KSA0               GACGTCCTTAAGCTATA-
                   KSA3                 GTCCTTAAGCTATAGTT-
                   KSA6                   CTTAAGCTATAGTTCGA-
                   KSA9                     AAGCTATAGTTCGAATA-
                   KSA12                      CTATAGTTCGAATAGCT-
                   KSA15                       3'TAGTTCGAATAGCTATG-
                                                              ┌ G
                                                              | A
                                                        AvaI  | G
                                                              | C
                                                              | C
                                                              └ C
                                                                G
                                                                C
                                                                T
                                                                G
                                                                5'
```

Figure 1 PCR primers for generation of rotational variants of any protein-binding site. Shown in the centre is the sequence in the pBluescriptII KS+ polylinker flanking the *Sma*I site. Shown at the upper left and lower right are primers to be used in the PCR generation of rotational variants of a binding site cloned into the *Sma*I site. The horizontal sequence of each primer is complementary to the vector sequence, while the vertical sequence is a non-complementary tail which contains the directional *Ava*I site to allow head-to-tail oligomerization following *Ava*I cleavage of the PCR product. To relate these primers to the CAP study (1), primers SKA0 and KSA2 were used to produce the 83-bp repeat construct shown in *Figure 3*.

shows an example set of primers that includes those used for the CAP study (1).

The primers consist of sequences that hybridize to the vector sequences that flank the original cloning site at *Sma*I in pBluescript II, with the sites of hybridization proceeding outward in one to three base increments for successive primers. To the 5′ end of each primer is added an *Ava*I site of the appropriate asymmetry, followed by four additional bases. A set of monomer fragments is then generated by PCR using combinations of primers that produce fragments that increase in size by one to two base-pair increments through a range of greater than one duplex turn, or 11 bp. Thus, tandem repeats of these monomers will contain successive that sites are progressively rotated relative to one another through a range of greater than 360°. Each monomer PCR product is then cloned into T-vector (Promega) to generate a set of monomer clones. It is suggested that each clone be sequenced at this point for confirmation. Monomer constructs are then isolated from midsize cultures (500 ml) and the monomer fragment excised by digestion with *Ava*I. Each fragment is then isolated by electrophoresis in low melting point agarose followed by elution from the gel slice. It should be noted that once synthesized, the same set of primers can be used to generate a rotational variant set of fragments from *any* protein-binding site once that site is blunt-cloned into the *Sma*I site of pBluescript or, alternatively (see below), once the site is synthesized into a suitable oligonucleotide containing the primer hybridization sites.

Protocol 1

PCR generation of rotational variants

Reagents

- 2.0 μl starter clone (pBluescript + insert, 50 μg/ml)
- 2.0 μl primer 1 (180 μg/ml)
- 2.0 μl primer 2 (180 μg/ml)
- 0.75 μl *Taq* polymerase (Promega, 5 U/ml)
- 10 μl 10x Taq buffer (Promega)
- H_2O to 100 μl

Method

For the primers shown in *Figure 1* we used the following thermal cycling protocol:

1. Step 1: 94 °C, 5 min.
2. Step 2: 96 °C, 1 min.
3. Step 3: 54 °C, 2 min.
4. Step 4: 72 °C, 3 min.
5. Step 5: Go to step 2, 29 times
6. Step 6: 72 °C, 7 min.
7. Step 7: End (4°C for duration)

PCR reactions can be analysed by electrophoresis of a 10 μl sample on a 7.5% polyacrylamide/TBE gel.

Protocol 2

Gel isolation of vector fragments and monomer inserts

1 Digest pTAV or pCLAV with *Ava*I and *Alw*NI; Digest T-vector, rotational variant constructs with *Ava*I.

Typical preparative-scale digests of pTAV or pCLAV are of the order 5–10 μg of DNA. For rotational variant monomer isolation, 10–50 μg of plasmid clone DNA is digested. Restriction enzymes (New England Biolabs) are used at 1 (manufacturer defined) unit/μg of DNA in a 1-h digestion at 37 °C. *Ava*I and *Alw*NI share compatible buffer conditions.

2 Isolate DNA fragments by agarose gel electrophoresis and band excision.

Vector fragments are isolated by electrophoresis through a 0.7% agarose/TBE gel. A longitudinal slice of the sample lane is stained using ethidium bromide. The stained portion of the gel lane is illuminated with UV light and migration of the two fragments measured with a ruler. Vector fragments are sliced out of the unstained portion of the gel. The slices are then frozen in liquid N_2 and centrifuged in microfuge tubes at 19000 **g** (12 500 r.p.m. in a Sorvall SS-34 rotor) for 15–30 min. To the supernatant is added NaCl to 150 mM and two volumes of ethanol, followed by centrifugation. Isolation of insert fragment is performed in the same manner except that 1.8–2.0% agarose is used for gel electrophoresis since the insert size will be ≈ 50–100 bp long.

Protocol 3

Tandem repeat cloning using dual initiator fragments

1 Assemble ligation reactions including ligase, buffer and vector fragment. Begin ligation reactions by adding an equimolar amount of rotational variant insert to each vector fragment reaction.

Insert is added in separate reactions to each vector half. Because only half the vector is present in each reaction, the predominant ligation event is that of insert to vector. Since the *Alw*NI restriction site is asymmetric, neither vector half can ligate to itself. Molar ratios of 1 : 1 are used to reduce the potential of insert–insert ligation and concomitant insert-only minicircle formation. Molar ratios can be determined by electrophoresing the vector and insert fragments in a 2.0% agarose gel and performing densitometry on the stained gel. For analysis gels were photographed using a digital camera (Ultra Violet Products) and the images saved as, e.g., bitmap files. The relative band weights can then be analysed using densitometry software available from several manufacturers (e.g. SigmaScan or SigmaGel, SPSS Science).

2 At 1/2–1 h intervals add monomer insert to each reaction.

When possible, dilute insert and/or vector fragment in ATP-containing ligase buffer. This affords incremental supplementation of ATP to the reaction. Insert is added at each step in equimolar amounts to vector fragment. Depending upon the number of ligation steps, supplement reaction with ligase.

Protocol 3 continued

3 After a sufficient number of insert additions, combine the two ligation reactions.

A rough estimate of the number of addition steps required can be based upon the target number of inserts desired in the final construct. Since insert is being added concomitantly to each vector half, the number of steps theoretically should equal 1/2 of the number of inserts desired in the final construct. In practice, we recommend more additions than this calculated number. By combining the separate ligation reactions, the original vector (now containing a number of inserts at the *Ava*I cloning site) is reconstituted.

Example ligation protocol

- 0.5 μl vector fragment (≈ 25 ng of DNA)
- 0.5 μl insert (dilute in ligase buffer)
- 1.0 μl T4 ligase (NEB 40 U/ml; dilute 1:5 in reaction buffer)
- 0.5 μl 10 × buffer (NEB)
- 1.0 μl 0.25 mg/ml BSA
- 1.5 μl H_2O

Incubate at room temperature, with hourly additions of 0.5 μl aliquots of insert over a total of 8 h, plus a 1.0 μl aliquot of T4 ligase (diluted 1:5) at 5 h. After 8 h, combine reactions (total volume 19 μl) and add 1 μl 10 mM ATP and 1 μl T4 ligase (diluted 1:5), followed by an overnight incubation. Transform 200 μl of competent bacterial cells (≈ 10^8 μl cfu/μg DNA) with 10 μl of this incubation and plate 1/10th and 9/10 of the transformation.

4 Transform bacteria. Screen 3 ml bacterial cultures from individual colonies. Alternatively, single colonies may be screened by PCR.

The transformation protocol which we have adopted is based on methodologies outlined elsewhere in the *Practical Approach Series* (4). A first approximation of the size of the insert included in a given clone can be determined by digestion of the construct with an appropriate enzyme. The amount of DNA isolated from 3 ml cultures is sufficient for initial restriction enzyme characterization. Addition of insert results in a progressive increase in size of the 364-bp fragment of a *Hin*FI digestion of pTAV. In order to reduce the amount of time required in screening we also implemented a PCR approach (6) for determining insert size. This method involves amplifying the insert sequence present in individual colonies directly and hence precludes the necessity of growing overnight cultures and extracting DNA from many cultures. Using primers which tightly flank the insert sequence, PCR products can be sized on 2.0% agarose gels.

5 Determination of the number of inserts by *Ava*I partial digestion

Midsize cultures (250 ml) are grown of clones determined to have multiple inserts and DNA extracted using a commercially available kit (Qiagen). DNAs resuspended in TE at a final concentration of 0.6–0.9 μg/ml were used in the following reaction for subsequent electrophoresis on an analytical gel:

- 1 μl DNA (15–20 μg)
- 1.5 μl AlwNI (10 U/ml)

Protocol 3 continued

- 0.5 μl *Ava*I (10 U/ml) - see below
- 4 μl 10 x reaction buffer
- 0.5 μl BSA (10 mg/ml)
- H_2O to 40 μl

After allowing *Alw*NI to digest the plasmid for 1 h, *Ava*I is added and the digestion continued for another 20–30 min. This time may need to be adjusted to obtain a suitable *Ava*I partial digestion. Reactions are stopped by adding EDTA to a final concentration of 12.5 mM. Digests are then fractionated by electrophoresis in a 0.7% agarose gel. A digest of a pTAV-derived plasmid results in a two-tiered restriction fragment ladder. This ladder is well fractionated when electrophoresed on a 0.7% TBE–agarose gel for 8–10 h at 2.8 V/cm. A typical loading for a gel well of 7.5 × 10 × 1.5 mm is on the order of 3–5 μg of DNA or 15–20 μl of the reaction volume. As shown in *Figure 2*, the lower tier of the ladder has, as its first rung, a 0.88 kb fragment which is the result of *Ava*I restriction at the cloning site and *Alw*NI restriction. This fragment corresponds to the smaller half of the empty vector. Migrating at regular intervals above the empty vector fragment are bands that contain the 0.88 kb fragment plus increasing numbers of inserts. The upper tier series of restriction fragments begin with the 2.1 kb empty vector fragment and increases in size at intervals reflecting increasing number of insert.

2.4 Monomer oligomerization and cloning

Each isolated monomer fragment is oligomerized and cloned into a recipient vector containing the same *Ava*I termini. The recipient vector can be generated from any of the monomer clones above by removing the insert by *Ava*I digestion, followed by dilution and reclosure. Our recent study (6) employs a recipient vector termed pTAV (for *T*-vector-derived, *Ava*I-containing plasmid) derived in this manner from Promega T-vector. The CAP study (1) required further modification to remove a CAP-binding site already present in the original T-vector, generating pCLAV (CAP-Less, *Ava* I-containing plasmid).

At this point in the original study (1) the pCLAV was phosphatase treated, mixed with a tenfold molar excess of monomer fragment, incubated with ligase, and the preparation used to transform bacteria. Screening the colonies resulted in isolation of constructs containing from eight to 23 repeats for the respective repeat lengths (see Table 1 in ref. 1), with the repeat number determined from partial *Ava*I digestion ladders (see *Figure 2* in ref. 1 and *Figure 2* here). However, our experience with this direct ligation approach is that considerable screening is required to obtain high repeat clones. This is likely due to significant self-circularization of the monomer fragment after an initial amount of oligomerization, a concern that was recognized in the original Hartley and Gregori study (2). Their approach to this problem was to construct a vector-derived 'initiator' fragment onto which the monomer fragment could be added without circularizing the product. A shortcoming with this approach is that a number of enzymatic steps are needed to prepare the initiator fragment in order to complete the final circularization.

We have modified the initiator fragment concept to obtain an increased repeat number with fewer steps. In this modification, which we term the double initiator approach, the recipient vector is cleaved with *Ava*I as well as with a second restriction enzyme that cleaves at a single asymmetric site elsewhere in the vector. The two resulting fragments are isolated by gel electrophoresis and ligated in separate oligomerization reactions with repeated additions of monomer at equal or lower stoichiometry. Neither of the purified initiator fragment can self-ligate nor circularize when monomers are oligomerized onto the *Ava*I terminus. After repeated additions of monomer, the overall reaction is completed by mixing the two separate oligomerization incubations, which results in completion of circles containing oligomers that have accumulated on the ends of the two respective initiator fragments.

We have obtained a significant improvement in oligomer length with this double initiator modification. This strategy is a considerable simplification over the original initiator fragment approach (2) because it does not require multiple additional enzymatic steps. Instead it requires only one additional restriction digestion at a second single cut site in the vector that exhibits the appropriate asymmetry. Numerous classes of restriction enzymes can generate sticky ends with this asymmetry: (1) enzymes that have multiple recognition sequences (e.g. *Ava*I, *Sfc*I), (2) enzymes that recognize a site with an odd number of bases (e.g. *Eco*RII, *Mae*III), (3) enzymes that recognize interrupted palindromes (e.g. *Bgl*I, *Alw*NI) and (4) enzymes that recognize non-palindromic sequences (e.g. *Bbv*I, *Sfa*NI). Once the appropriate second asymmetric site is identified, a single large preparation of the two purified initiator fragments can used for all the rotational variant oligomer constructs.

Figure 2 illustrates an example plasmid constructed using this approach. It is a 10-repeat rotational variant constructed for our A-tract study (unpublished observations) and employs the recipient vector pTAV, which contains an asymmetric *Ava*I site cloned into the T-vector site at nucleotide 51 in the original pGEM-5Zf(+) parent vector. The second asymmetric site used is the *Alw*NI site [nucleotide 918 in pGEM-5Zf(+)]. Cleavage with *Ava*I and *Alw*NI generates fragments of 0.88 kbp and 2.1 kbp, which, when gel purified, serve as the initiator fragments for ligation with added monomer fragments in separate reactions. Shown in *Figure 2* is a partial *Ava*I/partial *Alw*NI digest that was used to determine the number of repeats in the plasmid. The band S0 corresponds to the small initiator fragment (0.88 kbp), and the ladder of 10 bands above that band correspond to the 10 repeats in the construct. The second ladder of bands higher in the gel represents the series beginning with the large initiator fragment L0 (2.1 kbp). The highest ladder represents the series of fragments based on vector that was not cut by *Alw*NI in this *Alw*NI partial digest.

Adjustment or increase in the number of repeats in the construct is easily achieved by a further modification of the double initiator method. Thus the digest of the 10-repeat construct described in *Figure 2* can be used to readily generate a

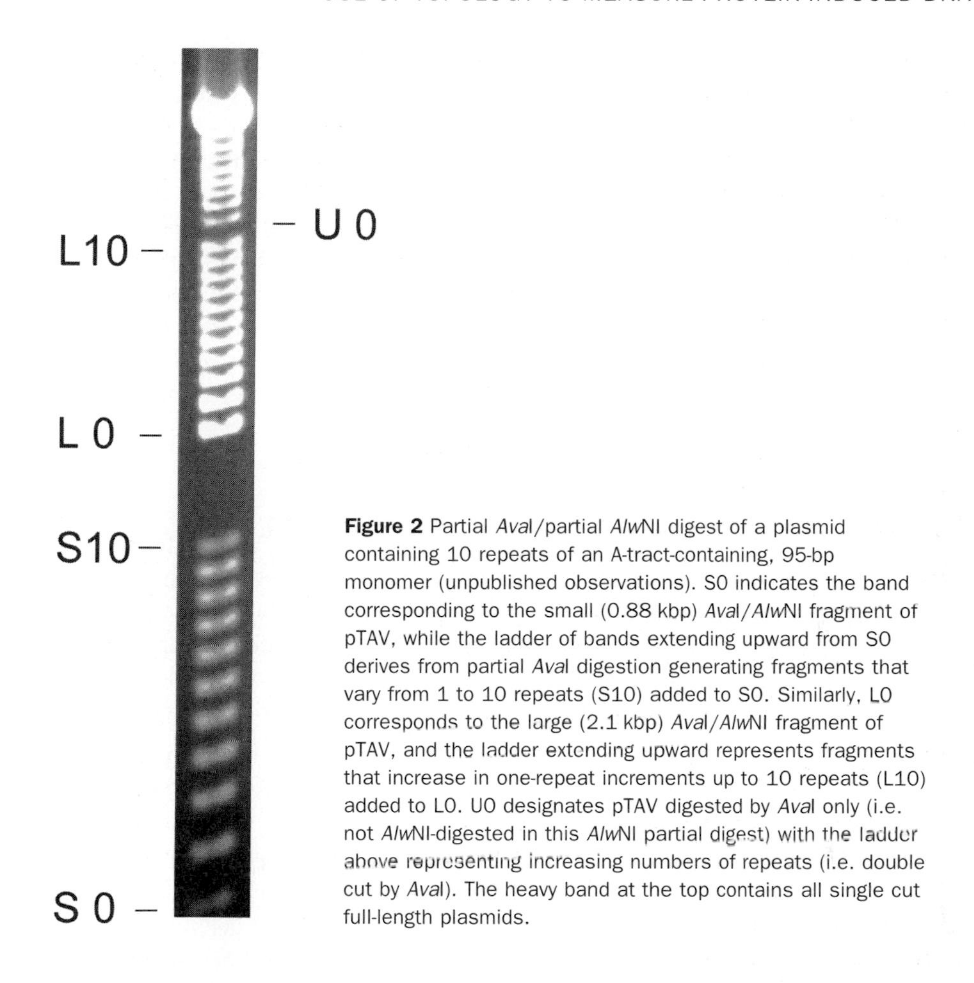

Figure 2 Partial *Ava*I/partial *Alw*NI digest of a plasmid containing 10 repeats of an A-tract-containing, 95-bp monomer (unpublished observations). S0 indicates the band corresponding to the small (0.88 kbp) *Ava*I/*Alw*NI fragment of pTAV, while the ladder of bands extending upward from S0 derives from partial *Ava*I digestion generating fragments that vary from 1 to 10 repeats (S10) added to S0. Similarly, L0 corresponds to the large (2.1 kbp) *Ava*I/*Alw*NI fragment of pTAV, and the ladder extending upward represents fragments that increase in one-repeat increments up to 10 repeats (L10) added to L0. U0 designates pTAV digested by *Ava*I only (i.e. not *Alw*NI-digested in this *Alw*NI partial digest) with the ladder above representing increasing numbers of repeats (i.e. double cut by *Ava*I). The heavy band at the top contains all single cut full-length plasmids.

20-repeat construct: the L10 and S10 fragments are eluted from the gel, after which they are mixed, ligated and transformed. Alternatively, to obtain specifically, e.g. a 14-repeat construct, respective fragments from the two ladders that add up to 14 repeats (e.g. S8 and L6) are isolated, mixed and ligated. Other combinations can be used to generate any desired repeat number up to and including one twice that of the original.

2.5 Strategies for further improvement in the process of plasmid construction

As this plasmid construction is the most labour intensive aspect of the method, a number of modifications are being developed to facilitate this stage. In the development of the original strategy used in the CAP study (1), the intent was to follow the PCR reaction directly with *Ava*I cleavage, followed, in turn, by oligomerization and cloning directly. This would avoid the separate for cloning the monomer into T-vector. Attempts during the original development of the method to use this strategy were unsuccessful owing to inadequate oligomerization, so the T-vector cloning step was introduced. However, the original abbreviated strategy is now being reinvestigated using the more highly purified primers available with current synthesis methodology.

An alternative procedure modification that avoids the original blunt cloning of the binding site involves the synthesis of an oligonucleotide containing the binding site as well as the sequences that hybridize to the primers at either end. This then serves as a template for use with the various combinations of PCR primers to generate the set of rotational variant monomers. If both of the above modifications prove successful, it should be possible to simply synthesize the binding site of the protein of interest and proceed directly to cloning the oligomerized rotational variants using the double initiator fragment oligomer cloning strategy.

3 Binding the protein and generating the ΔL

This stage involves binding the protein to each of the constructs generated above, followed by relaxation by topoisomerase of both bound and unbound DNAs to convert the alteration induced by the superhelix formation into a linking number change (ΔL). In addition to dealing with the conditions for the binding reaction, this section expands on two aspects of the binding reaction: non-specific binding and determination of saturation of sites.

3.1 Site saturation

The buffer conditions for binding of course will depend on the specific protein being analysed. For the CAP study (1), optimal binding conditions had been well established (see references therein), as had the dissociation constant (K_d) under those conditions. To determine conditions for full occupancy of the tandem sites in a construct, two measurement methods were used: bandshift in gels and topological plateau. To monitor occupancy by bandshift, a monomer fragment is excised from the tandem repeat, end-labelled, and used in a binding reaction to which increasing amounts of protein are added. The low end of the protein concentration range used should be about at the K_d and increase in threefold increments from there. A quantitative bandshift (see *Figure 2b*, lower, in ref. 1) indicates that the site is fully occupied. An advantage of this assay for occupancy is that the end-labelled fragment can be included in the same tube as the topoisomerase relaxation reaction and then analysed by polyacrylamide gel electrophoresis afterwards, thereby serving as an occupancy reporter in the same tube in which the relaxation is performed. A second measure of occupancy is the topoisomer distribution itself. If protein binding induces a topological change in the particular rotational variant plasmid studied, then there will be a progressive topological change as occupancy increases with added protein. This change will reach a plateau when full occupancy is achieved (see *Figure 2b*, upper, in ref. 1). The shortcoming of this assay is that it cannot be used if the particular rotational variant does not exhibit a ΔL upon protein binding. An example that approximates such a situation is the p89-8 CAP site construct (1), where the superhelix contribution to ΔL is largely cancelled by a roughly equal and opposite duplex unwinding contribution. In a case of such an apparent negative topological result, the internal bandshift occupancy reporter described above is important to establish occupancy.

3.2 Non-specific binding

A second consideration is non-specific binding to the plasmid. All DNA-binding proteins exhibit a non-specific affinity for DNA, but for the topological measurement performed here this binding must, in general, have a topological effect in the concentration range of interest to be relevant. This is best assessed by performing a relaxation reaction with the parent vector (e.g. pCLAV) over the protein concentration range used to determine if there is any topological change. No significant effect was observed in the CAP study (1), but if one is observed, that value can be used to correct the experimental data.

4 Gel electrophoresis

The topoisomerase relaxation can be terminated in a variety of ways, but we find the most convenient is simply to add heparin to 1 mg/ml. The heparin removes DNA-binding proteins and does not affect the electrophoretic separation of topoisomers, so the sample can be applied directly to the agarose gel. Electrophoresis was in a 0.7% agarose gel containing TBE plus 0.65 μg/ml chloroquine diphosphate (Sigma) for the 3–4 kbp plasmids used in the CAP study (1). This level of chloroquine causes plasmids that are in a relaxed state in normal buffer to migrate as well-separated, positively supercoiled topoisomers in the gel. An example of an electrophoretic separation of plasmid p83–23 relaxed in the presence and absence of CAP is shown in *Figure 3*.

Digitized gel images are recorded using a CCD-based documentation system for storing the ethidium bromide fluorescence image as, for example, a bitmap file. Any one of a number of programs can be used to display this file. We use SigmaScan (SPSS Science), which can be used to generate a densitometric scan from a gel lane in the image (*Figure 3*). From this is determined the topoisomer distribution centre, which corresponds to the mean linking number of the sample. The difference in the mean linking numbers of the bound and unbound samples is the ΔL for that construct. This will be dealt with in detail below.

5 Data processing

A sizeable investment is required to reach the point of actually having data in hand. This creates two conflicting desires in the experimenter. First is the urge to get some numbers as rapidly as possible and second, because each data point requires a significant investment, the desire to get the maximum information

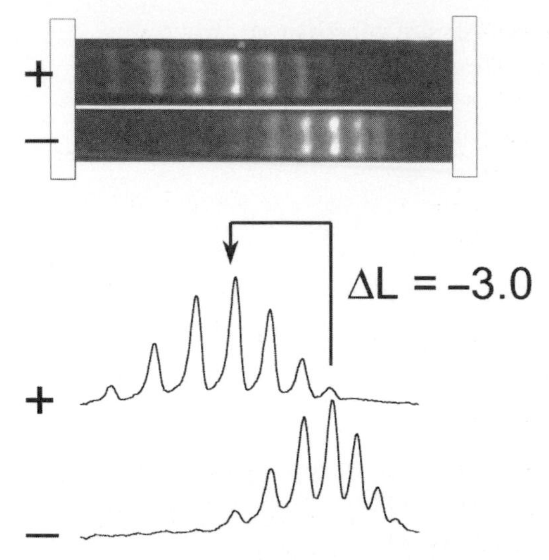

Figure 3 Topological analysis of the p83-23 CAP construct from (1). Shown at the top is a photograph of the ethidium bromide-stained agarose gel, while densitometric scans are shown below. Electrophoretic migration is from left to right. The presence of 0.65 μg/ml chloroquine diphosphate in the gel causes the DNA to migrate as a series of positively supercoiled topoisomers, so linking number increases from left to right. DNA samples were relaxed in the presence (+) and absence (–) of CAP. If the major peak in the straight (–) distribution is given an index of 0, the centre of that distribution, corresponding to the mean linking number of the sample, is at –0.2. The correspondence of individual peaks in the two scans is clear, so the major peak in the bent (+) DNA has an index of –3 and the value for the mean linking number is –3.2. The difference in mean linking numbers (ΔL) is then –3.0. It is only a coincidence that ΔL appears to be an integer.

from each experiment. Both issues are addressed, but with a bias towards the most quantitative treatment.

5.1 Capturing the data

5.1.1 Mean linking number

The experimental result is an electrophoretic gel, with DNA bands revealed by fluorescence, absorbance, autoradiography or some other means. With practice, it is possible to estimate the position of the centre of the topoisomer distribution directly from the image to within a few tenths of a topoisomer index (linking number). Although this practice discards information about the breadth of the distribution, recording this estimate can check against errors in subsequent steps of the analysis.

The image is digitized and entered into a commercial computer program (SigmaScan). The full width of each lane is integrated to provide a trace of integrated intensity as a function of migration distance. An estimate of the location of the centre of the distribution can be derived under the assumptions that each topoisomer band has the same bandwidth (full width at half-height) and that the trace returns to the baseline between the bands (the topoisomers are fully resolved). If these two assumptions hold, each band is assigned an index, j, and the height of each band h_j is recorded. The centre of the distribution is given by Equation 1.

$$\langle j \rangle = \Sigma(jh_j)/\Sigma(h_j) \quad 1$$

It is also possible to evaluate the breadth or standard deviation of the distribution by collecting a sufficient number of points.

$$\sigma^2 = \Sigma((j - \langle j \rangle)^2 h_j)/\Sigma(h_j) \quad 2$$

The breadth of the distribution reflects the energetics of statistical (thermal) fluctuations and thus can be used to study physical properties of the DNA. With σ expressed in turns, the torsional stiffness C is

$$C = Nhk_BT/(2\pi\sigma^2) \quad 3$$

where N is the total number of base pairs, h is the axial spacing of base pairs (0.34 nm), k_B is the Boltzmann constant, and T is the temperature (Kelvin).

A more satisfactory approach is to replace the height h_j with the area a_j of each peak. This avoids the restrictive assumptions of analysis by heights and permits the analysis of partially overlapping peaks of differing widths. It requires fitting an empirical function to the shape of the peaks. Several commercial programs designed for the analysis of chromatographic data would be suitable for this purpose, but we have developed our own. The advantages of this approach lie in the ability to deal with incomplete resolution and in a reduced chance for human error. The disadvantages of this approach arise from the large number of parameters. One should include all peaks that are significantly above the noise level in the trace and the smallest of these will be quite small relative to the largest. Each peak, regardless of size, needs three independent parameters for its description (area, bandwidth, and position) and the more poorly determined small peaks can introduce numerical instability into the fitting procedure. The workaround is to avoid fitting too many parameters simultaneously. The primary advantage of this approach over a technique to be described next, is that it is capable of analysing a distribution contaminated by a linear band.

Ordinarily, a Gaussian distribution of peak areas is fitted to the gel data, with an empirical function for the shape of the individual peaks, a common bandwidth for the peaks and a peak position on the gel that is a smoothly varying function of the peak number. The primary motivation for this approach is to reduce the complexity of the numerical analysis. Fitting the data with n independent peaks requires $3n$ parameters (individual areas, widths, and positions), but fitting with a Gaussian distribution of areas requires only six parameters (the total area, mean position (peak index), standard deviation of the Gaussian distribution of topoisomers plus the peak width (in gel coordinates) and two parameters to describe the relation between peak position and peak index). In addition to the advantage of searching in a parameter space of lower dimension, this approach also offers the convenience of automatically generating as many individual peaks as are needed to account for the total area.

One justification for this approach is derived from the theory of thin elastic rods. The energy of deformation increases quadratically with the size of the deformation (linking number change); this leads to a Gaussian distribution for the populations of the species. However, it is a standard statistical result that *all* distributions tend to the Gaussian with increasing sample size (10^{12} molecules of DNA).

One drawback to this approach is that it becomes impossible to search for asymmetry in the distribution of species. We have searched for possible asymmetry, using independent peaks, and found that vagaries in the assignment of a baseline, particularly with a curved baseline, made it impossible to obtain consistent results. In fact, finding a statistically significant third moment became a good indicator of an improperly assigned baseline. Another point that might be considered a drawback is that the fits are not quite as 'good', primarily because the relation between topoisomer and gel position is not smooth. Further, the bandwidth, as judged by fitting to independent peaks, is not constant. However, these parameters are of no real intrinsic interest and do not significantly influence the value of the linking number.

We use computerized fitting for the following reasons: it eliminates or markedly reduces human operator error and subjective bias; it permits collection of information otherwise not so readily available (e.g. standard deviation of the distribution); and it permits collection of information otherwise not available at all (e.g. standard errors or confidence regions for the parameters such as mean linking number).

A final point here is that the sign of the differential migration must be determined independently. The scan of the gel does not reveal if the most rapidly moving peak is more overwound or underwound than the slowest band (see Section 4).

5.1.2 Mean linking number difference (ΔL)

The procedure of the previous section is applied to a sample of the DNA in its bent state to obtain L(b). This is repeated with straight DNA to get L(s). The difference in mean linking numbers is then just $\Delta L = L(b) - L(s)$. The numbering of species must be the same in the two experiments. This is a step that can introduce serious error. If the linking number difference is small, the two samples can be run in adjacent lanes and the patterns (distributions) will overlap, enabling the establishment of register.

It is worthwhile devoting some time (at least initially) to determining the reliability of a new experimental technique. This issue has been explored by analysis of variance (ANOVA) on repeated determinations of L, investigating the influence of linking conditions, length of the repeat unit (geometry of the superhelix), electrophoretic gel and date. No identifiably dominant origin for the variability was found, suggesting that variability in the gel electrophoresis, which is common to all replications, is the primary cause of variability in measured values in L.

If the repeated measurements can be considered pairwise, calculate the mean of the differences and the standard deviation of the differences. Otherwise, take the difference between the means of the two groups and calculate the variance of the difference as the sum of the standard errors of the means. By studying repeated measures reproducibility of ΔL is found to be about 0.1 turn.

5.2 Taming the data

A data set comprises several linking number differences corresponding to different spacings between the bends. If these linking number differences are plotted against the length of the interbend segment they should show a regular variation about some common value. If there is any difference in bend between the two DNA states there will be a 'signal'. Because the screw of the DNA (bp/turn) is regarded as an adjustable parameter, the data should span one complete cycle of dihedral angle (say, 12 bp).

The general appearance of a data set is shown in *Figure 4*. The figure also shows the theoretical curve based on a geometric analysis of the superhelix. The derivation can be found in Lutter *et al.* (1); the result is presented here with some minor changes in notation. The angle of bend is B (degrees). The length of a repeat RL is the interbend distance in base pairs. Because of the rotation intrinsic

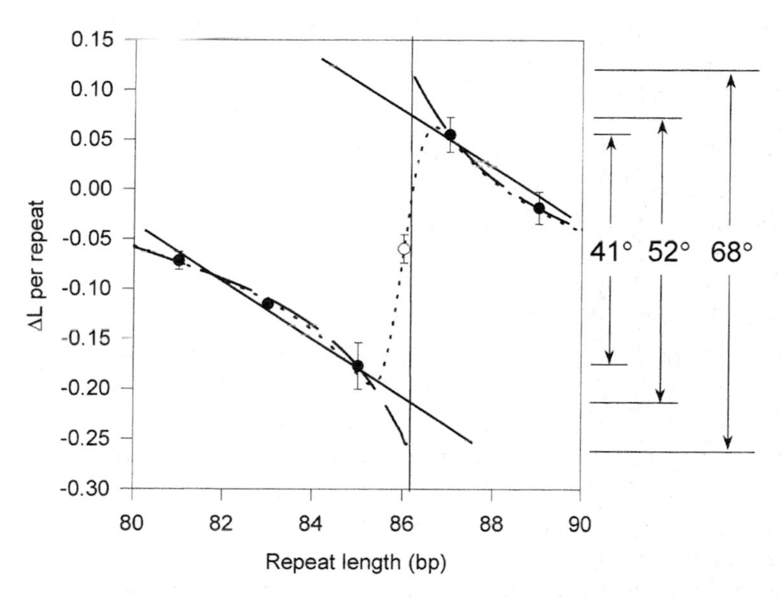

Figure 4 Linking number difference (ΔL) as a function of the repeat length (interbend spacing). The data are the discrete points. The results of estimating the bend B in different ways are shown projected on a vertical line at the crossover length X = 86.2 bp, which, for this example, was derived from the rigid form fit described below. The 41° bend value is calculated from the ΔΔL of the two data points (85 and 87) spanning the crossover. The 52° value is calculated from the ΔΔL determined from two parallel linear fits (solid lines) of the filled data points on either side of the crossover; the ΔΔL is the vertical displacement between the two lines. The 68° value is calculated from the best fit of *Equation 6* (rigid form). The dotted line represents the fit of *Equation 9* (torsional flexing), with the fitted data now including the 86 bp point (open circle). The bend value determined from this fit is 72°.

to the duplex DNA, the relative orientation of the two bends at the ends of the insert changes with the number of intervening base pairs according to the screw S (bp/turn). Two neighbouring bends thus define a dihedral angle ϕ

$$\phi = 360°(RL - X)/S \quad 4$$

where X denotes a repeat length corresponding to a planar (*cis*) structure. The use of X for the crossover length (instead of C) avoids confusion with the established use of C to denote the torsional stiffness (see *Equations 3* and *8*).

The quantity of interest is the angular mismatch produced by forming the superhelix, ΔL_{sh}. For one interbend segment

$$\cos(180°\Delta L_{sh} + \phi/2) = \cos(B/2)\cos(\phi/2) \quad 5$$

defines the connection between the angles. Part of the total ΔL is independent of the interbend spacing. This invariant part, I, can be expressed n I (for n bends) if it sensible to consider that protein binding at a bend introduces a specific change

$$\Delta L = (n - 1)\Delta L_{sh} + nI \quad 6$$

5.2.1 Crude analysis

There is a natural desire to know the implications of the data as they are being collected. Preliminary estimates of the bend can be determined before the set is complete. Extremes in the linking number difference occur when the superhelix changes sign discontinuously (the crossover point). For a construct with n bends ($n - 1$ dihedral angles) the magnitude of the discontinuity is $2(n - 1)(B/2)$, where B is the bend angle. These approximations always underestimate the magnitude of the discontinuity, so the bend is always greater than the range of the data multiplied by $360°/(n - 1)$.

As more data become available, a better estimate can be obtained by extrapolating into the region of the discontinuity (see *Figure 4*). Note that only the vertical offset between two parallel slanting lines is involved; it is not necessary to know where the crossover is located.

5.2.2 Refined analysis

A more accurate assessment of the bend and the other parameters of the model requires non-linear least-squares computer analysis. The starting point is an equation for the angular mismatch produced by the repeated bends (*Equations 5* and *6*).

The equation, in this form, is not adequate to describe the data because it assumes completely rigid connections. It is possible to produce a construct that corresponds to the crossover. The strictly planar crossover is impossible physically because of self-excluded volume. The experimental gel pattern for this construct is not bimodal, corresponding to a mixture of two limiting non-planar forms, but unimodal, representing some average about planarity. The only way the DNA can escape the problem is by relaxing through torsional fluctuations. The bending flexibility of DNA is well known, but bending does not relieve the problem. We address this issue by assuming independent random angular fluctuations between successive base pairs, arising from thermal energy. We assume a quadratic potential, hence a Gaussian distribution of angular fluctuations, but other forms could be applied equally as well.

$$p(\alpha) = \frac{1}{\sigma\sqrt{2\pi}} \exp\left[-\frac{1\alpha^2}{2\sigma^2}\right] \quad 7$$

where

$$\sigma = \sqrt{(nhk_BT/C)}\,(360°/2\pi)(\text{deg}) \quad 8$$

and n here is the product of the repeat length and the number of interbend links (number of bends − 1).

The expected effect is obtained from a convolution integral.

$$\Delta L = (1/2\pi)\int(n-1)\pi(\alpha)\Delta L_{sh}d\alpha \quad 9$$

(n here is the number of bends.) Because of the discontinuity at the crossover, the integral is evaluated by Monte Carlo integration. The parameters of the model are then adjusted to maximize the agreement, in a least-squares sense, between the numerical integrals and the data.

In studies of this kind it is not sufficient to say that the bend is, say, 80°. It is important to be able to assert that the bend is between 75° and 85° with some probability, or that it is between 40° and 120° with the same probability. A preliminary estimate of the confidence region for the parameters is provided by the non-linear least-squares analysis, but it needs to be accepted with caution. The mathematical form is distinctly non-linear and, typically, there are few degrees of statistical freedom (excess of data points over the number of parameters). Accordingly, we have adopted an additional Monte Carlo procedure, akin to the jack-knife or bootstrap. In a universe of all possible experimental outcomes, we could equally have collected a data set with each point perturbed by some random amount of its (experimental) standard error. Analysis of such a data set would result in slightly different values for the parameters. We generate several hundred such synthetic data sets, analyse them, and then determine the variability of the set of parameters. Because the problem is non-linear, the mean values of the parameters for the set of synthetic data do not correspond to the best parameters for the actual experimental data. The synthetic data sets are used only to determine the variability of the parameters.

5.2.3 Advanced analysis

In the preceding section it has been tacitly assumed that the difference being measured when forming ΔL is solely the bend. (If one of the DNA states is straight, changes in screw are immaterial.) It is easy to imagine situations where this is not the case. For example, if the protein-binding site has an intrinsic bend the straightforward analysis will underestimate the true bend and correspond more closely (but not exactly) to the change in bend that occurs when the protein binds.

The modified experiment is to take the difference between two states of differing screw, where temperature or an intercalator (ethidium bromide) has been used. In effect, the resulting difference is a double difference (ΔΔLk) where the virtual straight reference has cancelled out. Visually, the result looks like the derivative of *Figure 1*. The relation between repeat length RL and dihedral angle ϕ is now different for each data set, complicating the analysis. Alternatively, it is possible to vary this perturbation almost continuously, greatly increasing the amount of information available from a limited number of constructs. If there exists any uncertainty about possible existence of a bend in the putatively straight state for the DNA, a comparison with the effect of an intercalator is a powerful test. This topic is currently under active investigation.

Acknowledgement

This work was supported by grants from the Institute of General Medical Sciences at NIH.

References

1. Lutter, L. C., Halvorson, H. R., and Calladine, C. R. (1996). *J. Mol. Biol.* **261**, 620.
2. Hartley, J. L. and Gregori, T. J. (1981) *Gene* **13**, 347.
3. Maniatis, T., Fritsch, E. F., and Sambrook, J. (1982) *Molecular Cloning: a laboratory manual.* Cold Spring Harbor Laboratory Press, Cold Spring Harbor.
4. *DNA cloning 1: core techniques* (1995). Oxford University Press, Oxford.
5. *Current Protocols in Molecular Biology.* (1996) John Wiley and Sons, Inc., New York.
6. Gussow, D. and Clackson, T. (1989). *Nucl. Acids Res.* **17**, 4000

Appendix

Programs and instructions are available upon request by e-mail to L. C. Lutter (llutter@cmb.biosci.wayne.edu).

Chapter 5

Bandshift, gel retardation or electrophoretic mobility shift assays

Louise Fairall, Memmo Buttinelli and Gianna Panetta
MRC Laboratory of Molecular Biology, Hills Road, Cambridge CB2 2QH, UK

1 Introduction

In this chapter we describe how to perform a bandshift experiment, also known as the gel retardation assay or electrophoretic mobility shift assay, and some of the uses of the bandshift assay. The basis of the bandshift assay is that protein–DNA complexes remain intact when gently fractionated by gel electrophoresis and migrate as distinct bands, but more slowly than the free DNA fragment. The assay is simple and quick, and the use of radioactive binding-site DNA makes it highly sensitive. Originally, this method was primarily used in the detection of DNA-binding proteins in crude cell extracts, but as more and more proteins have become available in a pure form it is largely used to study the binding activity of purified DNA-binding proteins. It can be used for both highly sequence-specific (1, 2) and non-specific proteins, such as histones (3, 4). Bandshift assays can be used quantitatively to estimate dissociation constants for protein–DNA complexes. Bandshift experiments can also be used to visualize protein–protein interactions between a DNA-binding protein and other non-DNA-binding proteins. The binding of a second protein to a protein–DNA complex to form a triple complex is visualized by a further retardation of mobility that is called a supershift. Supershift experiments can also be used to assay the binding of a second DNA-binding protein to the DNA or the binding of a second molecule of DNA to the protein.

2 Some general considerations

When studying sequence-specific DNA-binding proteins it is helpful to keep in mind a few simple points.

(1) The binding constants of sequence specific DNA-binding proteins vary over a wide range (between 10^7 and 10^{14} M^{-1}). Consequently, the molarity of protein or binding-site containing DNA required for complex formation is dependent

on the binding constant and is likely to be different for different DNA-binding proteins. Therefore, when setting up a preliminary binding experiment, it is advisable to use a relatively high concentration of binding-site DNA (10^{-9}–10^{-7} M) to facilitate protein–DNA complex formation, even if the amount of protein is limiting.

(2) Cell extracts contain many DNA-binding proteins that are capable of interacting with DNA in a sequence-independent manner and thus may compete with the protein of interest for binding-site DNA. This problem can be minimized by the inclusion of competitor or carrier DNA in binding reactions.

(3) It is possible that one protein may bind specifically to more than one related sequence and, conversely, that a particular sequence may be recognized by more than one protein.

(4) Different proteins may require different conditions for optimal complex formation and stability. These may vary considerably in terms of pH, ionic strength and metal ion content. For example, proteins containing the 'zinc-finger' motif require zinc as an essential cofactor, and their binding activity may be inhibited by the presence of EDTA in buffers.

3 Gel systems

Differences in the size, aggregation state and pI of protein–DNA complexes affect the choice of conditions used for the bandshift assay. The process of electrophoretic separation may destabilize protein–DNA complexes. Hence, the ionic strength and composition of binding and electrophoresis buffers should be adjusted in accordance with the known or suspected stability of the protein–DNA complex. It is a good idea to keep the electrophoresis buffer similar to the binding buffer. The fractionation system most frequently used is a non-denaturing (or nucleoprotein) polyacrylamide gel, but agarose gels can also be convenient. The size of a protein–DNA complex (or complexes) may vary considerably, and is often unknown at the beginning of a study, so both gel systems should be tried to assess which gives the better separation and better binding. In general, protein–DNA complexes of high molecular weight are best fractionated in agarose gels, whereas small or closely-related complexes are resolved better in polyacrylamide gels. Furthermore the two gel matrices may affect the stabilities of various complexes differently.

Bandshift gels are adapted from gels developed for the separation of different length double-stranded DNA fragments. The construction of various types of apparatus, preparation of acrylamide and agarose gels, electrophoresis conditions, radioactive labelling of DNA, detection of radioactive DNA and recovery of DNA from gels have been described in great detail in (5), and will not be repeated here. Instead, their application to the fractionation and identification of protein–DNA complexes is described.

3.1 Electrophoresis buffers

The most common buffer used in both acrylamide and agarose bandshift gels is TBE (90 mM Tris base, 90 mM boric acid, 2 mM EDTA, pH 8.3). For the analysis of metal-binding proteins, a buffer without EDTA such as TB (90 mM Tris base, 90 mM boric acid, pH 8.3) is used. In some cases the use of low ionic strength electrophoresis buffers such as TE (10 mM Tris/HCl, 0.1 mM EDTA) or HE (20 mM HEPES/NaOH, 0.1 mM EDTA, pH 7.5) can improve the stability of the protein–DNA complexes and the resolution between bands.

3.2 Sample buffers

Protein–DNA complexes are loaded on to bandshift gels (acrylamide or agarose) in the presence of 4–5% glycerol. This can either be in the binding buffer or added just prior to loading the gel. Bromophenol blue (0.02% w/v) can also be added to the sample as a electrophoresis marker, but it sometimes disrupts protein–DNA complexes so it is better to add it to a separate lane. Careful application of the sample is required to prevent dilution.

3.3 Agarose gels

Agarose gels for bandshifts (0.7–1.5% w/v) are cast using high-temperature gelling agarose (BRL Ultrapure) with 0.25–1 × TBE or TB used as the buffer in the gel and for electrophoresis (Section 3.1). The use of low-temperature gelling agarose may at times be convenient to facilitate the recovery of protein or DNA. Optimal compositions for gels and electrophoresis buffers should be established empirically, as they may dramatically affect the quality of separation. The size of the gel can vary, but commercially available flat bed apparatus usually has a bed size of about 10 × 10 cm (called 'mini-gels') or the larger 20 × 25 cm. The gel is cast ≈ 0.5 cm deep (which would require either 50 ml of agarose for mini-gels or 250 ml of agarose for the 20 × 25 cm apparatus). Mini-gels are most frequently used to assay complex formation because they require shorter electrophoresis times. The larger bed size is used when greater separation between complexes is required. To prevent heating and the destabilization of protein–DNA complexes, a smaller electric current is applied than that used for the separation of DNA fragments: 20–30 mA for mini-gels and 50–60 mA for larger gels. Electrophoresis can be carried out at room temperature or in the cold-room and the same temperature is generally used for the binding reaction. It is customary to stop electrophoresis when the bromophenol blue marker dye has reached the bottom of the gel, but the electrophoresis time has to be optimized for the complex studied and the separation required. Following electrophoresis, the gel is dried on Whatman DE81 paper with a backing sheet of Whatman 3MM paper under vacuum at 60 °C. The gels are then visualized by autoradiography or by using a phosphorimager (Section 3.5). Gels may also be exposed wet for autoradiography, if subsequent elution of the DNA or protein–DNA complex is required.

3.4 Polyacrylamide gels

Polyacrylamide gels for bandshifts (3.5–6% w/v) are cast using an acrylamide to bisacrylamide weight ratio of 19 : 1 or 37.5 : 1, generally using 0.25–1× TBE or TB as the buffer in the gel and for electrophoresis. Optimal compositions for gels and electrophoresis buffers should be established empirically, as they may dramatically affect the quality of the separation (Section 3.1). For example, the stability of complexes in gels may be improved by the addition of 5% glycerol to the gel, although glycerol can affect the ability to separate the various components (6). Gels sized 20 × 20 × 0.3 cm are convenient and can be run at 5–15 V/cm at room temperature or 4 °C, depending on the stability of the complex. Following electrophoresis, the gels are placed on Whatman 3MM or DE81 paper without fixing and dried under vacuum at 80 °C. The gels are then visualized as described above for agarose gels.

3.5 Visualization of gels

After gels have been dried, they can be visualized by either autoradiography or using a phosphorimager. To autoradiograph a gel, the gel is exposed to X-ray film (such as Kodak X-AR and Fuji RX) with an intensifying screen at −70°C (HI speed X, GRI), or without a screen at room temperature or 4 °C. The intensifying screen reduces the exposure time by a factor of 5–10. Generally, 2–3 × 10^3 c.p.m. per track are sufficient to be visualized with an overnight exposure using a screen. The X-ray film should be preflashed if the signal is weak and if the autoradiograph is going to be quantified, since this brings the film into its linear range. Phosphorimaging is much more sensitive than autoradiography, nevertheless accurate quantification of Phosphorimaging requires a long exposure (16–24 h for 2–3 × 10^3 c.p.m. per lane). Another advantage of the phosphor storage plate is that the dynamic range is much greater than that of X-ray film, which means that only one exposure is required; when using X-ray film different exposure times are often needed to quantify the weak and strong bands. Scanning of the phosphor storage plate is carried out using specialized equipment such as the Molecular Dynamics Phosphorimager. Yet another advantage of the phosphorimaging technology is that the data are already digitized for quantitative analysis (see chapter 16 in this book).

4 Uses of the bandshift assay

4.1 Fractionation of transcription factor binding to differently positioned nucleosomes

The high resolution obtained with polyacrylamide bandshift gels can be used to detect more than one complex within the same binding reaction. Lane N in *Plate 1* shows an experiment in which nucleosomal complexes formed between histone octamer and a 249 bp DNA fragment, containing the *Xenopus borealis* 5S rRNA gene, are fractionated in a nucleoprotein polyacrylamide gel. The various bands correspond to differently positioned histone octamers on the DNA fragment, which have distinct electrophoretic mobilities owing to differences in conformation (7, 8). This gel system can also be used to visualize further retardation of specific nucleosomal bands after transcription factor binding in a supershift experiment., Lane +TFIIIA in *Figure 1* shows an example in which nucleosomal complexes are incubated in the presence of transcription factor IIIA (TFIIIA). Upon TFIIIA binding the different nucleosome bands run with lower mobility resulting in a changed pattern of bands (9). The nucleosome reconstitution, TFIIIA binding to the nucleosome and gel conditions are described in *Protocol 1*.

Protocol 1

TFIIIA binding to reconstituted nucleosomes on the somatic 5S rRNA gene by dialysis and analysis in acrylamide gels

Equipment and reagents

- Dialysis buttons (Pierce and Warriner, UK), small collodion bag (Sartorius) or analogous apparatus for microdialysis.
- Stock solutions: 1 M HEPES/NaOH pH 7.5, 5 M NaCl, 0.1 M EDTA pH 7.4
- Peristaltic pump
- Binding buffer: 20 mM HEPES/NaOH pH 7.5, 50 mM NaCl, 0.05% (v/v) Nonidet P-40 (BDH), 150 μg/ml poly [d(I-C)]

Method

1. Mix 60 nM radiolabelled DNA, 5 μM recombinant histone octamer and 10 μM mixed sequence nucleosomal DNA in 40 μl total volume containing 20 mM HEPES/NaOH pH 7.5, 2 M NaCl.
2. Dialyse for 2 h at 4°C against 200 ml of the same buffer supplemented with 1 mM EDTA.
3. Reduce the salt concentration stepwise by dialysing against 20 mM HEPES, 1 mM EDTA with the following NaCl concentrations (each for 2 h): 0.8, 0.65, 0.5, 0.0 M NaCl.
4. Finally dialyse overnight against 20 mM HEPES and 0.01% Nonidet P-40.
5. Centrifuge for ≈ 10 min at 4°C and transfer the supernatant to a fresh tube to remove any precipitated material.
6. Prepare a 30 μl binding reaction containing binding buffer, 5 nM 5S rRNA gene and 400 nM reconstituted nucleosomes (a fraction of the initial reconstitution mixture). Finally, add 300 nM RNase-treated 7S particle.
7. Incubate the binding mixture at 4°C for 30 min.
8. After addition of 5% (v/v) glycerol, analyse an aliquot ($2-3 \times 10^3$ c.p.m.) of reconstituted nucleosome and of TFIIIA-nucleosome complexes in a 5% nucleoprotein polyacrylamide (acrylamide-bis, 37.5 : 1) gel in buffer containing 20 mM HEPES-NaOH pH 7.5, 0.1 mM EDTA. For the 249 bp fragment, carry out electrophoresis out for about 6 h at 250 V at 4°C with buffer recirculation until the xylene cyanol marker has travelled down 2/3 of the gel.

4.2 Fractionation of protein–DNA complexes prior to further analysis

Polyacrylamide or agarose bandshift gels can be used as a general technique to fractionate mixtures of protein–DNA complexes prior to further analysis, as described in Section 4.1. In the footprinting assay (described in chapter 3 in this book) the region of DNA occupied (bound) by a protein is found by probing the accessibility of the DNA with a nuclease or chemical reagent. Consequently, the clarity of the footprint is dependent on the extent of occupancy of the binding site. When either 100% binding cannot be achieved or more than one complex is present in the binding reaction, the fractionation on bandshift gels can be used, after incubation with the nuclease or reagent of choice (10), to resolve the complex of interest from the naked DNA or other complexes (see *Plate 1*). The regions of DNA bound by protein in different complexes can be identified by cutting each band out of the gel, extracting the DNA and analysing in denaturing polyacrylamide gels.

4.3 Estimation of dissociation constants

When just one protein and its binding site are present in the binding reaction, the bandshift assay can be used simply to estimate dissociation constants. In the methods described here the first method requires that the protein concentration is known accurately and the second method requires the protein and the DNA concentration be known. In both methods the amount of free DNA and the amount of DNA bound in the protein–DNA complex have to be estimated either by measuring the intensity of a band in an autoradiograph using densitometry or by using phosphorimaging. It should be remembered that dissociation constant measurements obtained from analysis in bandshift gels may not reflect the absolute dissociation constant since the gel electrophoresis is likely to effect the equilibrium. Hence bandshift analyses are best used for comparative studies.

The first method involves using a low DNA concentration (below the dissociation constant) with increasing amounts of the protein. Then, as shown in *Equation 1* (11, 12), if the DNA concentration is much lower than the dissociation constant, then when the concentration of DNA is equal to the concentration of the protein–DNA complex (50% complex formation) the dissociation constant is equal to the concentration of the protein.

$$K_D = \frac{[P][D]}{[PD]} \qquad 1$$

where P = protein and D = DNA-binding site; when $[D] << K_D$ then $[P]_{free} \cong [P]_{total}$, so

$$K_D = \frac{[P]_{total} \times [D]}{[PD]}$$

This first method has been used to estimate the dissociation constants for the *Drosophila* transcription factor Tramtrack (TTK) to a number of different-length

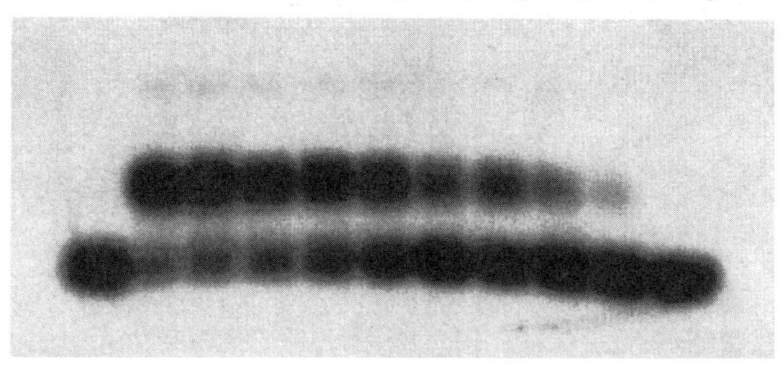

Figure 1 A titration of increasing amounts of TTK protein with a 14 bp oligonucleotide. The purpose of the titration is to find the concentration of protein required to obtain 50% binding. This value was then used to calculate the K_D of the protein to this oligonucleotide. The DNA concentration was 1×10^{-8} M and the protein concentration varied as shown. Samples were run in a 0.7% agarose gel run in 45 mM Tris-borate.

binding sites (13). TTK is a member of a large family of proteins that use the 'zinc-finger' motif to bind to DNA (14). Because removal of zinc inactivates these proteins, the binding buffer may need to be supplemented with Zn^{2+} (this is necessary if DTT is used) and care must be taken to omit EDTA from all buffers. For the binding reaction described below (*Protocol 2*), pure TTK (7 kDa) and a 14 bp oligonucleotide containing the TTK binding site were used. *Figure 1* shows a bandshift gel of a titration with increasing amounts of TTK to determine the concentration of protein required for 50% complex formation. This concentration corresponds to the dissociation constant for TTK binding to the 14 bp oligonucleotide.

Protocol 2

Estimation of the dissociation constant of the TTK–DNA complex

Equipment and reagents

- Binding buffer (10 μg/ml poly[d(I–C)], 100 μg/ml BSA, 0.1% NP40, 2 mM $MgCl_2$, 40 mM NaCl, 20 mM MES pH 6.5, 10% glycerol)
- 0.5 or 1.5 ml siliconized microcentrifuge tubes

Method

1 For each protein concentration and a naked DNA control prepare a 20 μl binding reaction mixture containing binding buffer, DNA at 1×10^{-8} M (^{32}P labelled 14 bp oligonucleotide containing the TTK-binding site) and, finally, varying amounts of protein

Protocol 2 continued

from 0 to 1×10^{-6} M. The high protein concentration is needed because of the low DNA-binding affinity of TTK (the dissociation constant is 4×10^{-7} M to a longer binding site).

2. Incubate the binding mixtures at room temperature for 30 min.
3. Analyse complex formation in a 10×10 cm 0.7% agarose gel using $0.5 \times$ TB as the gel and electrophoresis buffer. Run the gel at 30 mA until the bromophenol blue marker is 2/3 of the way down the gel (about 40 mins).

The second method for determining dissociation constants requires the preparation of a range of different concentrations of a one-to-one molar mix of protein and DNA. The proportion of DNA in the complex at the different concentrations depends upon the ratio between the total DNA concentration and the dissociation constant. The most convenient way to perform this measurement is to titrate (labelled) DNA at high concentration (at least 1000 times the expected dissociation constant) with increasing amounts of protein. Analysis in a bandshift gel is first used to determine the point at which a one-to-one complex is formed. A complex, formed under these conditions, is then diluted serially and the analysis carried out again in a bandshift gel to estimate the percentage of complex at different DNA concentrations. The amount of complex can then be used to calculate the dissociation constant using *Table 1*. See (15) for measurements of differences of the binding affinity for the oestrogen receptor DNA binding domain to different sequences using this method. From *Equation 2* it can be seen that the concentration of complex is dependent upon the dissociation constant and the total concentration of DNA, permitting the values in *Table 1* to be calculated.

Assuming equilibrium: $P + D \longleftrightarrow PD$

$$K_D = \frac{[P]_{free}[D]_{free}}{[PD]} \qquad 2$$

$[D]_{total} = [PD] + [D]_{free}$.

Required condition: $[P]_{total} = [D]_{total}$, then $[P]_{free} = [D]_{free}$ and

$$K_D = \frac{([D]_{total} - [PD])^2}{[PD]}$$

so

$$[PD]^2 - (2[D]_{total} + K_D)[PD] + [D]^2_{total} = 0$$

then

$$[PD] = \frac{(2[D]_{total} + K_D) - \sqrt{(2[D]_{total} + K_D)^2 - 4[D]^2_{total}}}{2}$$

In the presence of competitor DNA the dissociation constant for binding to the competitor DNA is also a consideration as in, for example, the nucleosome reconstitution experiments described in *Protocol 1*. In these experiments, a large

Table 1 Relationship between percentage complex and [DNA]/K_D

% Complex	[DNA]/K_D
96.9	1000
90.5	100
73	10
38.2	1
8.4	0.1
1	0.01

amount of bulk nucleosomal DNA is present as well as the labelled restriction fragment. *Equation 3* shows how the binding affinities for different binding sites can be compared in the presence of competitor DNA. The main requirement is that the competitor DNA concentration is far higher than the protein concentration and the protein concentration is far higher than the specific DNA-binding site concentration. The differences are usually reported as a free energy variation.

Assuming the equilibrium $P + D + C \longleftrightarrow PD + PC$ 3

$$K_D = \frac{[P][D]}{[PD]} \quad \text{and} \quad K_{Dc} = \frac{[P][C]}{[PC]}$$

where C = competitor DNA (without the specific binding site of interest). Then, (substituting [P] from K_{Dc} to K_D equation)

$$\frac{[D]}{[PD]} = \frac{K_D[C]}{K_{Dc}[PC]}$$

since $[P]_{total} = [P] + [PD] + [PC]$ when $[C]_{total} >> [P]_{total} >> [D]_{total}$, then $[P]_{total} \cong [PC]$ so

$$\frac{[D]}{[PD]} \cong \frac{K_D[C]}{K_{Dc}[P]_{total}}$$

When comparing two different specific sites, D1 and D2, the equation can be greatly simplified since, at the specified conditions, the terms $[C]/K_{Dc}$ and $[P]_{total}$ are independent of the specific binding site used. Then the value of the ratio between K_{D1} and K_{D2} depends directly on the values of [D] and [PD] at equilibrium, which can be obtained by quantifying bandshift experiments. So that

$$\frac{K_{D1}}{K_{D2}} \cong \frac{[D_1]/[PD_1]}{[D_2]/[PD_2]}$$

and (in terms of free energy variation)

$$\Delta\Delta G_{D1\text{-}D2} = -RT\,(\ln K_{D1} - \ln K_{D2}) \cong -RT\left(\ln\frac{[D_1]}{[PD_1]} - \ln\frac{[D_2]}{[PD_2]}\right)$$

where R = gas constant and T = absolute temperature.

Using the above equation to analyse a competitor DNA titration experiment a logarithmic curve will be obtained by plotting the ΔG values as ordinate and the competitor DNA concentration as abscissa. This curve will reach a plateau when the concentration of competitor DNA far exceeds that of the protein. The ΔG values at the plateau can be used to calculate the difference in ΔG ($\Delta\Delta G$) between two DNA-binding sites of the same length.

The equation can be rearranged so that a plot of [D]/[PD] on the ordinate and [C]total − [P]total/[P]total on the abscissa should be a straight line. The slope of the line corresponds to K_{DC}/K_D. The main advantage of calculating K_D values in this way is that each point of the titration curve is taken into account.

4.4 Some other applications for the bandshift assay

Bandshift experiments can be used to determine the sequence of the DNA-binding site of a protein complex using a technique known as SELEX (systematic evolution of ligands by exponential enrichment) (16–18). This involves binding the protein of interest to a random DNA sequence, contained within a longer oligonucleotide template, and separating the bound and unbound DNA in a bandshift gel. The bound DNA is then amplified by PCR. After several rounds of selection and amplification, high-affinity binding sites should be obtained. This procedure is particularly useful when there is protein available but the binding site is unknown or just a consensus binding site is known.

Bandshift experiments can also be used to visualize protein–protein interactions between a DNA-binding protein and other non-DNA-binding proteins in what is called a 'supershift assay'. In a supershift assay the binding of a second protein to a protein–DNA complex is visualized by a further retardation of mobility. An example of this application is the use of antibodies to confirm the identity of a protein in a protein–DNA complex, as in the identification of a basic helix–loop–helix protein which binds to the promoter of the insulin gene (19). The constituents of transcriptional complexes have also been identified using bandshift gels, such as in a study of a multicomponent complex which includes TFIID, TFIIA, TFIIB, ZEBRA (a non-acidic activator of the Epstein–Barr virus lytic cycle) and TATA-binding protein associated factors (20).

Individual components of protein–DNA complexes containing multiple transcription factors can also be identified by using bandshifts and immunoblotting (called 'Shift-Western blotting') (21). The components of protein–DNA complexes, after separation in agarose or acrylamide gels, can be transferred simultaneously onto two membranes: one (nitrocellulose) for detection of protein by immunoblotting and one (anion-exchange membrane) for detection of DNA by autoradiography. Antibodies to each component can be used to detect more than one protein in a complex.

Despite the fact that this chapter has focused on the use of bandshift assay in the study of protein–DNA complexes, it has been used extensively in the study of protein–RNA complexes. For example, bandshift gels using mutant RNA, have been used to study the interaction of U1, a small nuclear ribonucleoprotein, with U1 RNA (22) and bandshift gels have been very useful in a study of the cooperative binding of the HIV protein rev onto the Rev Response element RNA (23).

References

1. Garner, M. M. and Revzin, A. (1981). *Nucl. Acids Res.* **9,** 3047.
2. Fried, M. and Crothers, D. M. (1981). *Nucl. Acids Res.* **9,** 6505.
3. Linxweiler, W. and Horz, W. (1984). *Nucl. Acids Res.* **12,** 9395.
4. Rhodes, D. (1985). *EMBO J.* **4,** 3473.
5. Rickwood, D. and Hames, B. D. (ed.) (1982). *Gel electrophoresis of nucleic acids: a practical approach.* IRL Press, Oxford.
6. Pennings, S., Meersseman, G., and Bradbury, E. M. (1992). *Nucl. Acids Res.* **20,** 6667.
7. Duband-Goulet, I., Carot, V., Ulyanov, A. V., Douc-Rasy, S., and Prunell, A. (1992). *J. Mol. Biol.*, **224,** 981.
8. Meersseman, G., Pennings, S. and Bradbury, E. M. (1992). *EMBO J.*, **11,** 2951.
9. Panetta, G., Buttinelli, M., Flaus, A., Richmond, T. and Rhodes, D. (1998). *J. Mol. Biol.* **282,** 683.
10. Topol, J., Ruden, D. M., and Parker, C. S. (1985). *Cell* **42,** 527.
11. Riggs, A. D., Suzuki, H., and Bourgeois, S. (1970). *J. Mol. Biol.* **48,** 67.
12. Johnson, A. D., Meyer, B. J., and Ptashne, M. (1979). *Proc. Natl. Acad. Sci. USA* **76,** 5061.
13. Fairall, L., Harrison, S. D., Travers, A. A., and Rhodes, D. (1992). *J. Mol. Biol.* **226,** 349.
14. Schwabe, J. W. R. and Klug, A. (1994). *Nature Struct. Biol.* **1,** 345.
15. Schwabe, J. W. R., Chapman, L., and Rhodes, D. (1995). *Structure* **3,** 201.
16. Blackwell, T. K., Kretzner, L., Blackwood, E. M., Eisenman, R. N., and Weintraub, H. (1990). *Science* **250,** 1149.
17. Tuerk, C. and Gold, L. (1990). *Science* **249,** 505.
18. Irvine, D., Tuerk, C., and Gold, L. (1991). *J. Mol. Biol.* **222,** 739.
19. Naya, F. J., Stellrecht, C. M. M., and Tsai, M.-J. (1995). *Genes Dev.* **9,** 1009.
20. Chi, T., Lieberman, P., Ellwood, K., and Carey, M. (1995). *Nature* **377,** 254.
21. Demczuk, S., Harbers, M., and Vennstrom, B. (1993). *Proc. Natl. Acad. Sci. USA* **90,** 2574.
22. Jessen, T.-H., Oubridge, C., Teo, C. H., Pritchard, C., and Nagai, K. (1991). *EMBO J.* **10,** 3447.
23. Mann, D. A., Mikaelian, I., Zemmel, R. W., Green, S. M., Lowe, A. D., Kimura, T., Singh, M., Butler, P. J. G., Gait, M. J., and Karn, J. (1994). *J. Mol. Biol.* **241,** 193.
24. De Santis, P., Fua, M., Palleschi, A., and Savino, M. (1995). *Biophys. Chem.* **55,** 261.
25. Nolte, R., Conlin, R., Harrison, S., and Brown, R. (1998). *Proc. Natl. Acad. Sci. USA* **95,** 2938.

Chapter 6

Practical aspects of fluorescence resonance energy transfer (FRET) and its applications in nucleic acid biochemistry

Frank Stühmeier and Robert M. Clegg

Department of Physics, Laboratory for Fluorescence Dynamics, University of Illinois at Urbana–Champaign, Loomis Laboratory of Physics, 1110 West Green Street, Urbana, IL 61801-3080, USA

Alexander Hillisch and Stephan Diekmann

Institute of Molecular Biotechnology, Beuten bergstraße 11, D–07745 Jena, Germany

1 Introduction

Fluorescence spectroscopic methods can be used to detect substances (which contain fluorescent chromophores) at very low concentrations and are especially useful for studying biomolecular structures and their dynamics in solution (see (1–3) for introductions into fluorescence spectroscopy). FRET, which is a non-radiative exchange of excitation energy between two chromophores, is strongly dependent on the distance between the two chromophores. If two such molecules are attached to a single biopolymer at different and well-defined positions, FRET can be used to estimate the distance between the two chromophores, and these data can be used to study structural aspects of the macromolecule. The distances that can be estimated are in the range of 0.5 to ≈ 10 nm, which are in the same range as many distances in macromolecules. Because well-defined biopolymers can be synthesized and labelled at defined sites, FRET has become an important tool for structural studies in biophysics.

The comparison of FRET efficiencies within a series of experiments where two dyes are either covalently attached to different positions of identical biopolymers or to similar positions of structurally related biopolymers is an elegant approach for deducing information about the global structure of these molecules. The orientation of the double helical arms in the DNA four-way junction (4), the geometry of the hammerhead ribozyme (5) and the structure of the DNA three-way junctions (6, 7) were determined with this method. The helical geometry of double-stranded DNA (8) and the kinking of bulged DNA and RNA molecules (9) were detected using FRET. Furthermore, the bending of the DNA helix upon the binding of proteins has been detected and quantified (10–12).

This chapter describes practical aspects of FRET that are useful for determining the global geometry of DNA-structures in solution. We first give an overview of the basic aspects of FRET, highlighting potential problems in the interpretation of FRET experiments (Section 2). Methods for determining FRET efficiencies from steady-state fluorescence measurements are discussed (Section 3) and then the preparation and characterization of DNA structures, which are labelled covalently with chromophores, is described (sections 4 and 5). Finally we discuss the interpretation of the FRET efficiencies (section 5), emphasizing the use of computer molecular modelling to determine the stereochemistry of the DNA molecules and to define the interaction of the dyes with the DNA.

2 Basic principles of fluorescence resonance energy transfer

FRET is a photophysical process by which energy is transferred non-radiatively from a fluorophore (the energy-donor D) in an excited state to another chromophore (the energy-acceptor A) by means of intermolecular long-range dipole-dipole coupling (13–18). The energy transfer is non-radiative: thus D does not emit a photon and A does not absorb a photon.

There are four essential requirements for an effective energy transfer over distances in the range from 1 to 10 nm.

(1) The fluorescence spectrum of D and the absorption spectrum of A must overlap adequately.

(2) The quantum yield of D in the absence of A, ϕ_D, should be sufficient (e.g. $\phi_D \geq 0.1$).

(3) The absorption coefficient of A should be sufficient (e.g. $\approx$ 1000/M.cm).

(4) The transition dipoles of D (emission dipole) and A (absorption dipole) must be orientated favourably relative to each other during the lifetime of D to allow effective dipole–dipole interactions between the two chromophores.

It can be shown that the quantum yield of energy transfer – i.e. the FRET efficiency, E – depends on the distance between the dyes (see *Figure 1*) and is given by $E = 1/(1 + (R/R_0)^6)$, and $E = 1/2$ when $R = R_0$. R_0 depends on the spectroscopic parameters of the dyes and their mutual orientation, it is in the range of 1–7 nm for commonly used D–A pairs (see refs. 14 and 19 for lists of common donor–acceptor pairs). The practical use of FRET in biology is usually related to the strong diminution in E as R increases. A good estimate of E can often be determined from steady-state (non-dynamic) fluorescence spectroscopic measurements (see below). Once E has been determined, it can be used for estimating the distance between an energy donor and an energy acceptor, and if these chromophores are securely associated to defined positions of a DNA/RNA struc-

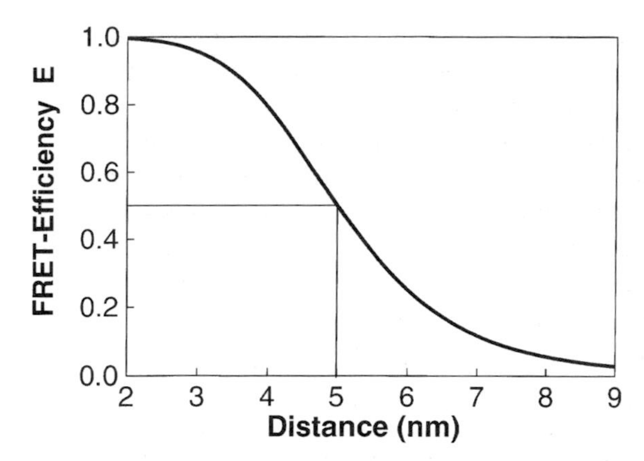

Figure 1 Distance dependence of the FRET efficiency, E, calculated according to $E = 1/(1+(R/R_0)^6)$ with $R_0 = 5.0$ nm. E is a very strong function of the distance R for distances from 4 to 6 nm. At $R = R_0$ the FRET efficiency is 0.5: 50% of the energy of the donor is transferred to the acceptor.

ture, these 'FRET distances' can be used to probe the geometry of the macromolecules.

There are a few potential pitfalls when interpreting the experimentally determined FRET-efficiencies between two dyes covalently attached to defined positions of a nucleic acid molecule. Although many problems can arise owing to defective preparation and insufficient characterization of the dye-labelled DNA structures (see chapters 13.3 and 13.4), certain physical aspects of FRET should be kept in mind before interpreting the experimental data.

2.1 Orientation effects of the dyes

The FRET efficiency, E, depends on the orientation of the dipoles of the energy-donor D and the energy-acceptor A (for an introduction to this problem see refs. 13 and 14). In most cases no detailed information is available about the orientational distribution of the dye molecules relative to each other or relative to the biopolymer.

If the rotational diffusion of both dye molecules is isotropic and rapid compared with the fluorescence lifetime of D, the orientational factor κ^2 (which is included in R_0) is 2/3. Dyes covalently attached to a biopolymer almost never show complete isotropic rotational freedom (they usually interact to some extent with the biopolymer) and, consequently, the assumption $\kappa^2 = 2/3$ may not be appropriate in many circumstances (20); this unwarranted assumption can produce significant errors in the calculation of the distances. However, even if one dye molecule remains essentially statically orientated in space while the other dye molecule undergoes complete rapid dynamic rotation (this is approximately true for many of the donor–acceptor pairs that have been used in studies of nucleic acid structures) it can be shown that the maximum possible error of R_0 calculated on the basis of $\kappa^2 = 2/3$ is 12% (13, 20, 21).

The assumption that $\kappa^2 = 2/3$ is nevertheless frequently used in the literature. The good agreement between FRET and other experimental approaches (7) indicates that the distributions and dynamics of the relative orientations between the A and D transition dipoles are often such that κ^2 is approximately a constant within a series of samples under investigation and that the approximate value of 2/3 for κ^2 is satisfactory (to see why this is true, see ref. 13). The exact value of κ^2 is less critical when estimating global structures if a large number of similar but structurally slightly modified samples are compared.

2.2 Influence on the spectroscopic properties of the dyes by other mechanisms than FRET

The FRET efficiency, E, depends on the value of the quantum yield of D in the absence of A. E is independent of the quantum yield of A—even a non-fluorescent molecule can be used as an energy acceptor—but E depends on the molar absorbance of A. Spectroscopic properties of chromophores often depend on their molecular surroundings; therefore, it is important to hold the molecular environment constant in a series of measurements where FRET efficiencies of several doubly-labelled nucleic acid structures are to be compared. Some common problems associated with this are listed here.

(1) The spectroscopic properties of some commonly used chromophores are sensitive to changes in the pH of the solution (lists of the useful pH ranges for many dye molecules are given in ref. 22). Therefore, all spectroscopic measurements should be carried out at the same pH for all samples, or the pH dependence of the chromophores must be known so that the data can be corrected.

(2) In general the spectroscopic parameters of chromophores in non-interacting solvents are temperature independent; however, the interactions of the chromophores with the solvent (or with the macromolecules themselves) are often very temperature dependent. The temperature dependence of a commonly used acceptor TMRh conjugated to the 5′-end of DNA has been investigated in detail (23); this is an example of an extreme effect of temperature. If the temperature dependence of the spectroscopic parameters of a dye is not known, all measurements within a series of measurements to be compared should be performed at the same temperature.

(3) The quantum yield of a dye often depends on the concentration of compounds which can act as quenchers of the excited state of the dye (3). In addition, the interaction of fully or partly charged dyes with the negatively charged backbone of the nucleic acid may vary with the presence of additional monovalent and divalent ions; therefore, the concentration of the components in the buffer should be kept constant. If one wants to follow structural transitions of nucleic acids induced by ions with FRET, it is necessary to investigate the influence of the ions on the spectroscopic parameters of the dyes.

(4) Most dyes used for labelling of nucleic acids interact with the neighbouring base pairs, and the formation of several distinct complexes of dyes with DNA has been reported (23). The excited state of a dye can therefore be quenched by the interactions of the dye with the nucleotides of the DNA backbone; these interactions, in general, may be sequence dependent. Different sequences in the immediate environment of the dye may cause the formation of discrete dye–DNA complexes with completely different spectroscopic and stereochemical properties. The neighbouring base pairs should always be the same for all dye-labelled samples to enable a direct quantitative comparison of the experimentally determined FRET efficiencies.

2.3 The influence of donor–acceptor distance distributions

The conformation of the DNA or RNA molecules can change with time, and the orientation of the dyes attached to the nucleic acid will also change during the lifetime of the excited state of the dye owing to thermal fluctuations. Steady-state fluorescence measurements do not provide direct information on the flexibility of the nucleic acid (the FRET efficiencies determined from steady-state measurements are related to an average distance between the dyes). If the DNA/RNA structure is so flexible that the distances between D and A form a broad distribution (this could be the case especially for large DNA structures with structural elements that contribute efficiently to the flexibility), those conformers with shorter D–A distances would transfer more effectively than the conformers with larger distances; in this case, structural differences would be more difficult to detect and the interpretation of the data is more complex. Such pronounced effects of flexibility are not serious limitations provided that the macromolecules are relatively small (in the range of about 10–30 bp even for bulged molecules with up to seven unpaired nucleotides in the bulge (9)). The determination of structures using steady-state FRET experiments for molecules of this size have been in very good agreement with the results from other experimental techniques. Distance distributions between an energy donor and an energy acceptor are usually measured with time-resolved fluorescence spectroscopy in the nanosecond time-scale. These dynamic experiments are experimentally more difficult to perform and analyse (see ref. 24 for an introductory overview of this field) and are not discussed here.

3 Experimental determination of FRET efficiencies

The FRET efficiency, E, can be determined from fluorescence spectroscopic experiments in several ways (see refs. 13, 14, 25, 26 for an introduction and for extensive references to the literature concerning FRET). Here, two steady-state methods are described which have been applied successfully in the study of DNA/RNA structures using a conventional steady-state fluorimeter with polarizers.

3.1 Normalizing the enhanced steady-state fluorescence of the energy acceptor

The most clear-cut observation unequivocally demonstrating that energy transfer has taken place is to measure the enhanced production of the excited state of the acceptor in the presence of the donor, and the easiest way to do this is to chose an acceptor that fluoresces and monitor its fluorescence. The efficiency of energy transfer can be determined by normalizing the enhanced emission of the energy acceptor, A (after exciting primarily the donor D), by its fluorescence intensity excited at a wavelength where only it absorbs. This method has proven to be useful for D–A pairs covalently attached to DNA/RNA structures (26). It is especially simple to apply provided that two prerequisites are fulfilled:

(1) If A can be excited independently from D one can normalize the enhanced emission to a fluorescence signal that emanates only from A;

(2) If the fluorescence spectra of D can be measured over a wavelength range without a fluorescence contribution from A, it is very easy to determine accurately the fluorescence intensity of A in wavelength regions where both D and A contribute to the signal.

The following measurements and analysis are performed: The sample labelled with D and A is illuminated at a wavelength ν' which excites mainly D (see *Figure 2A*). The fluorescence emission spectrum is divided into two parts: one part of the emission spectrum where only D contributes to the fluorescence signal, and another part where the fluorescence signal has contributions from both D and A. The excited state of A is populated due to direct excitation at ν' and also energy-transfer from molecules of D that were also excited at the ν' wavelength. The spectrum in the range where only D contributes to the fluorescence is fitted to a spectrum of a DNA sample which is labelled only with D. The donor spectrum can then be subtracted from the total spectrum, even in the regions where D and A contribute to the fluorescence; thus one can obtain the pure emission spectrum of A (called the extracted acceptor signal). This extracted acceptor spectrum contains the fluorescence from acceptor molecules that are directly excited, and which have become excited through energy transfer. A is then excited at a wavelength ν'' (where D is not excited). The signal from the emission spectrum of A at a wavelength ν_2 can be used to normalize the extracted acceptor spectrum at the wavelength ν_1 (integrals over the spectrum can also be used; see ref. 26). The ratio of the extracted acceptor fluorescence signal due to excitation at ν' and the acceptor fluorescence signal due to excitation at ν'' is called $(\text{ratio})_A$ and is related to the FRET efficiency E for $\nu_1 = \nu_2$ (identical emission wavelength of acceptor and extracted acceptor) by:

$$(\text{ratio})_A = E \cdot d^+ \cdot (\varepsilon^D(\nu')/(\varepsilon^A(\nu'')) + (\varepsilon^A(\nu')/(\varepsilon^A(\nu''))$$

The FRET efficiency E can be calculated from $(\text{ratio})_A$ if the percentage labelling with D ($d^+ = 1$ for 100% labelled samples) and the values for the spectroscopic constants are known. $\varepsilon^A(\nu')/\varepsilon^A(\nu'')$ can be determined from the $(\text{ratio})_A$ value of a

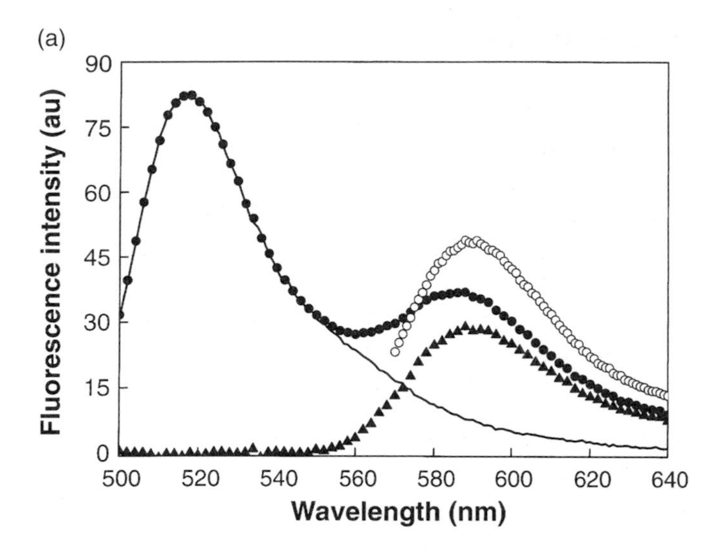

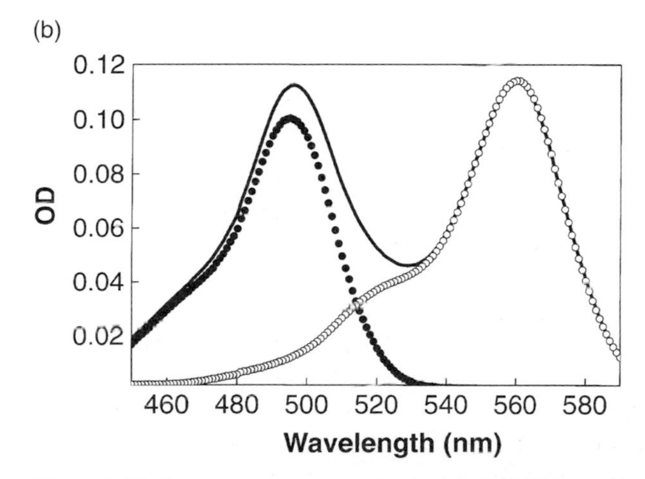

Figure 2 (A) Fluorescence spectra of a dye-labelled DNA sample. The donor-dye (6-carboxyfluorescein) and the acceptor-dye (5-TMRh) are attached to the 5′-ends of the DNA. After illumination at $\nu' = 490$ nm the spectrum is acquired from 500 to 640 nm (solid circles). The region where only the donor emits (from 500 to $\approx$ 540 nm) is fitted to the spectrum of a DNA sample labelled with the donor-dye only (solid line) and then subtracted from the entire spectrum to generate the 'extracted acceptor spectrum' (solid triangles). The extracted acceptor spectrum is normalized with the spectrum of the acceptor, excited at $\nu'' =$ 560 nm and acquired from 570 to 640 nm (open circles). (B) Absorption spectrum of a DNA sample labelled with 6-carboxyfluorescein and 5-TMRh (solid line). The absorption of 5-TMRh at its maximum (560 nm) can be measured without interference from 6-carboxyfluorescein. The absorption of 6-carboxyfluorescein (in the range 460–520 nm) has to be corrected for the contribution of TMRh. The absorption spectrum in the range 550–580 nm is fitted to the absorption spectrum of a DNA sample labelled with 5-TMRh only (open circles) and then subtracted from the entire spectrum to extract the 6-carboxyfluorescein contribution in the range 460–520 nm (solid circles).

sample labelled only with A ($d^+ = 0$) or from an excitation spectrum. $\varepsilon^D(\nu')/\varepsilon^A(\nu'')$ can be determined from the absorption spectra of the samples labelled with D and A (see *Figure 2B*).

Using the $(\text{ratio})_A$ method, very low FRET efficiencies can be determined accurately (as low as 0.01–0.05). Structural interpretations based on the FRET efficiencies determined with the $(\text{ratio})_A$ method have been in a excellent agreement with other experiments. The main advantages of this method are (26):

(1) All fluorescence measurements are made on the same solution; additional absorption measurements and concentration determinations are not necessary.

(2) $(\text{ratio})_A$ is independent of the fraction of labelling with A and depends only on the fraction of labelling with D.

(3) The quantum yield of A does not enter into the determination of $(\text{ratio})_A$.

(4) The quantum yield of D does not influence the method for determining E, it only influences the actual value of E because the actual value of R_0 is a function of the quantum yield of the energy donor.

3.1 Measuring changes in the fluorescence anisotropy of the energy donor

FRET can be determined by measuring the anisotropy of the donor fluorescence (for an introduction into the steady-state theory of the anisotropy see refs. 3 and 27). The anisotropy of a fluorescent molecule depends on the angle between the absorption and the emission dipole of the dye and on the extent of rotational diffusion of the dye during the lifetime of the excited state. The lifetime of the donor τ_D in the absence of A is longer than the lifetime of the donor in the presence of an acceptor τ_{DA}; that is, $\tau_D > \tau_{DA}$. Therefore, the extent of rotational diffusion of D (while D is in the excited state) in the presence of A will be less than in the absence of A. The FRET efficiency E can be calculated from the anisotropy r_{DA} of D (in the presence of A) if the anisotropy r_D for D (in the absence of A) and the limiting anisotropy r_0 is known (9):

$$E = 1 - \{(r_0/r(\nu,\nu')) - 1\}/\{(r_0/r_D(\nu,\nu')) - 1\}.$$

The FRET efficiency calculated according to this equation is independent of the amount of labelling with D but depends on the amount of labelling with A (the situation is reversed when determining E from the normalized sensitized emission of the acceptor). This method can only be applied if the fluorescence of D can be observed independently of the fluorescence of A. The sample labelled with D only and the sample labelled with D and A must be excited at the same wavelength.

In general, the FRET efficiencies calculated from the normalized sensitized emission of the acceptor and from the anisotropy of the donor are in good agreement for a given sample if all spectroscopic constants in the two equations are known and if the sample is virtually 100% labelled with D and A. The FRET efficiencies determined from the normalized enhanced emission of the acceptor are, according to our experience, a better indicator for small structural changes when structurally similar but slightly different samples are compared.

One must remember that the determination of E observing either the fluorescence of D, or the fluorescence of A, may not agree if there are several conformers of the donor–DNA or acceptor–DNA complexes with different spectroscopic parameters. For example, there are several conformers of TMRh–DNA, and for one conformer the quantum yield is almost zero (23); therefore, by observing the fluorescence of A we will not observe the energy transfer process to this conformer of TMRh–DNA, and we would only be sensitive to the energy transfer to the TMRh molecules that fluoresce. In general, this has not been a problem because the FRET measurements for structural studies have been made under conditions where the population of TMRh–DNA which does not fluorescence is small.

The original papers referenced should be consulted for more information on both of these methods and for discussions and references concerning alternative methods of determining the FRET efficiency, E.

4 Preparation of dye-labelled DNA structures

The determination of the FRET efficiency between two dyes covalently attached to a nucleic acid is simplified if the labelling efficiency is virtually 100%. It is therefore preferable to prepare the samples such that the labelling is complete for both dyes, although it is possible to make corrections for incomplete labelling (26). The labelling of DNA and RNA structures has been reviewed several times (28, 29). Here we will give some general comments on the labelling and purification strategies and describe some of the potential problems in the preparation of dye-modified nucleic acid structures.

4.1 Labelling with an amino-reactive compounds

Only a few dyes are available as phosphoramidite compounds which can be attached to oligonucleotide during solid-phase synthesis. Most dyes are conjugated post-synthetically to the amino-modified oligonucleotide using succinimidester or isothiocyanate derivatives of the dye (22, 30). Incomplete reaction of amino-reactive dye derivatives with the aminolinker of an HPLC-purified oligonucleotide can have several causes:

(1) Commercially available dye derivatives are generally of high quality. Degradation of the amino-reactive group of the dye derivative results mainly from improper storage of the compounds – dyes with amino-reactive groups are degraded in the presence of small amounts of water and should therefore be stored dry (preferably in the complete absence of water).

(2) The reaction of amino-reactive groups with an aminolinker is highly pH-sensitive. The yield of the reaction is strongly decreased at lower pH-values (< 8.5).

(3) A relatively small excess of the amino-reactive dye is sufficient for complete reaction. A 5- to 10-fold excess of the succinimidester of TMRh is sufficient for the complete reaction with a C6-aminolinker attached at the 5′-end of an oligonucleotide.

(4) The reaction yield depends on the length of the aminolinker and on the nature of the dye. The reaction is less effective for short linkers and for linkage reactions at internally modified bases. Amino-derivatives of dyes with a tendency for a more or less strong non-covalent interaction with DNA or RNA (this is the case for rhodamine and cyanine dyes) show a high reactivity in the reaction with an amino group.

(5) Incomplete reaction of amino-reactive compounds is, in our experience, very often caused by non-reactive aminolinkers. The chemical quality of the aminolinker can be controlled by derivatization with fluorescamide (22).

4.2 Purification by denaturing and native PAGE

After the removal of unreacted dye by conventional or spin-down chromatography on Sephadex, further purification of the labelled oligonucleotides by preparative denaturing PAGE (31) is the method of choice for removing unreacted oligonucleotides. This method is time-consuming but extremely effective, even for longer oligonucleotides (30–70 bases). Oligonucleotides should be denatured completely before applying the samples to the gel (31). The samples can be removed from the gel by electroelution.

The dye-modified oligonucleotides are hybridized to form double-stranded DNA or RNA molecules. Unreacted dye-labelled single-stranded oligonucleotides can be removed conveniently by native PAGE: the presence of dye-labelled single-strands would otherwise interfere with the analysis of the spectroscopic measurements.

5 Characterization of dye-labelled DNA and RNA structures

The biochemical and spectroscopic characterization of the dye-labelled nucleic acid molecules is the most important step for accurate determination and interpretation of the FRET efficiencies. There are six important steps for characterizing the samples:

(1) The purity of the oligonucleotides which are used to construct the DNA or RNA structures can be estimated by their migration behaviour during the denaturing PAGE. The migration of the oligonucleotide should be in accordance with its length when compared with a standard oligonucleotide. The amount of decomposition products should be small compared with the amount of the main product.

(2) The quality of the double-stranded dye-labelled DNA or RNA structure can be controlled by the migration behaviour during the PAGE under native con-

ditions. Only two bands should appear on the native gel: one 'slow band' (the hybridization product) and one 'fast band' (the excess oligonucleotide). The appearance of several "slower bands" is a clear indication for the formation of several different structures. FRET measurements on a mixture of structures cannot be interpreted clearly with confidence.

(3) The extent of labelling with D and A can be estimated by comparison of the relative optical densities for a larger number of samples with a different number of bases. A plot of the OD(UV)/OD(VIS) values (the ratio of the optical density of the DNA or RNA—OD(UV)—with the optical density of the dye - OD(VIS)—at the absorption maxima) versus the number of bases of the structure should be linear.

(4) The OD(donor)/OD(acceptor) values (ratio of the optical density of the energy donor D—OD(donor)—with the optical density of the energy acceptor A - OD(acceptor)—at their absorption maxima) are very sensitive indicators for incomplete labelling and should be the same for all samples within a 2–5% error range. Larger deviations indicate either incomplete labelling or a change in the extinction coefficient of the dye caused by different interactions of the dye with the DNA or RNA structures.

(5) The maxima of the absorption spectra and of the fluorescence spectra of the dyes should be at the same wavelength for all samples within ≈ 1 nm. The shape of the absorption and fluorescence spectra should be the same for all samples. Stronger shifts and different spectral shapes indicate that the close molecular environments of the dyes for the different samples are different.

(6) The fluorescence anisotropy of A (excited at a wavelength where D is not excited) should be constant for all samples. Different anisotropies of A may be caused by differences in the direct molecular environment of A or by different deactivation pathways of the excited state of A, and can also indicate the presence of single-stranded dye-labelled oligonucleotides or free dye. In general, the anisotropy of a dye is less for single-stranded molecules or free dye than when the dye is attached to the 5′-end of an intact duplex structure.

6 Structural interpretation of FRET efficiencies

The comparison of FRET efficiencies between two dyes covalently attached to a series of similar nucleic acid structures is an elegant approach for deducing information about the global structure of these molecules. A series of measurements with an increasing number of base pairs in the stem regions were carried out in several of these studies. Thus, the location of the dyes are different on the macromolecule and their relative positions and orientations are different. Structural information can then be deduced by quantitatively estimating the intramolecular distances within the series experiments from the FRET measurements.

For a detailed interpretation of the data, information should be available about the positions of the dyes at the DNA helix ends. A population representing discrete spatial distributions of dye positions at the biopolymer can be calculated with a systematic conformational search method. Here, we describe this procedure for dyes attached to the 5′-end of the DNA molecule. Molecular modelling can be employed for assembling molecular building blocks from known X-ray and NMR-structures, taking into account the measured dye-to-dye distances. Starting from one molecular model, several alternative structures are generated with local conformational search methods or molecular dynamic simulations. Only those models that fulfil all experimental constraints are accepted. When two (or more) totally different structural models are in agreement with the data, one cannot draw any structural conclusions with confidence.

6.1 Modelling the dye positions at the DNA helix ends

FRET measurements can be used for estimating apparent singular distances between the donor and the acceptor, even where there is a distribution of positions of one or both of the dyes. Such discrete positions are convenient to use in model building. The singular positions of the dyes must represent averages of the actual distributions of the dyes. The location of a dye molecule relative to the DNA molecule depends mainly on their steric and electrostatic interactions with DNA. Two dyes commonly used for FRET measurements are 6-carboxyfluorescein and 5-carboxytetramethylrhodamine (5-TMRh). At a pH in the range of 7.5–9, which is common for the FRET measurements discussed here, 6-carboxyfluorescein is twofold negatively charged and 5-TMRh is neutral. In general, these differences in charge and shape lead to different interactions of the two dyes with DNA. Anisotropy measurements indicate that 6-carboxyfluorescein rotates freely in solution, whereas 5-TMRh is relatively immobile owing to strong interactions with the DNA. Positions that represent the distribution of conformations of the dyes at the helix ends can be suggested by molecular modelling. Systematic conformational searches with force-field methods were applied to find low-energy conformers where the dyes interact most favourably with the DNA or to calculate a distribution of conformations with lower energy.

Models of the small organic dye molecules are built and subsequently energy optimized with force-field methods. These models can be refined by applying semi-empirical or *ab initio* optimizations. The partial charges of the dye molecules should be calculated with semi-empirical or *ab initio* methods. Programs available for the calculation of charges are, e.g., MOPAC (32), GAMESS (33) and GAUSSIAN (34). The dyes are then attached to the 5′-end or a modified base in the DNA with an extended linker conformation. B-DNA structures from X-ray data are used to represent the nucleic acid molecule. Torsion angles in the linkage are corrected by a short molecular mechanics minimization of the DNA-dye complex. It is advisable to use force fields that are parameterized for biopolymers, e.g. AMBER (35), CHARMM (36) or GROMOS (37), for these purposes. Conformers of the dyes from all regions of the conformational space are then generated by systematically varying the torsion angles in the linkers. Such algorithms are implemented in commercially available modelling programs such as InsightII (38), MacroModel (39) or SYBYL (40). For 6-carboxyfluorescein with C6-linkers there are 12 rotatable bonds. If all torsion angles are varied with increments of 60°, a number of $(360/60)^{12} = 2.2\ 10^{9}$ conformations are generated. All conformers with steric clashes are excluded, drastically reducing the number of contending conformations. The potential energy of the resulting conformations is then calculated with force-field methods. No computer-time consuming minimization is needed in this step because only the single point energy of each conformer is evaluated. It is not yet possible to include explicit water molecules in this calculation because the complexity of the conformation space generates an enormous number of conformations. In order to approximately simulate the effects of a solvent, the electrostatic interactions should be damped by a factor of 2–4. The conformers are ranked by their internal energy values and a subset of 1000–10 000 low-energy conformers is selected for further refinement. A database of the conformers is generated and an energy optimization is performed using force-field methods. The atoms of the DNA molecule are kept fixed during this minimization.

The two dyes, 6-carboxyfluorescein and 5-TMRh with C6-linkers at the DNA ends, demonstrate disparate behaviour. For most of the 6-carboxyfluorescein conformers, the linker is close to fully extended, and is directed away from the DNA helix (see *Figure 3*). This is presumably due mainly to the electrostatic repulsion between the twofold negatively charged fluorescein molecule and the polyanionic DNA. The force-field energy values of the 1000 optimized conformations with 6-carboxyfluorescein differ by only ≈ 5 kcal/mol. Since 6-carboxyfluorescein does not interact by direct contact with the DNA, a representative position of this dye can be calculated by averaging the coordinates of a central atom in the dye over a sufficient number of low-energy conformations. This central atom should be chosen to be part of the chromophore's ring system. The average position for 6-carboxyfluorescein is located in a virtual elongation of the phosphate backbone ≈ 10 Å away from the 5′C-atom of the DNA. In the case of 5-TMRh, most of the chosen conformers interact with the DNA (see *Figure 3*). The energy difference between the 5-TMRh conformers with the lowest and the highest energies of interaction (≈ 20 kcal/mol) are much greater than for 6-carboxyfluorescein. In the case of 5-TMRh, a well-defined energy minimum was found. The 5-TMRh conformers with low energies all interact very similarly with the DNA. The preferred location for the 5-TMRh molecule on the DNA molecule is found to be in the major groove, next to the phosphate backbone of the unlabelled strand, and in a stacked conformation on top of the DNA double helix. The uncertainties for the average positions of 6-carboxyfluorescein and the low-energy conformer of 5-TMRh are 3 Å and 4 Å, respectively. The results of the calculations are in accord with the available experimental data (9, 23, 41). The anisotropy measurements indicate that 5-TMRh interacts mainly with the DNA, whereas 6-carboxyfluorescein freely rotates in solution. The dye-to-dye distances from a series of seven double-helical DNA molecules determined with FRET (8) correlate well with the theoretically calculated distances. These

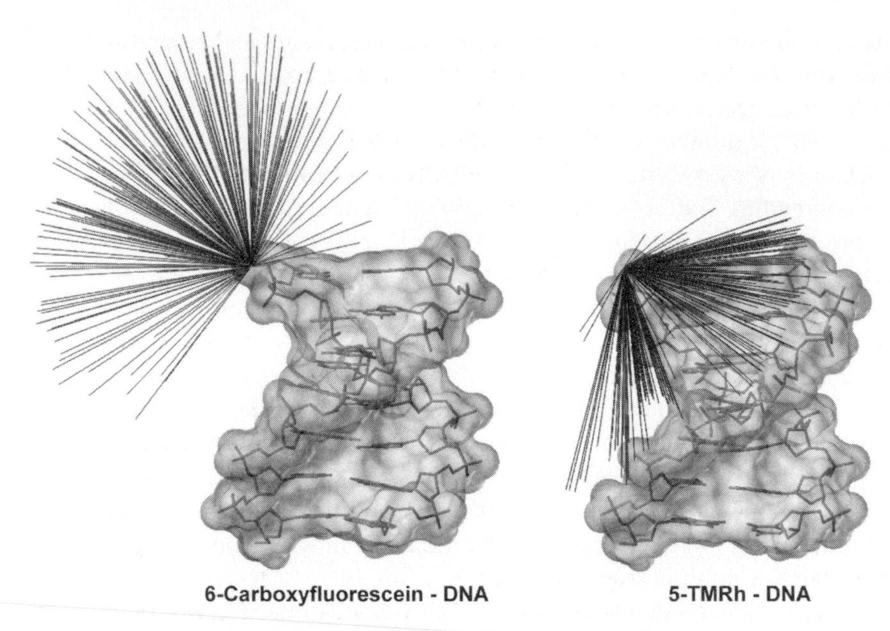

Figure 3 Distribution of the 200 energetically most favourable conformers for 6-carboxyfluorescein and 5-TMRh at the DNA helix ends. A central atom in the 200 superimposed conformations is connected with atom C5′ of the DNA through straight lines. For clarity, the other atoms of the dyes are not displayed. The dyes adopt a different position with respect to the DNA is shown.

approximate dye positions at the DNA helix-ends can now be used to transform measured FRET efficiencies into low-resolution structural constraints for DNA model building.

6.2 Modelling nucleic acid conformations and protein–DNA interactions

The measured FRET efficiencies are used to determine intra- or inter-molecular distances, and these distances are then used to construct molecular models. The approximate positions of the dyes that are attached to the DNA molecules (see above paragraph) are used when adjusting the nucleic acid geometry to fit the experimentally observed dye-to-dye distances. Owing to the flexibility of the dyes and the remaining uncertainty in the dye positions it is not possible to use dye-to-dye distances as exact constraints, as is done with NOE-constraints in NMR spectroscopy. However, by completing a series of FRET measurements with different labelling positions, valuable information can be acquired concerning the global orientation of subparts of the nucleic acid macromolecule. Other experimental data, such as mobilities in gel electrophoresis, footprinting and crosslinking data can contribute to the success of the model building and reduce the number of viable alternative structures. Models are constructed from known structural elements derived from X-ray and NMR data. These fragments are manually assembled on a graphics terminal to satisfy all low-resolution constraints. A wide variety of nucleic acid structures, such as double-helical DNA and RNA, hairpin, bulged and internal loops are available in the Brookhaven Protein Databank (42) and the Nucleic Acid Database (43). Examples of bent or kinked DNA conformations are available from DNA–protein complexes (for a review see ref. 44). These structures of shorter DNA fragments can be extended to longer lengths by fitting overlapping parts of DNA to normal helical duplex DNA. Useful tools for generating unusual nucleic acid conformations are the programs NAMOT (45) and JUMNA (46). After assembling the molecular fragments, the energy of the model is optimized to remove steric clashes and deviations from standard bond lengths and angles.

Since the interactive modelling procedure is always intuitive and biased from the point of view of the user, it is advisable to refine the models by searching the local vicinity of the conformational space for alternative solutions using systematic or random methods. Two principle methods for doing this are described here. In a first step, successive molecular mechanics optimizations applying internal coordinates with an implicit solvent model are used to search for low-energy conformers from larger regions of the conformational space. More detailed questions can then be answered by carrying out molecular dynamic simulations taking into account explicit water molecules and counter ions.

A powerful molecular mechanics program for searching for low-energy conformations of nucleic acid molecules is JUMNA (46). The principal characteristic of this program is that the energy optimization of nucleic acids is carried out in internal coordinates. Unlike other molecular mechanics programs, much larger conformational changes can be obtained during the JUMNA energy optimization. The tendency of all molecular mechanics programs to get trapped in local energy minima is lower using JUMNA. This program is therefore appropriate for systematic searches of nucleic acid conformations. Three force fields parameterized for nucleic acids are currently available within JUMNA. The program employs an implicit model of a sigmoidal distance-dependent dielectric function to represent the environment of the nucleic acid, and the phosphate charges are reduced to 50%, in order to mimic the surrounding water and counter ions. Regions of local flexibility such as bulge-, internal- and hairpin-loops can be studied with this method. Dye-to-dye distances from FRET measurements can help to reduce the number of alternative structure models, along with other experimental data and the internal force field energy values from JUMNA.

Because these molecular mechanics calculations are carried out with electrostatic models that only rudimentarily describe the solvent environment, it is advisable to test the stability of the conformations with molecular dynamic (MD) simulations. It has been shown that an explicit representation of the solvent and ionic environment of nucleic acids is necessary for a precise description of local structural changes in nucleic acids (47). If the molecules are not too large, the particle-mesh Ewald method (48) can be used to calculate all non-bonded interactions in a periodic water box without truncations. Stable MD trajectories in the nanosecond time-frame have been carried out for DNA molecules (49),

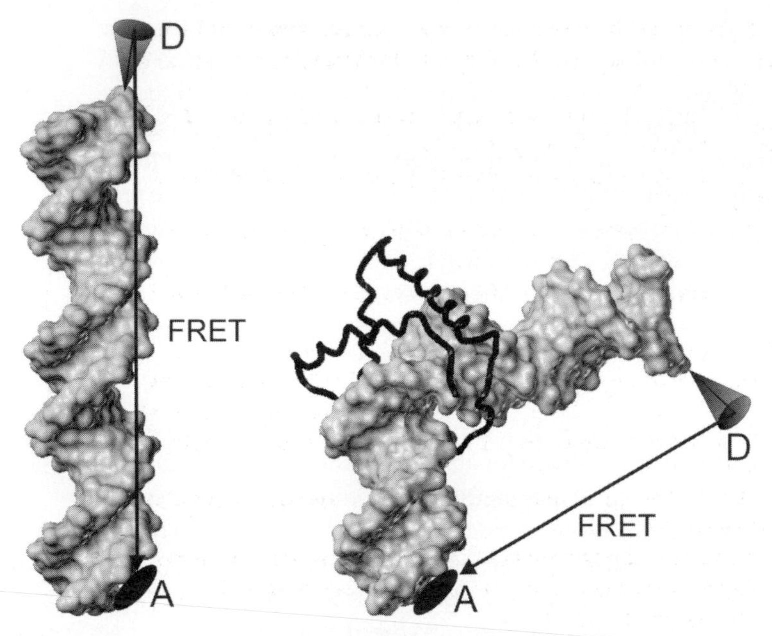

Figure 4 Principle of structural interpretations of FRET measurements. The proposed positions for the donor (D) and acceptor (A) dyes with their positional uncertainties are indicated. The linear B-DNA (left) is kinked upon binding of a protein (right, protein displayed in black tube representation). The differences in the dye-to-dye distances of several FRET measurements can be used to build models of the kinked DNA and to estimate bending angles.

proteins (50) and DNA–protein complexes (A. Hillisch, unpublished data). Very large protein–DNA complexes can be simulated in an environment consisting of several shells of water molecules (51). The accurate treatment of long-range electrostatic interactions is one advantage of molecular dynamics simulations. These methods can also provide information about the flexibility of the molecules studied by FRET and can, therefore, help interpret the experimental data on a structural level. A drawback of most molecular dynamics calculations is that the conformational space is sampled only locally because of the limited time scales that can be computed.

The principle of the structural interpretation of FRET measurements is depicted in *Figure 4*. A DNA molecule is kinked because of the binding of a protein. This leads to a significant reduction of the dye-to-dye distance and an increase of the FRET efficiency. Kinking angles can be assessed from molecular models which are built by including e.g. FRET, NMR and additional low-resolution experimental data (11, 12).

References

1. Galanin, M. D. (1996). *Luminescence of molecules and crystals*. Cambridge International Science Publishing, Cambridge.
2. Turro, N. J. (1991). *Modern molecular photochemistry*. University Science Books, Sausolito.

3. Lakowicz, J. R.(1983). *Principles of fluorescence spectroscopy*, Plenum Press, New York.
4. Murchie A. I., Clegg R. M., von Kitzing E., Duckett D. R., Diekmann S., and Lilley D. M. (1989). *Nature* **341**,763.
5. Tuschl, T., Gohlke, C., Jovin, T. M., Westhof, E., and Eckstein, F. (1994). *Science* **266**, 785.
6. Yang, M. and Millar, D. P. (1996). *Biochemistry* **35**, 7959.
7. Stühmeier, F., Welch, J. B., Murchie, A. I. H., Lilley, D. M., and Clegg, R. M. (1997). *Biochemistry* **36**, 13530.
8. Clegg, R. M., Murchie, A. I. H., Zechel, A., and Lilley, D. M. (1993). *Proc. Natl. Acad. Sci. USA* **90**, 2994.
9. Gohlke, C., Murchie, A. I.H., Lilley, D. M., and Clegg, R. M. (1994). *Proc. Natl. Acad. Sci. USA* **91**, 11660.
10. Parkhurst, K. M., Brenowitz, M., and Parkhurst, L. J. (1996). *Biochemistry* **35**, 7459.
11. Payet, D., Hillisch, A., Lowe, N., Diekmann, S., Travers, A. A. (1999). *J. Mol. Biol.* **294**, 79.
12. Lorenz, M., Hillisch, A., Goodman, S. D., Diekmann, S. (1999). *Nucleic Acids Res.* **27**, 4619.
13. Clegg R. M. (1996). In *Fluorescence imaging spectroscopy and microscopy*, p. 179, Wiley-Interscience, New York.
14. Wieb van der Meer, W. B., Cooker, C., and Chen, S. Y.(1998). *Resonance energy transfer—theory and data*. VCH-Verlag, New York.
15. Cantor, C. R. and Schimmel.P. R.(1980). *Biophysical chemistry II*. Freeman, San Francisco.
16. Kuhn, H. (1977). In *Biophysik*, Springer Verlag, Berlin, Heidelberg, New York.
17. Förster, T. (1949). *Z. Naturforsch.* **4a**, 321.
18. Förster, T. (1948). *Ann. Phys.* **2**, 55.
19. Wu, P. and Brand, L. (1994). *Anal. Biochem.* **218**, 1.
20. Dale, R. E. and Eisinger, J. (1974). *Biopolymers* **13**, 1573.
21. Dale, R. E., Eisinger, J., and Blumberg, W. E. (1979). *Biophys.J.* **26**, 161.
22. Haugland, R. P.(1996). In: *Handbook of fluorescent probes and research chemicals* (ed Larison, K. D.) Molecular Probes, Eugene.
23. Vamosi, G., Gohlke, C., and Clegg, R. M. (1996). *Biophys. J.* **71**, 972.
24. Millar, D. P. (1996). *Curr. Opin. Struct. Biol.* **6**, 322.
25. Yang M. and Millar D. P. (1997). p. 417. Academic Press, London.
26. Clegg, R. M. (1992). *Methods Enzymol.* **211**, 353.
27. Pesce, A. J., Rosen, C. G., and Pasby, T. L.(1971). In *Fluorescence spectroscopy*. Marcel Dekker Inc., New York.
28. Waggoner, A. (1995). *Methods Enzymol.* **246**, 362.
29. Mergny, J. L., Boutorine, A. S., Garestier, T., Belloc, F., Rougee, M., Bulychev, N. V. *et al.* (1994). *Nucl. Acids. Res.* **22**, 920.
30. Connolly, B. A. (1987). *Nucl. Acids. Res.* **15**, 3131.
31. Ellington A. (1993). In *Current protocols in molecular biology*. John Wiley & Sons, New York.
32. Stewart J. J. (1990). *J. Comput.-Aided Molec. Des.* **4**, 1.
33. Schmidt, M. W., Baldridge, K. K., Boatz, J. A., Elbert, S. T., Gordon, M. S., Koseki, S. *et al.* (1993). *J. Comput. Chem.* **14**, 1347.
34. Frisch, M. J., Trucks, G. W., Schlegel, H. B., Gill, P. M.W., Johnson, B. G., Robb, M. A. *et al.* (1995). *GAUSSIAN 94 (revision D.1). User's reference*. Gaussian Inc., Pittsburgh PA.
35. Pearlman, D. A., Case, D. A., Caldwell, J. W., Ross, W. S., and Cheatham III, T. E. (1995). *AMBER 4.1*. University of San Francisco, San Francisco.
36. Brooks, B. R., Bruccoleri, R. E., Olafson, B. D., States, D. J., Swaminathan, S., and Karplus. M, (1983). *J. Comput. Chem.*, **4**, 187.
37. van Gunsteren, W. F. and Berendsen, H. J. C. (1996). *GROMOS 96*. BIOMOS, University of Groningen, Groningen.
38. Molecular Simulations Inc. (1997). *InsightII*. Molecular Simulations Inc., San Diego.

39. Mohamadi, F., Richards, N. G. J., Guida, W. C., Liskamp, R., Lipton, M., Caufield, C. *et al.* (1990). *J. Comput. Chem.* **11**, 440.
40. Tripos Inc. (1997). *SYBYL 6.4*. Tripos Inc., St Louis.
41. Clegg, R. M., Murchie, I. H. M., Zechel, A., Carlberg, C., Diekmann, S., and Lilley, D. M. J. (1992). *Biochemistry* 31, 4846.
42. Bernstein, F. C., Koetzle, T. F., Williams, G. J., Meyer, E. F. J., Brice, M. D., Rodgers, J. R. *et al.* (1977). *Eur. J. Biochem.* **80**, 319.
43. Berman, H. M., Olson, W. K., Beveridge, D. L., Westbrook, J., Gelbin, A., Demeny, T. *et al.* (1992). *Biophys. J.* **63**, 751.
44. Allemann, R. K. and Egli, M. (1997). *Chem. Biol.* **4**, 643.
45. Carter E. and Tung C. S. (1994). *NAMOT 2.1*. National Laboratory, Los Alamos.
46. Lavery R. (1988). In *DNA bending and curvature* (ed. Olson, W. K., Sarma, M. H., Sarma, R. H., and Sundaralingam M.), p. 191. Adenine Press.
47. Westhof E., Rubin-Carrez C. and Fritsch V. (1995). In *Computer modelling in molecular biology* (ed. Goodfellow J. M.). VCH, New York.
48. York, D. M., Darden, T. A., and Pederson, L. G. (1993). *J.Chem.Phys.*, **99**, 8345.
49. Cheatham, T. E. and Kollman, P. A. (1996). *J. Mol. Biol.* **259**, 434.
50. York, D. M., Wlodawer, A, Pedersen, L. G., and Darden, T. A. (1994). *Proc. Natl. Acad. Sci. USA* **91**, 8715.
51. Harris, L. F., Sullivan, M. R., and Popken-Harris, P. D. (1997). *J. Biomol. Struct. Dyn.* **15**(3), 407.

Chapter 7

Determination of DNA–ligand interactions by fluorescence correlation spectroscopy

J. Langowski and M. Tewes
Division of Biophysics of Macromolecules, German Cancer Research Centre, Im Neuenheimer Feld 280, D–69120 Heidelberg, Germany

1 Introduction

The quantitative characterization of protein–DNA interactions is of fundamental importance for our understanding of the mechanism of transcription regulation. A number of methods are applied to measure the thermodynamics and kinetics of such interactions, but many of those commonly used, such as gel shift, nitrocellulose filter binding, or surface plasmon resonance are restricted to systems that are bound to or confined within some kind of matrix.

To measure binding parameters with an accuracy that allows to distinguish between different binding mechanisms, cooperativity, etc., one needs to determine these quantities free in solution. Classical titration methods that have been applied to this problem, which monitor the change in an optical parameter such as absorbance, fluorescence intensity or depolarization, or circular dichroism, have the disadvantage that often large quantities are needed and binding constants greater than 10^8/M can be measured only with great difficulty because of the limited sensitivity.

FCS is a method that has recently gained importance for the measurement of interactions between biomolecules in solution. It is used to determine the concentrations and hydrodynamic properties of fluorescent molecules by analysing their number fluctuations.

This technique and its theoretical foundations were described some time ago (1–3), but routine measurements of biomolecular interactions have become possible only through recent improvements (4, 5): these include the use of confocal optics for excitation and detection, and avalanche photodiode detectors that offer a quantum efficiency > 50% in the red range of the visible spectrum (a factor of 10 over that of typical photomultipliers).

The principle of the method is shown schematically in *Figure 1*. Concentration fluctuations—which can arise from Brownian motion or chemical reactions – are

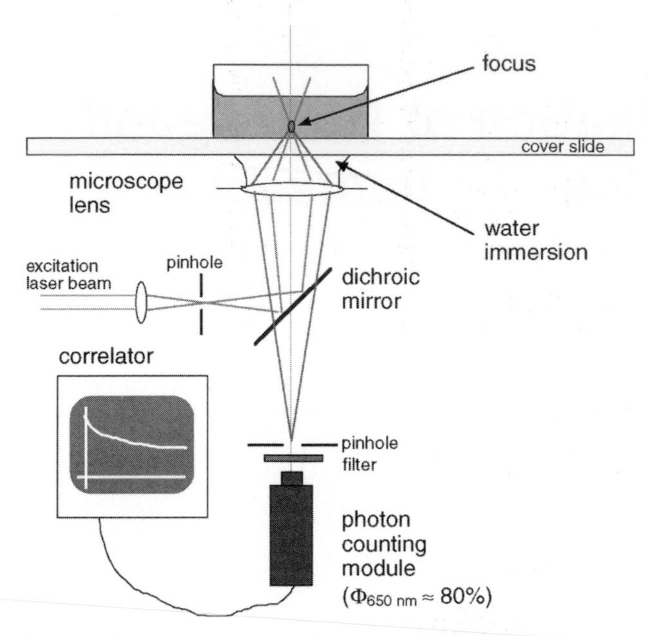

Figure 1 Schematic principle of the FCS method.

detected by exciting the fluorescent molecules in a very small detection volume ($\approx$ 1 fl) with a laser beam focused through a microscope lens. The fluorescence emitted is detected through the same optics; excitation and emission wavelengths are separated by a dichroic mirror and filters.

Very small concentrations (< 1 pM) may be detected because individual fluorescent particles will give clearly distinguishable bursts of fluorescence intensity above the background arising from detector noise, Raman scattering and optical imperfections. The amplitudes and characteristic time scales of the fluctuations measured are connected with macroscopic properties such as particle concentration or diffusion constants by fundamental laws of statistical physics; therefore FCS can be used to measure unknown concentrations and sizes in solutions of fluorescent molecules, or mixtures thereof.

As an example, interactions between molecules may manifest themselves in a decrease of the measured concentration: when dimers are formed, only half as many independently moving particles are present in the solution. In parallel, the mean residence time of the molecules in the observation volume will increase because of their slower diffusion. Thus, an association can be readily quantified, even in the presence of other fluorescent particles which can be distinguished through their different diffusion time.

Owing to the small focus of the laser beam, measurements inside biological objects also become possible. A typical high-resolution microscope lens has a focal spot of 300 nm diameter and 1.5 μm length, so that diffusion processes inside cells or organelles can be probed in a position-dependent manner. FCS has, for example, been used to probe chromatin in the cell nucleus (6).

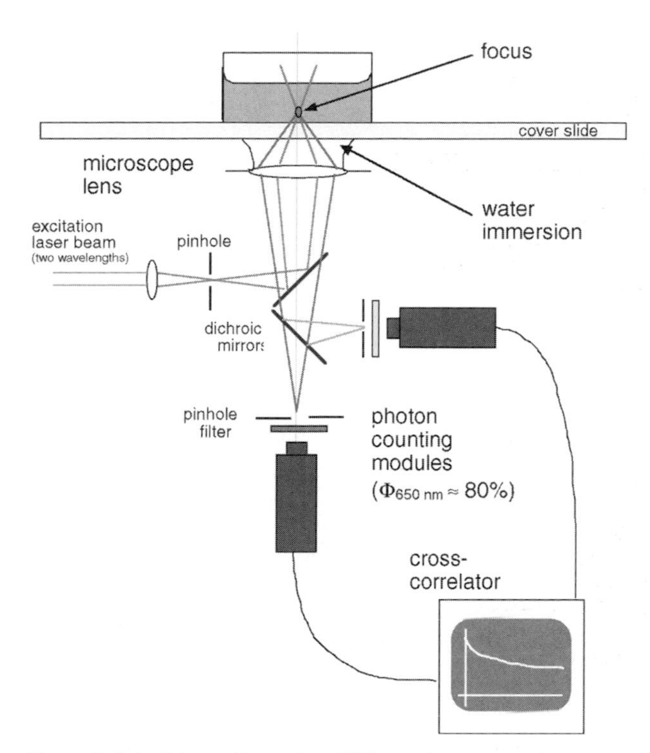

Figure 2 Principles of two-colour FCS.

Another recent development in FCS is the use of two-colour detection with cross-correlation (FCCS; *Figure 2*) (7, 8). Here, emitted light from the same focal volume is detected at two wavelengths. In this case, particles which fluoresce at both wavelengths will give simultaneous bursts of intensity in the two channels. This correlated emission is detected by computing a cross-correlation function. FCCS is a convenient means to show binding between two ligands labelled with different fluorophores because the complex will show correlated fluorescence at the two wavelengths.

2 Theoretical foundation of FCS

2.1 Concentration fluctuations in small systems

In a solution of concentration c, the number of solute molecules N in a given volume element V fluctuates such that $\langle\delta N^2\rangle = \langle N\rangle$, where $\langle N\rangle = cV$ is the average number of molecules in V and $\langle\delta N^2\rangle = \langle(N - \langle N\rangle)^2\rangle$ the mean squared fluctuation. Furthermore, the time-dependence of the fluctuations is directly related to the diffusion coefficient of the molecule, as will be worked out in more detail below. Therefore, by observing the concentration fluctuation of a solute in a very small volume of known size, one can determine its concentration and its diffusion coefficient.

Table 1 Number fluctuations in a 1 nM solution as a function of volume

Size (mm)	Volume (l)	No. of particles	ΔN	ΔN/N (%)
10	10^{-3}	6.023×10^{11}	776080	0.00013
1	10^{-6}	6.023×10^{8}	24541	0.0041
0.1	10^{-9}	6.023×10^{5}	776	0.129
0.01	10^{-12}	602.3	24.5	4.075
0.001	10^{-15}	0.6023	0.776	128.9

Let us assume that we measure the fluorescence of a 10^{-9} M rhodamine solution in volume elements of various sizes. *Table 1* shows the absolute and relative number fluctuations for this case. The typical observation volume in a fluorescence spectrometer is 1 ml. It is easily seen that, at this sample, size no observable fluctuation is expected. If, however, one measures the fluorescence of the same solution in a smaller volume, the fluctuations become increasingly important until they reach the size of the fluorescence signal itself at a sample size of 1 fl; here, less than one molecule is present, on average, in the observation volume. The characteristics of the fluorescence fluctuations and their relation to molecular properties are summarized in the following text.

2.1.1 Autocorrelation, one species

The primary data obtained in an FCS measurement is the time-dependent fluorescence intensity $F(t)$, which is proportional to the number of particles in the observation volume at time t. The autocorrelation function of $F(t)$ contains all relevant information relating to the diffusion of the fluorophores. The normalized autocorrelation function $G(\tau)$ is computed as

$$G(\tau) = \frac{\langle F(t)F(t+\tau)\rangle}{\langle F(t)\rangle^2} \qquad 1$$

To obtain quantities such as diffusion coefficients, concentrations or reaction rate constants, one has to fit a theoretical correlation function to the measured $G(\tau)$ which is based on a model that contains these quantities as free parameters. For a solution of a single fluorescent species with diffusion coefficient D and molar concentration c and for Gaussian profiles for the excitation intensity and detection efficiency, $G(\tau)$ becomes (5):

$$G(\tau) = \frac{1}{cV_{eff}}\left(1+\frac{4D\tau}{w_0^2}\right)^{-1}\left(1+\frac{4D\tau}{z_0^2}\right)^{-1/2} + 1 \qquad 2$$

Here, V_{eff} is the effective observation volume which depends on the geometry of the focus for excitation and emission, w_0 and z_0 are the half-widths of the focus in the x–y plane (the observation plane of the lens) and in the z-direction, respectively.

V_{eff}, w_0 and z_0 can be measured independently by calibration with a solution of a fluorophore of known concentration and diffusion coefficient. If only relative

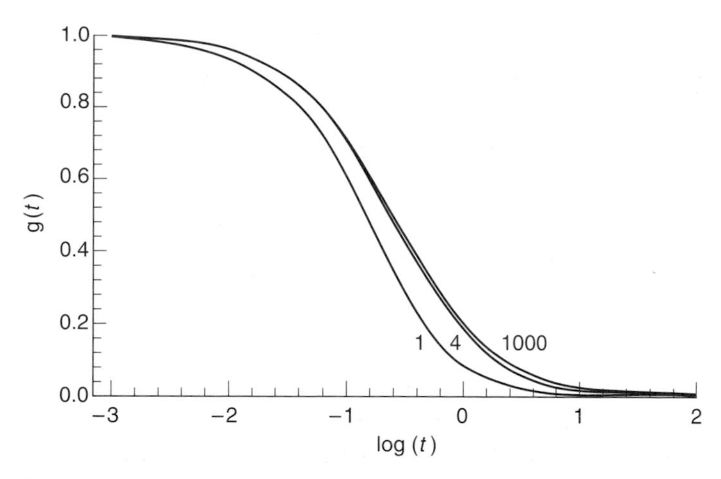

Figure 3 Influence of the structure factor κ on the FCS autocorrelation function. Three curves are displayed for the same diffusion time, τ, and κ = 1, 4, 1000. Since κ = 4 for typical confocal optics, the relevant range in practice is between the two furthest-right curves.

changes are of interest, one can use the average particle number $N = cV_{\text{eff}}$ and an effective diffusion time $\tau_{\text{diff}} = w_0^2/4D$ as parameters:

$$G(\tau) = \frac{1}{N}\left(1 + \frac{\tau}{\tau_{diff}}\right)^{-1}\left(1 + \frac{\tau}{\tau_{diff}\,\kappa^2}\right)^{-1/2} + 1 \qquad 3$$

κ (also called the structure factor) is the axial ratio of the observation volume, z_0/w_0.

For a high aperture lens (NA = 1.2) at optimal alignment, κ typically ranges between 4 and 6. As can be seen in *Figure 3*, the influence of κ on the shape of the correlation function is rather small in this region, and errors on the measured $G(\tau)$ that are caused by insufficient statistics or slowly diffusing components such as dust or aggregates may easily lead to a wrong estimate for κ. In a typical FCS experiment one would therefore determine κ on a monodisperse solution of a known fluorophore and keep its value fixed for the measurements of the unknown sample.

2.1.1.1 Concentration determination

The intercept of the FCS autocorrelation function $G(\tau)$ is inversely proportional to the number of particles in the focal volume, and thus to their concentration. In practice, deviations from this ideal behaviour are found at very high and very low concentrations. At low concentrations these deviations are due to the background, which becomes comparable to the fluorescence signal and which is caused by incomplete suppression of the excitation light, detector dark counts and background fluorescence. At a particle concentration c the measured particle number N in the presence of background is then

$$N = cV_{\text{eff}}\left(1 + \frac{\nu}{c}\right)^{-2} \qquad 4$$

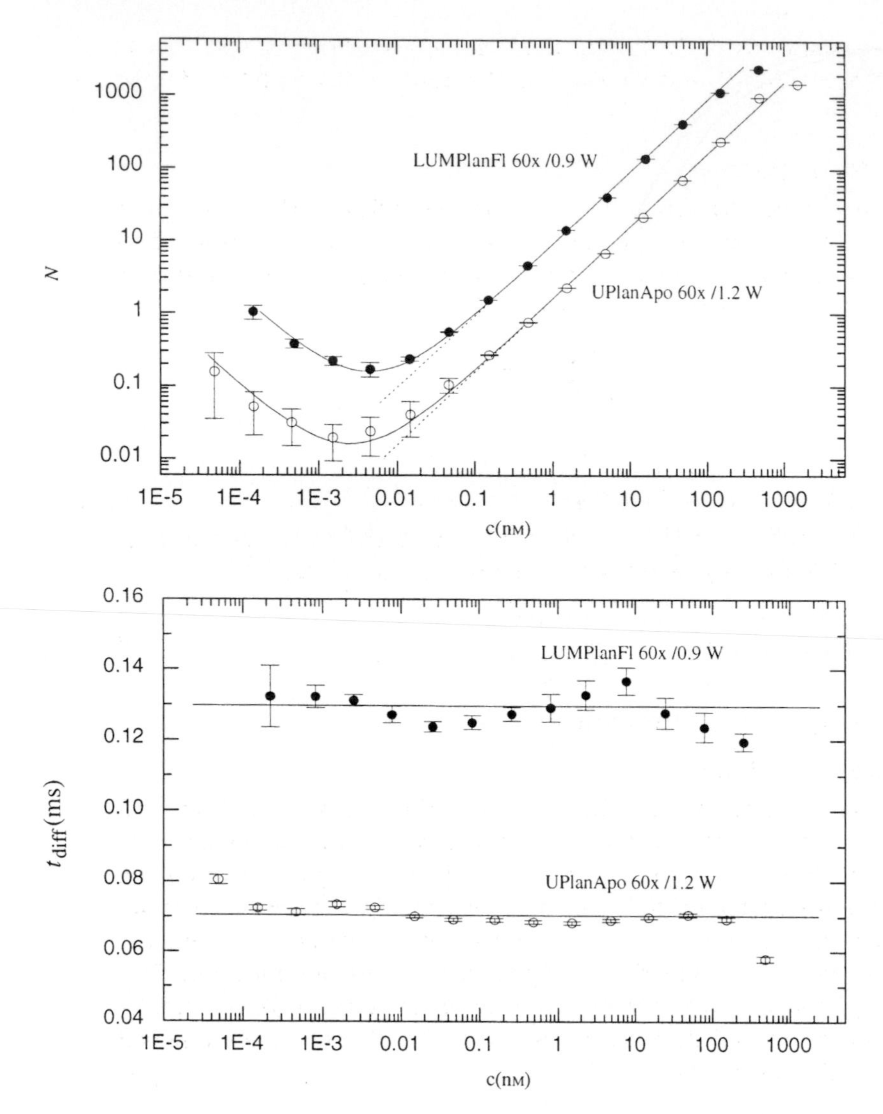

Figure 4 Observed particle number N and diffusion time t_{diff} as a function of sample concentration for a solution of Texas Red-cysteine. Two lenses with different numerical apertures were used. Reliable diffusion times are obtained down to sample concentrations of 200 fM.

where $\nu = \langle U \rangle / \tilde{F}$ is the ratio of the background signal to the normalized fluorescence intensity of the fluorophore. *Figure 4* shows this effect on a solution of Texas Red-cysteine in 10 mM Tris/HCl pH 8.0, 0.1 mM EDTA and 0.02% Tween 20 over a concentration range of 50 fM to 1 µM. Deviations at high concentrations are caused by slow fluorescence intensity fluctuations which can arise from slow adsorption of the dye to the cuvette walls or to larger particles (e.g. dust). If the concentration of the dye is known, the effective detection volume V_{eff} and the relative background v can be obtained from a fit to the curve in *Figure 4*. With the two different lenses used, these parameters are $V_{eff} = 0.26$ fl and $\nu = 2.4 \times 10^{-10}$ M for the 60×/1.2 W lens and $V_{eff} = 1.21$ fl and $\nu = 5.5 \times 10^{-10}$ M for the 60×/0.9 W lens.

2.1.2 Multiple species

In a mixture of molecules with different diffusion coefficients the fluorescence intensity autocorrelation function is a sum of the contributions of the individual species. The general form of $G(\tau)$ for a mixture of m different fluorescent species with diffusion times $\tau_{diff,i}$ is then given by

$$G(\tau) = \frac{1}{N}\sum_{i=1}^{m} \rho_i g_i(\tau) + 1 \qquad 5$$

with

$$g_i(\tau) = \left(1 + \frac{\tau}{\tau_{diff,i}}\right)^{-1} \left(1 + \frac{\tau}{\tau_{diff,i}\,\kappa^2}\right)^{-1/2}$$

The ρ_i are the relative amplitudes corresponding to molecules with distinct diffusion coefficients; they are related to their concentrations c_i by

$$\rho_i = \frac{\phi_i^2 c_i}{\sum_{i=1}^{m} \phi_i^2 c_i} \qquad 6$$

where ϕ_i is the quantum yield of species i.

2.1.3 Triplet contribution

Up to now only number fluctuations in the detection volume have been considered to contribute to the fluctuations of the light intensity at the detector, under the simplifying assumption that an excited fluorophore will emit a constant light flux. Because of the quantum nature of light and the photophysics of fluorescent molecules this is not the case. The most important effect that has to be considered is a transition of the excited molecule into the triplet state. This will 'interrupt' the stream of photons for approximately the triplet lifetime of the fluorophore and add another contribution to the autocorrelation function which—in good approximation—is then (9):

$$G(\tau) = (1 + \beta e^{-\lambda\tau})\left(\frac{1}{N}\sum_{i=1}^{m} \rho_i g_i(\tau)\right) + 1 \qquad 7$$

The amplitude of the triplet term β and its relaxation time λ increase with the excitation light intensity up to a limit given by the excitation, emission and intersystem crossing probabilities of the fluorophore. In practice, β can reach amplitudes higher than the number correlation function itself. Since relaxation time of the triplet term is of the same order as the diffusion times of small molecules (some μs), it is important to conduct the FCS experiment with a laser intensity that keeps β as small as possible.

2.1.4 Two-colour cross-correlation

The detection of specific binding between biomolecules by FCS depends on a change in molecular size: when the diffusion time changes sufficiently upon binding, the complex can be distinguished in $G(\tau)$ as a second species and its concentration determined (*Equations* 5 and 6). However, in cases when the diffusion time changes only very slightly or not at all, i.e. when a non-fluorescent ligand binds to a larger fluorescent particle, or in the case of exchange reactions (see below), this approach is no longer practicable.

Recently, Schwille *et al.* (8) presented a device for two-colour fluorescence cross-correlation spectroscopy (FCCS). In this method the fluorescence is detected at two distinct wavelengths simultaneously in the same detection volume (*Figure* 2). The signals from the two detectors are analysed by computing their cross-correlation function. It is easily seen that in a mixture of two fluorescent molecules emitting at the two wavelengths but not interacting with each other the particles will diffuse independently and the amplitude of the cross-correlation function will be zero. On the other hand, when the particle is labelled with two dyes and emits simultaneously at the two detection wavelengths, the cross-correlation function is equal to the autocorrelation function for single-colour FCS (assuming equal detection efficiencies and exact overlap of the detection volumes for the two channels). This latter case occurs when the two fluorescent species form a complex.

In FCCS, therefore, the amount of complex formation between two fluorescently labelled biomolecules can be obtained simply by measuring the cross-correlation amplitude.

2.2 Construction of a typical FCS instrument

2.2.1 The confocal setup

An inverted microscope with confocal optics attached represents a very convenient means to measure fluorescence fluctuations in a very small volume. *Figure* 5 shows a picture of a setup developed in our laboratory (unpublished data). The laser source in this case is an argon/krypton laser which is coupled to a fibre which is attached to the box containing the confocal optics at G. The beam emitted from the end of the fibre is collimated and focused in D; the position of the laser focus can be adjusted in three dimensions by micrometer screws. A is a filter/dichroic mirror combination which selects a laser wavelength and reflects it into the video port of the microscope (H). The microscope lens focuses the laser beam into the measuring cell (outside the picture), and the emitted fluorescence is imaged by the same lens through the dichroic mirror A on the pinhole E. The laser exit point in D can be adjusted such that its image in the measuring cell coincides with that of the pinhole E (confocal condition). B and C are dichroic mirror/filter combinations which select the fluorophore emission wavelengths and image the pinhole on the active area of the avalanche photodiode single-photon detectors F. The photon pulse stream is sent to the correlation electronics, where the autocorrelation function is formed and analysed.

2.2.2 Lenses

The microscope lenses used in FCS should be of very high numerical aperture (at least 0.9) to minimize the size of the focal volume and therefore maximize the

Figure 5 Photograph of a confocal FCS attachment to an inverted microscope (explanation see text).

fluctuation amplitude for a given fluorophore concentration. Some lenses used in FCS and their characteristics are described in *Table 2*.

The sample is generally present in aqueous solution and is observed in an inverted microscope through a standard cover slide (0.13–0.17 mm thick). In most cases, a water immersion lens is used because an oil immersion lens will lose focus very close (some 4–6 μm) above the inner surface of the cover slide. The water immersion lens has a focal spot of very high quality even at working depths of 200 μm above the cover slide surface.

The high-aperture water immersion lenses all are adjustable for varying cover slide thickness. This adjustment is important as *Figure 6* shows that while the total fluorescence intensity varies only slightly with the setting of the cover slide adjustment of the lens, this parameter has a strong effect on the number of molecules in the observation volume and on the measured diffusion time.

Some lenses can be adjusted for immersion and sample fluids of varying refractive index, such as water-glycerol-sucrose solutions; such lenses have been applied successfully in FCS (5).

Table 2 Some microscope lenses used in FCS

Type	Power	NA	Characteristics
Zeiss C-Apochromat	40	1.2	Correction collar for cover glass thickness
Olympus	60	1.2	Correction collar for cover glass thickness
Olympus	40	0.9	Large working distance (2 mm)

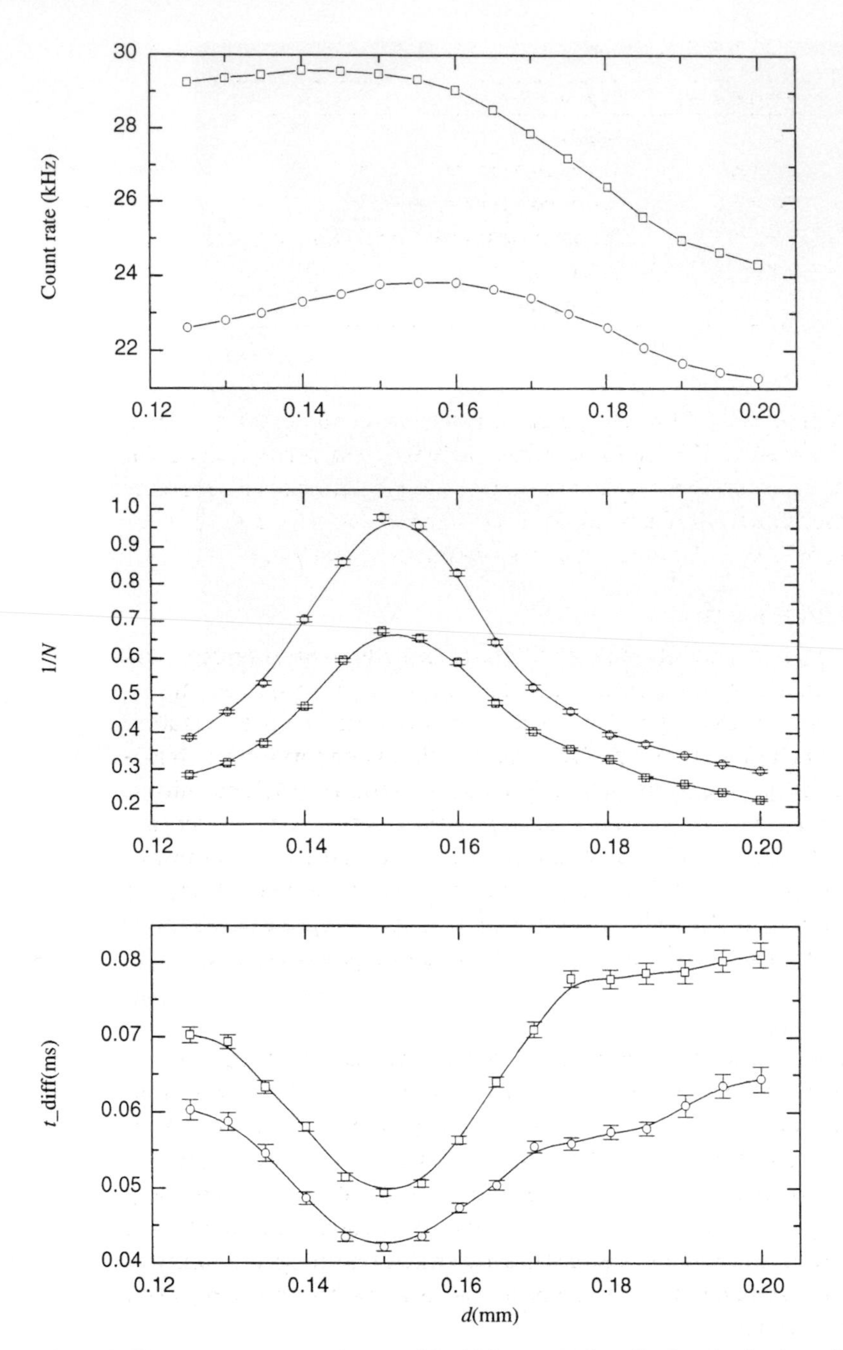

Figure 6 Effect of adjustment of cover-slide thickness on the effective focal volume in FCS. The samples were 5 nM solutions of fluorescein–cysteine (circles) or Texas Red–cysteine (squares). The upper graph shows fluorescence intensity measured (photon count rate in kHz), the middle graph shows the reciprocal of particle number in observation volume, and the lower graph shows measured diffusion time.

Table 3 Lasers applicable in FCS, their powers and wavelengths

Wavelength (nm)	Power (mW)	Type of laser
442	10	He/Cd
457, 488, 514	10-200	Air-cooled argon ion
488, 514, 568	10–50	Air-cooled argon/krypton ion
510	10–200	Diode-pumped frequency-doubled Nd:YAG
543.5	1–5	Green He/Ne
632.8	5–50	Red He/Ne

2.2.3 Laser

Generally, a CW laser is used for excitation. Emission wavelengths of typical CW lasers that can be used in FCS are summarized in *Table 3*. The correct alignment of two excitation lasers for two-wavelength excitation into the same focal spot is a formidable mechanical problem (8) Multi-wavelength lasers, such as Ar or Ar/Kr ion lasers, offer the advantage that this alignment is avoided.

2.2.4 Optics and filters

Dichroic beam splitters and interference filters are usually used to separate the excitation and emission wavelengths in FCS. For multi-wavelength lasers the excitation wavelength is selected with a suitable bandpass filter in the excitation pathway. Even for single-wavelength lasers such a filter is important for obtaining good results, since some parasitic light is usually present. The dichroic mirror, on the other hand, is not essential and might be replaced by a beam splitter that directs only 10% of the excitation light into the sample, because the laser power is usually not a limiting factor. The fluorescence emission is detected through a second filter which can be of the low-pass or bandpass type depending on whether the Raman scattering from water needs to be suppressed or not.

2.2.5 Detector

One of the essential components of the FCS device is a detector that registers the emitted photons with very high efficiency. Most of the dyes used in FCS emit in the yellow to red range of the spectrum, where the quantum efficiency of even red-enhanced photomultipliers is of the order of a few per cent. However, recent avalanche photodiode detectors, such as the SPCM series (EG&G Optoelectronics), have a quantum yield of up to 70% at 600 nm, with dark count rates of 50 c.p.s. or lower. The advent of these devices has greatly enhanced the practicability of FCS because, very often, count rates can be as low as a few hundred c.p.s. even with avalanche detectors.

2.2.6 Autocorrelator

The computation of the autocorrelation function (ACF) of the fluorescent light intensity is central to the FCS experiment. Generally, the ACF is constructed from the detected photon pulses by an electronic autocorrelator. This device multiplies the number of pulses $n(t)$ counted during a time interval δt with the number $n(t - \tau)$ counted during the same interval at an earlier time, building the average $<n(t)n(t - \tau)>$. This process is done simultaneously for a large number of different time delays τ, accumulating the ACF $G(\tau)$ in real time.

Modern autocorrelators will allow one to measure the ACF simultaneously over a range of delay times of 10^{-8} s to > 1000 s, with a choice of either auto- or cross-correlation mode (e.g. ALV-5000, ALV GmbH).

2.3 Sample requirements

The measured sample should fulfil some basic conditions for a successful FCS experiment. First, care has to be taken in the choice of the labelling dye. An obvious criterion is that its excitation and emission maxima should be compatible with the laser light source and filters used. Furthermore, many systems (such as living cells) exhibit intrinsic fluorescence which can cause artefacts; in such cases, dyes that can be excited in the red part of the visible spectrum (such as Cy5) are recommended. A large selection of fluorescent dyes and their derivatives can be found in the Molecular Probes catalogue.

Self-association of the dye or non-specific binding to sample impurities can lead to the formation of fluorescent aggregates. Their presence interferes with the measurement because another, usually slower component will be present in the ACF as an artefact. Self-association is especially critical with large hydrophobic fluorophores, such as Texas Red, and when the sample is labelled to a high degree, as in cross-correlation experiments where labelling of 100% of the sample molecules is necessary. In those cases, care has to be taken that the association state of the biomolecule is the same as for the unlabelled sample. This can be verified by other methods such as light scattering or analytical ultracentrifugation.

3 Some examples from current research

3.1 Triplex formation

DNA triple helices can be formed by binding a single-stranded homopyrimidine sequence to a complementary homopurine-homopyrimidine duplex (10, 11). The direct observation of this complex formation in solution has been difficult because of the lack of an optical signal which could be used for its detection. FCS provides a very convenient means of studying the binding thermodynamics and kinetics of triplex formation, because the complex has a significantly smaller diffusion coefficient than the free ligand. Thus, two relaxation times can be distinguished in a mixture of complex and free ligand and the relative amounts quantified. Because FCS is relatively fast, the kinetics of complex formation can be observed on a time-scale of minutes.

In a recent study (C. Pfannschmidt and J. Langowski, unpublished data), triplex formation was studied on a 2.5 kbp superhelical plasmid which contained the 27 bp sequence 3′-TTCCTCCTTCCTTCCTTCCTTCCTCCC-5′. The complementary triplex-forming oligonucleotide (TFO) had the same sequence in opposite direction and was rhodamine-labelled at the 5′-end. The measurements were done in 10 mM sodium acetate, 50 mM $MgCl_2$, 0.01% NP-40 at pH values between 4.0 and 7.0.

Here we used a Zeiss/Evotec Confocor FCS spectrometer with an excitation wavelength of 488 nm. The diffusion times of the free oligonucleotide and of the complex were sufficiently different that the two components of the autocorrelation function could be separated. The free TFO had a diffusion time between 140 and 200 μs and the complex between 3.56 and 4.25 ms. These variations resulted from alignment of the instrument, which had to be repeated before each set of measurements; the ratio of the diffusion times for the TFO and for rhodamine was constant within ± 3%. The change in quantum yield upon binding was determined separately to $Q_{bound}/Q_{free} = 0.35$.

The binding kinetics of the rhodamine-TFO to the plasmid DNA are shown in *Figure 7*. It is evident that the quantum yield of the bound TFO has a large influence on the measured degree of binding and must be considered in the evaluation of the data. Association rate constants k_1 were determined by a fit to the initial slope of the plot and the binding constant K_{ass} from its plateau for large times; the dissociation rate constant k_{-1} was calculated from K_{ass} and k_1. We found $k_1 = 3.3 \times 10^3$ l/mol.s, $K_{ass} = 1.54 \times 10^8$ l/mol, and $k_{-1} = 2.14 \times 10^{-5}$/s at pH 7.0.

3.2 NtrC protein

NtrC from enteric bacteria is a transcription factor that activates a variety of genes involved in nitrogen utilization by contacting simultaneously a binding site on the DNA and RNA polymerase complexed with the σ^{54} sigma factor at the promoter. The NtrC-binding sites found *in vivo* are several-hundred base pairs upstream from the promoter, and activation requires looping of the intervening DNA for interaction with RNAP·σ^{54}.

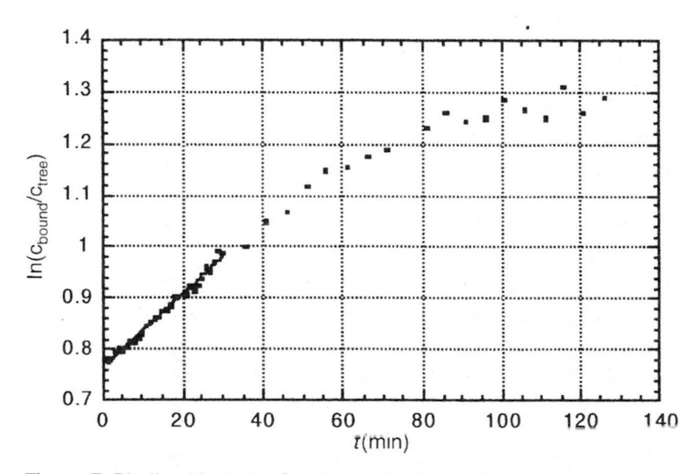

Figure 7 Binding kinetics of a rhodamine-labelled triplex-forming oligonucleotide to a plasmid DNA containing the complementary sequence. The association rate constant is obtained from the initial slope of a plot of $\ln(c_{bound}/c_{free})$ vs. time.

NtrC is a dimer in solution and binds as a dimer to a single binding site, as demonstrated recently by analytical ultracentrifugation (12). While the latter technique is a useful tool for determining the binding stoichiometry, the binding constant is too high under physiological conditions to be determined at the protein and DNA concentrations used in ultracentrifugation. FCS has been used recently (13) to measure the binding constant of NtrC to a fluorescently-labelled oligonucleotide at low ionic strengths where the binding is very strong and consequently has to be measured at very low concentrations.

An oligonucleotide was used that contained the NtrC-binding site and was labelled with tetramethylrhodamine at the 5′ end of one strand. First, the diffusion time τ_1, the structure factor κ, the triplet relaxation time λ and the triplet amplitude β were determined for free DNA in the absence of protein ($\theta = 0$). At saturating protein concentrations ($\theta = 1$) the diffusion time τ_2 of the protein–DNA complex could be measured. The ratio of diffusion times τ_1 and τ_2 of the two species is equal to the ratio of their translational diffusion constants D_1 and D_2:

$$\frac{\tau_2}{\tau_1} = \frac{D_1}{D_2} \qquad 8$$

Knowing the ratio τ_1/τ_2, which is constant, and the diffusion time of the free DNA, τ_1, which can vary from experiment to experiment owing to differences in optical alignment, enables one to determine the relative amplitude of the two components with diffusion times τ_1 and τ_2 in the measured autocorrelation function, corresponding to free and complexed DNA. *Figure 8* shows a plot of the relative amount of bound DNA as a function of free protein concentration from which the binding constant was determined by fitting a standard binding curve.

We found $K_{ass} = (7.1 \pm 2.5) \times 10^{10}$/M at 15 mM KCl and $K_{ass} = (1.4 \pm 0.4) \times 10^{8}$/M at 600 mM KCl. This result shows that the binding constant is strongly salt dependent and can be interpreted by the formation of two ion pairs upon binding of an NtrC dimer to DNA (13).

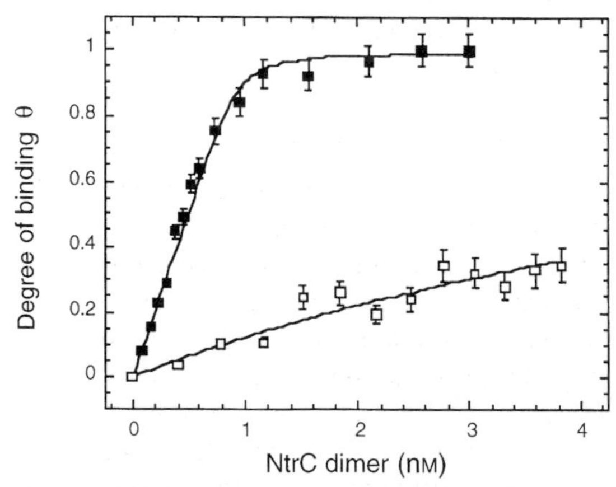

Figure 8 FCS titration curves for NtrC (filled squares) at 15 mM KCl, (empty squares) at 600 mM KCl. The lines are theoretical binding curves for a 1 : 1 complex fitted to the experimental data.

3.3 Vimentin oligomerization

Vimentin is one of the major protein components of the cytoskeleton of eukaryotic cells. One of its essential properties is the formation of intermediate filaments through self-association. The first step of this reaction is a dimerization where two protein monomers associate side by side to form an elongated rod; the formation of tetramers from two dimers is generally assumed to be the next step. Using FCS one can decide whether the protein forms a dimer or a tetramer in solution, and whether subunits can exchange between the complexes.

3.3.1 Stoichiometry of vimentin oligomers

The simplest method for determining the number of proteins per vimentin oligomer is to compare the mean count rate per particle for oligomers and monomers. However, vimentin forms monomers only at very high urea concentrations (8 M), and the FCS measurement becomes problematic because of the high refractive index of the urea solution. Comparison with free dye is imprecise because the quantum yields of the labelled protein and the free dye may be different. Therefore, we used as a reference a solution of vimentin oligomers that had been reconstituted by dialysis from a mixture of 98% unlabelled and 2% labelled protein in 8 M urea. In this solution, the majority of the labelled oligomers will carry only one labelled vimentin. The fluorescence intensity (count rate) per molecule can be computed for this reference and the fully labelled sample by normalizing the average fluorescence intensity $\langle F \rangle$ with the average number of particles in the detection volume, $\langle N \rangle$, and the oligomerization stoichiometry n is then obtained as

$$n = \frac{\langle F_{sample} \rangle / \langle N_{sample} \rangle}{\langle F_{ref} \rangle / \langle N_{ref} \rangle} \qquad 9$$

In the case of vimentin in 5 mM Tris·HCl, 1 mM EDTA, 0.005% Tween 20, pH 9.5 (low salt buffer), a stoichiometry of $n = 2.2 \pm 0.1$ was obtained, indicating that vimentin is a dimer under these conditions.

3.3.2 Exchange of vimentin monomers measured by FCCS

FCCS can be used very conveniently to detect the exchange of vimentin monomers between dimers in solution. The strategy of the measurement is to prepare vimentin samples labelled with either fluorescein (F-vimentin) or Texas Red (TR-vimentin), mix equal amounts of the two samples and measure the FCCS cross-correlation function (CCF). As outlined above, the amplitude of the CCF will be zero for non-interacting molecules (except for crosstalk between the detection channels, which cannot be completely avoided) because their diffusion is independent. When an exchange takes place, dimers will form which contain both fluorophores and the CCF amplitude will increase in proportion to their concentration.

In the experiment, a 1:1 mixture of F- and TR-vimentin was incubated for 10

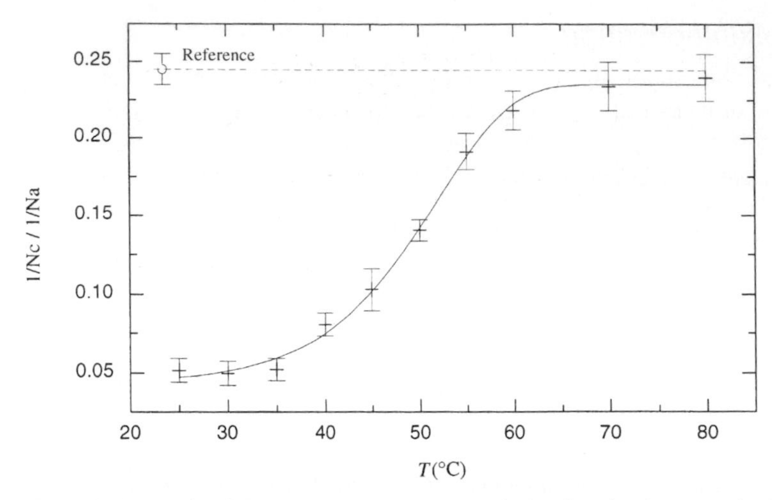

Figure 9 Amplitude of the fluorescence cross-correlation function between the fluorescein and Texas Red channels for mixtures of fluorescein- and Texas Red-labelled vimentin dimers which had been incubated for 10 min at various temperatures. The reference line gives the value expected for complete exchange, measured on a sample that had been mixed in 8 M urea and then dialysed to form dimers.

min at various temperatures, then cooled to room temperature and measured in the FCCS with detection channels for fluorescein and Texas Red emission. *Figure 9* shows the ratio of CCF to ACF amplitudes as a function of incubation temperature. The reference line is the ratio that is obtained from a 1:1 mixture of F- and TR-vimentin monomers in 8 M urea that was dialysed against low-salt buffer to form mixed dimers. It is clearly seen that the proportion of mixed dimer as detected in the CCF increases with temperature, and reaches the maximum value at 70 °C.

Details of this experiment, as well as thermodynamic parameters for the thermal dissociation of vimentin, will be published in a later paper (M. Tewes, H. Herrmann and J. Langowski, unpublished data).

Acknowledgement

This work was supported by BMBF (German Ministry for Education and Research) Grant No. 01 KW 9620/2 to J. L.

References

1. Elson, E. L. and Magde, D. (1974). *Biopolymers* **13**, 1.
2. Magde, D., Elson, E. L., and Webb, W. W. (1974). *Biopolymers* **13**, 29.
3. Webb, W. W. (1976). *Quart. Rev. Biophys.* **9**, 49.
4. Qian, H. and Elson, E. L. (1991). *Appl. Optics* **30**, 1185.
5. Rigler, R., Mets, Ü., Widengren, J., and Kask, P. (1993). *Eur. Biophys. J.* **22**, 169.
6. Sorscher, S. M., Bartholomew, J. C., and Klein, M. P. (1980). *Biochim. Biophys. Acta* **610**, 28.

7. Ricka, J. and Binkert, T. (1989). *Phys. Rev. A* 39, 2646.
8. Schwille, P., MeyerAlmes, F. J., and Rigler, R. (1997). *Biophys. J.* **72**, 1878.
9. Widengren, J., Mets, Ü., and Rigler, R. (1995). *J. Phys. Chem.* **99**, 13368.
10. Wells, R. D., Collier, D. A., Hanvey, J. C., Shimizu, M. and Wohlrab, F. (1988). *FASEB J.* 2, 2939.
11. Hélène, C. and Toulmé, J. J. (1989). In *Oligodeoxynucleotides: antisense inhibitors of gene expression* (ed. Cohen, J. S.), pp. 139ff. Macmillan Press.
12. Rippe, K., Mücke, N., and Schulz, A. (1998). *J. Mol. Biol.*, **278**, 915.
13. Sevenich, F. W., Langowski, J., Weiss, V., and Rippe, K. (1998). *Nucl. Acids Res.* **26**, 1373.

Chapter 8

DNA wrapping in *Escherichia coli* RNA polymerase open promoter complexes revealed by scanning force microscopy

Claudio Rivetti
Istituto di Scienze Biochimiche, Università di Parma, I–43100, Italy
Martin Guthold
Computer Science Department, Department of Physics and Astronomy, University of North Carolina, Chapel Hill, NC 27599–3255, USA
Carlos Bustamante
HHMI, Department of Physics and of Molecular and Cell Biology, University of California, Berkeley, CA 94720, USA

1 Introduction

Transcription is the first step in the expression of the genetic information encoded in DNA. In *Escherichia coli*, this process is carried out by RNAP, a single enzyme that can exist in two forms: the core enzyme, with a subunit composition $\alpha_2\beta\beta'$, M_r 380 000, is responsible for the actual RNA synthesis and the holoenzyme, which possesses an additional σ subunit that enables RNAP to recognize promoters and to initiate transcription. In prokaryotes, the majority of the promoters are recognized by the 70 kDa σ^{70} factor. σ^{70}-RNAP binds to the promoter to form a close promoter complex (CPC) which spontaneously isomerizes through various intermediates into an open promoter complex (OPC) containing an unwound DNA section (for reviews see refs. 1 and 2). Upon addition of nucleoside triphosphates RNAP begins RNA synthesis, σ factor is released, and the core enzyme moves in a highly processive manner along the DNA until a terminator is reached (3, 4). During transcription, RNAP translocates on the DNA, unwinding the DNA ahead and rewinding it in its wake, thereby maintaining the transcription bubble with no net free energy cost.

The structure of the holoenzyme, obtained by electron crystallography at about 25 Å resolution, revealed an overall enzyme size of about 100 × 100 × 160 Å and the presence of a channel 25 Å in diameter and 55 Å long that could represent the DNA-binding site of the enzyme (5). Recent DNAse I and hydroxyradical DNA footprinting studies of σ^{70}-RNAP OPCs have shown that the DNA sequence protected by the RNAP is of about 70–95 bp (240–320 Å) (6–13) which is much larger then the length of the supposed DNA-binding channel and even larger than the longest axis of RNAP. It has, therefore, been proposed that the DNA may be wrapped around the RNAP in a nucleosome-like structure (1, 6). Footprinting data also revealed that most of the DNA contacting the enzyme in OPCs is part of the upstream region (about 60 bp) whereas only about 25 bp of the downstream region is protected from cleavage. Although evidence for DNA wrapping has emerged from DNA supercoiling experiments (14), DNA footprinting (6) and microscopy analysis (5, 15), the hypothesis of DNA wrapping around the RNAP is not yet widely accepted.

This paper describes a Scanning Force Microscopy (SFM) study of *E. coli* RNAP open promoter complexes formed on the λ_{PR} promoter. The complexes were formed in solution, deposited onto freshly cleaved mica and imaged in air by SFM. Taking advantage of the high resolution and contrast that can be obtained with this technique, the DNA contour length of a large number of RNAP–DNA complexes has been accurately measured and compared with that of DNA alone. The present study demonstrates that high-resolution microscopy of transcription complexes, combined with accurate image analysis of a large number of molecules, reveals important structural features that would otherwise be undetected by a simple inspection of the images.

2 Scanning force microscopy

2.1 Imaging protein–DNA complexes with the SFM

In an SFM, a flexible cantilever carrying a sharp tip is scanned relative to a surface and its deflections, resulting from the interactions with the sample, are recorded to generate a topographic image of the surface (16). In addition to this *contact* mode of operation, the microscope can also operate in *tapping* mode where the cantilever is oscillated vertically at high frequency while scanning. In this case, the surface topography is reconstructed from the clipping of the oscillation amplitude of the cantilever (for reviews see refs. 17–21). The most common way to image DNA and protein–DNA complexes with the SFM is to deposit the molecules onto an atomically flat substrate such as mica. Good depositions are often obtained with a low-salt buffer containing Mg^{2+} or other divalent cations that favour the interaction between the negatively charged DNA and the negatively charged mica (21). The surface is then rinsed with ultrapure water and dried with a weak flux of nitrogen. Using DNA and 1–2 nM protein concentrations and a deposition time of 2–3 min yields good surface coverage. In the case of transcription complexes, which must be assembled in relatively high salt condition, it is advisable to dilute the reaction in a low-salt buffer just before the deposition.

To analyse quantitatively the structure of DNA and protein–DNA complexes, the deposition procedure should not distort the conformation of the molecules upon adsorption onto the surface. It has been shown that under the conditions used in this study, DNA molecules deposited onto freshly cleaved mica are able to move on the surface (22). This mechanism of adsorption allows the molecules to equilibrate thermodynamically in a two-dimensional system before they are captured and 'frozen' in a particular conformation on the substrate. Therefore, meaningful information about the structure of the molecules in solution can be inferred from their image on the surface (23).

2.2 Preparation and deposition of RNAP–DNA complexes

His-6 tagged RNA polymerase enzyme was purified as described in (24). Three different DNA templates were used in this study. Template A (1008 bp) contains a λ_{PR} promoter in which position +1 (the first base to be transcribed) is 439 bp from the upstream end and 569 bp from the downstream end (*Plate 2a*, top). The asymmetric location of the promoter with respect to the ends permits discrimination between the upstream and downstream DNA sections. Template B (1054 bp) is very similar to template A (not depicted in *Plate 2*). It contains the same λ_{PR} promoter in which position + 1 is 439 bp from the upstream end and 615 bp from the downstream end. Template C (1150 bp) contains two λ_{PR} promoters separated by 298 bp and orientated in the same direction. Using position +1 as a reference point, the first promoter is located 439 bp from the upstream end and the second promoter is located 413 bp from the downstream end (*Plate 2b*, top). The two promoters were separated by 298 bp to facilitate the recognition of the two bound RNAP molecules.

Protocol 1

Preparation of OPCs

Equipment and reagents

- Transcription buffer (20 mM Tris-HCl pH 7.9, 50 mM KCl, 5 mM $MgCl_2$)
- Deposition buffer (4 mM HEPES pH 7.4, 10 mM NaCl, 2 mM $MgCl_2$)
- TE buffer (50 mM Tris-HCL pH 7.4, 1 mM EDTA)

Method

1. Prepare clean DNA template with a suitable method (restriction digestion or PCR). The DNA fragment should preferably be gel purified. Clean the DNA solution by phenol and phenol-chloroform extraction followed by DNA precipitation in 95% ethanol. Wash the DNA pellet with 70% ethanol, dry and dissolve the pellet in TE buffer.
2. In 10 μl volume of transcription buffer mix 200 fmol of DNA template and 200 fmol of RNAP.
3. Incubate the reaction for 15 min at 37 °C.
4. Dilute 2 μl reaction in 18 μl deposition buffer and immediately deposit onto freshly cleaved mica (see *Protocol 3*).

The three DNA templates were excised by *Hin*dIII restriction digestion of the following plasmids: template A from pDE13; template B from pSAP; template C from pDSP. All plasmids carried the ampicillin resistance gene. The plasmid DNA was prepared using Qiagen plasmid purification kit.

OPCs on template C were made as described in *Protocol 1* but using 240 fmol of DNA template and 600 fmol of RNAP to increase the probability of finding DNA molecules with both promoters occupied. DNA solutions for SFM imaging of DNA alone were prepared following *Protocol 1*, except that no RNAP was added.

Protocol 2

Preparation of heparin-resistant complexes

Equipment and reagents

- Deposition buffer containing heparin (4 mM HEPES pH 7.4, 10 mM NaCl, 2 mM $MgCl_2$, 70 μg/ml heparin)

Method

1. Incubate RNAP stock solution with 70 μg/ml heparin for 10 min.
2. In 15 μl volume of deposition buffer containing heparin mix 10 fmol of DNA template and 40 fmol of RNAP preincubated with heparin.
3. Incubate the reaction for 15 min at 37 °C.
4. Deposit the 15 μl reaction onto freshly cleaved mica without further dilution (*Protocol 3*).

Protocol 3

Deposition of DNA and protein–DNA complexes onto mica

Equipment and reagents

- Deposition buffer (4 mM HEPES pH 7.4, 10 mM NaCl, 2 mM $MgCl_2$)
- Ruby mica (Mica New York)

Method

1. Prepare freshly cleaved mica disk by peeling off the top layer of mica with Scotch tape.
2. Dilute the protein-DNA complexes in deposition buffer to a concentration of about 1–2 nM molecules.
3. Deposit 20 μl of the diluted reaction onto the mica disk.
4. Incubate the solution drop onto the mica surface for about 2 min.
5. Rinse the surface with ultrapure water using a wash bottle.

Protocol 3 continued

6 Blot the excess of water with filter paper and dry with a weak flux of nitrogen. Perform the washing and drying steps by holding the mica disk with tweezers far from possible sources of dust.

7 Load the mica disk onto the SFM scanner and start imaging. If the surface appears too crowded or too scantly covered with molecules, adjust the deposition time at point 3 accordingly.

2.3 Image analysis: DNA contour length measurements

SFM images were acquired in air with a Nanoscope III (Digital Instruments) operating in tapping mode. Images analysis was performed with ALEX, an image analysis program written locally in the Matlab environment (ALEX can be obtained from the authors on request). The image integer values of the Nanoscope file were converted to nanometres using the relation given in the Nanoscope III documentation. The images were flattened by subtracting from each scan line the background fitted by a least-square polynomial with degree 2 or 3. No additional filters were applied to the images. The DNA path was digitized as follows: DNA ends and several points along the DNA were selected with the mouse. These points were then interpolated with one-pixel steps and around each pixel the highest intensity in a given window (5–7 pixels wide) was sought. In this way, all the coordinates of the pixels with the highest value, corresponding to the DNA trace, were found. To reduce the noise, the trace was smoothed by polynomial fitting. Because the DNA that makes contact with the RNAP cannot be seen under the microscope, in the case of RNAP–DNA complexes, the DNA trace was made to pass through the centre of the protein according to a simple model in which the DNA just runs through the RNAP. The position of the RNAP centre along the DNA trace was selected with the mouse and, together with the x, y, and z coordinates of the DNA trace, was stored for further analysis.

3 DNA contour length analysis

3.1 DNA alone

The mean values of the contour-length distributions obtained from the three DNA templates with no protein bound are shown in *Table 1*; also shown is the DNA rise per base pair determined from the ratio of the measured contour length and the number of base pairs for each DNA fragment. The rise per base pair determined by SFM is, to some extent, smaller than that determined by X-ray crystallography of B-DNA (0.338 nm). The discrepancy is probably due to the limited resolution of the microscope and to the smoothing routine applied to the DNA trace to reduce the noise. Indeed, in an image formed by 512×512 pixels, with a scan size of 2 μm, each pixel represents a DNA length of 3.9 nm, which corresponds to about 11 bp. Deviations from linearity within this number of bases are not resolved.

Table 1 DNA Contour length measurements

	Contour length (nm)	DNA rise (nm/bp)	Contour length reduction (nm)	Number of molecules
DNA alone, 1008 bp	329 ± 12	0.326	—	947
One OPC on 1008-bp DNA	297 ± 34	—	32	788
DNA alone, 1150 bp[a]	363 ± 8	0.316	—	458
One OPC on 1150 bp DNA	329 ± 15	—	34	216
Two OPCs on 1150 bp DNA	308 ± 20	—	55	173
DNA alone, 1150 bp[b]	363 ± 8	0.316	—	317
One artificial complex on 1150-bp DNA	363 ± 12		0	338
DNA alone 1054 bp	330 ± 13	0.313	—	302
One HRC on 1054-bp DNA	329 ± 20	—	1	221
Two HRCs on 1054-bp DNA	335 ± 22	—	+5	46

[a] Contains all the molecules in (b) plus 141 DNA molecules with no protein bound that were measured from images of OPCs on template C.

Since an accurate measure of the DNA contour length is crucial for this work, particular care has been taken with those experimental parameters that could bias the measurements: (i) all the images were collected with equal scan size, using the same microscope equipped with the same scanner and calibrated along the *x* and *y* directions using a silicon calibration standard with a periodicity of 200 nm (Nanosensors GmbH); (ii) all the depositions were made in the same buffer conditions, using the same kind of mica as a substrate; (iii) all the DNA contour length measurements were done using the same procedure by the same user; (iv) when possible, molecules of DNA alone and transcription complexes were measured from the same set of images.

3.2 Open promoter complexes

A typical SFM image of OPCs assembled with the 1008 bp DNA template (template A; *Protocol 1*) and deposited onto mica (*Protocol 3*) is shown in *Plate 2a*. In these conditions, most of the DNA molecules have one RNAP bound and the concentration of complexes on the surface allows easy scoring and contour-length measurements. Furthermore, the size of the DNA template and the knowledge of the exact location of the promoter permitted the separation of specific complexes from non-specific complexes that were not included in the analysis.

Figure 1a shows the distributions of the contour length measurements obtained with template A alone and bound to RNAP to form an OPC. The mean values, given by the Gaussian fitting of these distributions, are reported in *Table 1*. A significant shift in the contour length distribution of OPCs toward shorter values is notable in the histograms. The contour length reduction, expressed in terms of the difference between the mean values of the two distributions, is 32 nm. A similar result was obtained with the 1150 bp DNA template (template C) which contains two λ_{PR} promoters. The promoters have the same orientation

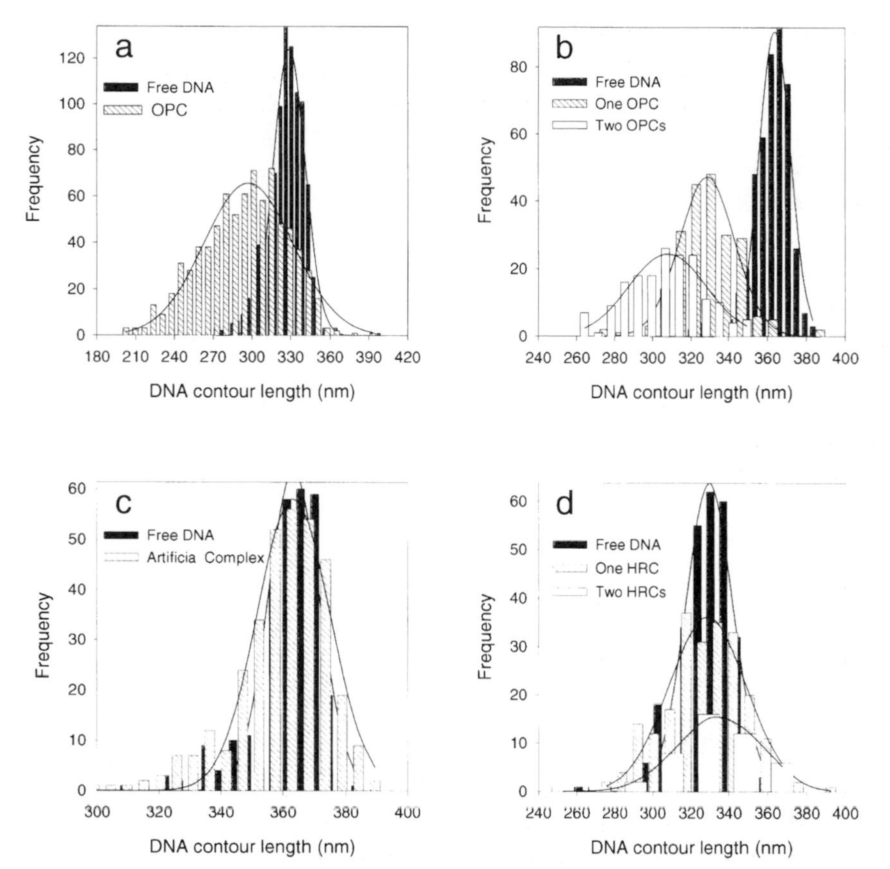

Figure 1 (a) Contour length distributions of DNA template A alone and in OPC; DNA molecules alone and OPCs were measured from two different set of images. (b) Contour length distributions of DNA template C alone, with one OPC and with two OPCs; 141 of the 458 DNA molecules of template C alone and all OPCs were measured from the same set of images. (c) Contour length distributions of template C alone and with artificial complexes created as described in the text. (d) Contour length distributions of DNA template B alone, with one heparin-resistant complex (HRC) and with two HRCs. All molecules were measured from the same set of images. In all graphs the lines represent the Gaussian fitting to the distribution.

and are separated by 298 bp. Using the conditions described above, it is possible to obtain DNA molecules in which one or both promoters are occupied by RNAP (*Plate 2b*). Usually, 50% of the DNA molecules in the image had a single RNAP bound, 30% had two RNAP bound and 20% were free of proteins. This particular situation made it possible to measure the contour length of DNA molecules alone and in complex with the RNAP from the same set of images. To increase the number of DNA molecules evaluated, some measurements were also done on a separate set of images of DNA alone. No difference in the mean contour length was observed between molecules measured from this set of images and those measured from images of OPCs. *Plate 2c* shows DNA molecules of template C alone and with one or two OPCs. All molecules are part of the same SFM image. The lines represent the DNA contours traced as described above. The small dots represent the nearest point along the contour to the centre of the RNAP that the user has chosen.

The contour length distributions obtained with template C are shown in *Figure 1b*. Similar to the observation made with template A, the presence of one OPC results in a reduction of the DNA contour length of 34 nm (*Table 1*). In addition, when two RNAP were bound to the two λ_{PR} promoters of template C, the reduction of the DNA contour length was nearly twice of that observed with one RNAP bound (*Table 1*).

The observed contour length reduction of OPCs with respect to DNA alone could be explained by a model in which the actual path of the DNA in OPC wraps around the surface of the protein to form a nucleosome-like structure. Indeed, the reduction of about 30 nm is in agreement with the circumference of a globular feature of about 10 nm in diameter, as is the case for *E. coli* RNAP. Because of the broadening effect of the tip, the RNAP in these images appears bigger than its real dimensions, hiding the DNA that makes contact with the protein. Therefore, the exact path of the DNA in an OPC could not be resolved. However, in a few cases, because of the particular orientation of the complexes on the surface, or especially good imaging conditions, the appearance of the images strongly suggest that the DNA is wrapped around the RNAP. A selection of such complexes is shown in *Plate 3*.

3.3 Analysis of the DNA arms

The asymmetric location of the promoter within the 1008 bp DNA template allowed discrimination between the upstream and the downstream regions of the DNA with respect to the RNAP. It is therefore possible to determine the partition of the DNA wrapping between the upstream and downstream arms of the template. To address this point, the position of the RNAP centre with respect to the two DNA ends was recorded for each complex. Considering the transcription starting site as the reference point, the ratio of the length in base pairs of the upstream arm with the length in bp of the downstream arm is 0.77 (439/569). However, the ratio of the contour lengths of the upstream and downstream arms measured, obtained from the SFM images, was 0.70. This reduced ratio is not due to the overall reduction in contour length of the OPC. Indeed, assuming an equivalent shortening of both DNA arms the expected ratio is:

$$\frac{439\ (\text{bp}) \times 0.326\ (\text{nm/bp}) - 32/2\ (\text{nm})}{569\ (\text{bp}) \times 0.326\ (\text{nm/bp}) - 32/2\ (\text{nm})} = 0.75$$

where 0.326 nm/bp is the rise per base pair determined from the SFM images. Therefore, the experimentally observed value of 0.70 indicates that RNAP occupies a larger portion of the upstream arm. The amount of upstream DNA (U_{DNA}) and downstream DNA ($32 - U_{DNA}$) wrapped around the polymerase can be obtained from the equation:

$$\frac{439\ (\text{bp}) \times 0.326\ (\text{nm/bp}) - U_{DNA}\ (\text{nm})}{569\ (\text{bp}) \times 0.326\ (\text{nm/bp}) - (32 - U_{DNA})\ (\text{nm})} = 0.70$$

from which a U_{DNA} value of 20.5 nm is obtained. Thus, $\approx 2/3$ of the total DNA length wrapped around the RNAP in OPCs can be attributed to the upstream arm and 1/3 to the downstream arm. This result is in agreement with footprinting (6) and crosslinking data (25), and with the high degree of conservation found in the -10 and -35 regions among prokaryotic promoters (26).

3.4 Heparin-resistant complexes as a control experiment

Two different control experiments were performed to further validate the data interpretation proposed above. The first addresses the question of whether or not the presence of a globular feature along the DNA path could have biased the performance of the tracing routine used to measure the DNA contour length. To this end, a set of images of DNA alone was analysed and the mean contour length was determined. Next, using software tools, the globular feature of an RNAP was 'copied' from an image of RNAP molecules and 'pasted' onto each of the DNA molecules to simulate RNAP–DNA complexes. Such artificial complexes were then re-measured and analysed with the same procedure used for DNA alone. The contour length distributions obtained in the two cases (*Figure 1c*; *Table 1*) were almost identical indicating that no artefacts were introduced in the contour length measurement by a globular feature along the DNA path.

The second control was designed to determine whether a non-specifically bound RNAP induces significant changes of the DNA contour length. This test was difficult to design because a DNA template could contain sequences that function as pseudo-specific binding sites for RNAP (27). A way to overcome this problem is to form non-specific complexes in the presence of inhibiting amounts of heparin (*Protocol 2*). Heparin inhibits transcription, presumably, by competing for the DNA-binding site of RNAP. Under these conditions it was observed that RNAP can, to some degree, still bind to the DNA and appears to have a higher affinity for the DNA ends (*Plate 2d*). These HRC must bind the DNA through some type of non-specific interaction. This assumption was supported by the random distribution of the RNAPs bound along the DNA. The analysis of these complexes shows that the presence of one or even two RNAPs bound to the DNA in the presence of inhibiting concentrations of heparin does not reduce the DNA contour length (*Figure 1d*; *Table 1*). DNA molecules with RNAP bound to one or both ends were not scored since the DNA contour length could not be accurately measured.

4 Concluding remarks

In the present study, high-resolution SFM was used to analyse *E. coli* RNAP open promoter complexes. A careful analysis of a statistically large number of molecules clearly shows that the DNA contour length in OPCs is reduced by about 32 nm, or about 95 bp, relative to the free DNA. This observation agrees with recent hydroxyl radical footprinting experiments on the λ_{PR} and rrnB P1 promoter complexes (6, 10) where it was found that RNAP covers a region of about 94 bp, ranging from position -70 to $+24$. Moreover, in the images, the RNAP molecule often appears adjacent to the DNA. These two observations strongly suggest that in open promoter complexes, the DNA wraps around the circumference of the RNAP. Furthermore, careful statistical analysis of the data revealed that 2/3 of the full DNA region wrapped around the RNAP is part of the upstream arm and 1/3 is part of the downstream arm of the DNA. When the same analysis was done on complexes treated with heparin, which is known to prevent the formation of specific RNAP/DNA interactions, no significant reduction in the DNA contour length was observed.

Several studies of *E. coli* RNAP CPC and OPC have suggested a model in which the DNA wraps around the surface of the protein:

(1) Wrapping was proposed to explain the extent of unwinding observed from DNA topology experiments (14, 28).

(2) In DNA footprinting experiments the extent of DNA protection by RNAP and the periodic pattern of DNA cleavage was explained as evidence of DNA wrapping (6).

(3) On the basis of the RNAP structure determined by electron crystallography it was suggested that DNA may wrap around the RNAP (5).

(4) DNA contour length reduction observed in SFM measurements of σ^{54}-OPCs and CPCs was interpreted as an indication of DNA wrapping around the polymerase (29).

Furthermore, recent crosslinking and EM studies on eukaryotic RNAP II indicate that in these complexes the DNA is also wrapped around the polymerase (30). It was found that the two large subunits of RNAP II (RPB1 and RPB2), which are homologous to the *E. coli* subunits β' and β, extensively crosslink to opposite faces of the promoter DNA. The authors infer that these subunits form a channel 240 Å long suitable for DNA binding. From EM images it was observed that the DNA contour length of RNAP II–DNA complexes was shorter than that of DNA alone. Finally, SFM studies of OPCs and stalled elongation complexes were previously interpreted as evidence that the polymerase bends the DNA by about 57° and 90°, respectively (31). It is possible that those observations corresponded to the wrapping of the DNA around the polymerase and that the deviation of the arms on opposite sides of the polymerase simply corresponded to the entry and exit angles of the DNA wrapped around the enzyme.

DNA wrapping around the polymerase in OPCs opens a number of possible relevant interactions between the enzyme and specific sequences near the promoter. These interactions may play an important role during promoter clearance and transcription regulation. Future SFM studies should provide information on the presence of DNA wrapping in OPCs formed with different promoters, as well as in elongation complexes. In particular, it may be possible to test the predictions of the model by introducing in-phase and out-of-phase pre-bent DNA sequences upstream and downstream of the promoter to investigate their effect on the stability of the OPCs. The results of this study show that high-resolution SFM, together with careful statistical analysis of the data can be a powerful tool in structural characterization of complex macromolecular assemblies.

Acknowledgements

This work was supported by NIH grant 5RO1 GM 32543 and NSF grants MCB 9631153 and BIR 9318145. C. R. was supported by a HFSPO fellowship.

References

1. de Haseth, P. L. and Helmann, J. D. (1995). *Mol. Microbiol.* **16**, 817.
2. Leirmo, S. and Record, M. T. (ed.) (1990). In *Nucleic acids and molecular biology* (ed. M. J. Lilley and F. Eckstein), Vol. 4. Springer-Verlag, New York.
3. Kassavetis, G. A. and Chamberlin, M. J. (1981). *J. Biol. Chem.* **256**, 2777.
4. Rhodes, G. and Chamberlin, M. J. (1974). *J. Biol. Chem.* **249**, 6675.
5. Polyakov, A., Severinova, E., and Darst, S. A. (1995). *Cell* **83**, 365.
6. Craig, M. L., Suh, W. C., and Record, M. T. (1995). *Biochemistry* **34**, 15624.
7. Kolb, A., Igarashi, K., Ishihama, A., Lavigne, M., Buckle, M., and Buc, H. (1993). *Nucl. Acids Res.* **21**, 319.
8. Kovacic, R. T. (1987). *J. Biol. Chem.* **262**, 13654.
9. Newlands, J. T., Ross, W., Gosink, K. K., and Gourse, R. L. (1991). *J. Mol. Biol.* **220**, 569.
10. Ross, W., Gosink, K. K., Salomon, J., Igarashi, K., Zou, C., Ishihama, A. *et al.* (1993). *Science* **262**, 1407.
11. Schickor, P., Metzger, W., Werel, W., Lederer, H., and Heumann, H. (1990). *EMBO J.* **9**, 2215.
12. Spassky, A., Kirkegaard, K., and Buc, H. (1985). *Biochemistry* **24**, 2723.
13. Carpousis, A. J. and Gralla, J. D. (1985). *J. Mol. Biol.*, **183**, 165.
14. Amouyal, M. and Buc, H. (1987). *J. Molec. Biol.* **195**, 795.
15. Darst, S. A., Kubalek, E. W., and Kornberg, R. D. (1989). *Nature* **340**, 730.
16. Binnig, G., Quate, C. F., and Gerber, C. H. (1986). *Phys. Rev. Lett.* **56**, 930.
17. Bustamante, C., Keller, D., and Yang, G. (1993). *Curr. Opin. Struct. Biol.* **3**, 363.
18. Bustamante, C., Erie, D. A., and Keller, D. J. (1994). *Curr. Opin. Struct. Biol.* **4**, 750.
19. Hansma, H. G. and Hoh, J. (1994). *Annu. Rev. Biophys. Biomol. Struct.* **23**, 115.
20. Bustamante, C. and Keller, D. J. (1995). *Physics Today* **48**, 32.
21. Bustamante, C. and Rivetti, C. (1996). *Annu. Rev. Biophys. Biomol. Struct.* **25**, 395.
22. Bustamante, C. J., Godsey, M., Guthold, M., Zhu, X., Rivetti, C., and Yang, G. (1996). *The Robert A. Welch Foundation conference on chemical research, XL. Chemistry on the nanometer scale*, p. 263. Robert A. Welch Foundation, Houston.
23. Rivetti, C., Guthold, M., and Bustamante, C. (1996). *J. Mol. Biol.* **264**, 919.
24. Kashlev, M., Martin, E., Polyakov, A., Severinov, K., Nikiforov, V., and Goldfarb, A. (1993). *Gene* **130**, 9.
25. Brodolin, K. L., Studitskii, V. M., and Mirzabekov, A. D. (1993). *Molec. Biol. Mosk.* **27**, 1085.
26. Hawley, D. K. and McClure, W. R. (1983). *Nucl. Acids Res.* **11**, 2237.
27. Kadesch, T. R., Williams, R. C., and Chamberlin, M. J. (1980). *J. Mol. Biol.*, **136**, 79.
28. Buc, H. (1986). *Trans. Biochem. Soc.* **14**, 196.
29. Rippe, K., Guthold, M., Hippel, P. H. v., and Bustamante, C. (1997). *J. Mol. Biol.*, **270**, 125.
30. Kim, T.-K., Lagrange, T., Wang, Y.-H., Griffith, J. D., Reinberg, D., and Ebright, R. H. (1997). *Proc. Natl. Acad. Sci. USA* **94**, 12268.
31. Rees, W. A., Keller, R. W., Vesenka, J. P., Yang, G., and Bustamante, C. (1993). *Science* **260**, 1646.

Chapter 9
Microcalorimetry of protein–DNA interactions

Alan Cooper
Chemistry Department, Glasgow University, Glasgow G12 8QQ, Scotland

1 Introduction

This chapter describes the application of current DSC and ITC calorimetry methods to the study of protein–DNA complexes and related protein–nucleic acid interactions, together with the minimal thermodynamic background necessary to interpret the experimental data.

Microcalorimetry is a potentially non-invasive technique for studying biomolecular stability and interaction in solution. Since almost all chemical or physical changes involve uptake or release of heat energy, calorimetric measurement of this heat effect can be used as a general, relatively non-specific analytical probe of the extent or magnitude of the particular process under investigation. In this respect, calorimetric methods are just like other analytical methods (fluorescence, UV/vis, etc.) that rely on intrinsic properties of the molecules concerned, and it is not necessary to understand the fundamental thermodynamics lying behind the heat effects. However, for those that need it, calorimetry also provides direct measurements of real thermodynamic quantities that relate to the fundamental forces at work in the process and which can give additional analytical information.

Heat effects in protein–DNA and other biomolecular interactions are relatively small for two basic reasons. First, since we are normally dealing with non-covalent interactions, the forces themselves are relatively weak, with heat (enthalpy change) contributions usually of order 5–100 kJ/mol (1–25 kcal/mol in old units; 1 cal = 4.184 J). Second, the molar amount of active macromolecular material available in a typical experiment is quite small even in relatively concentrated solutions – for example, a 1 mg/ml solution of a 50 kDa macromolecule corresponds to just 20 nmol (2×10^{-8})/ml. Consequently, anticipated heat effects for a typical 1 ml sample in a ligand-binding experiment for example might be of order 100 μJ or less, corresponding to temperature changes of order 20 millionths of a degree Celsius or less. At this level, such heat effects are easily masked by spurious effects arising from heats of buffer dilution, pH mismatch, and so forth; it is for this reason that sample preparation and equilibration are central to successful application and interpretation of calorimetric experiments, as described below. Background details of modern calorimetric techniques and their uses are given in refs. (1–5). Relevant recent work includes microcalorimetry of DNA–repressor complexes (6–9) and other protein–double-stranded (10–12) and single-stranded DNA interactions (13,14).

2 Thermodynamics: all you need to know...

Formal thermodynamics can be a daunting subject, prone to misunderstandings and misconceptions. But the basic concepts are quite straightforward and conform to common sense. This section reviews all the non-expert need know to allow satisfactory interpretation of calorimetric data.

2.1 Basics

Thermodynamic equilibrium represents the balance between two opposing tendencies in the molecular world: first, the tendency for all things to move to lower energy (ΔH or enthalpy effect – first law of thermodynamics), second, the tendency for thermal motion to disrupt things (ΔS or entropy effect – second law of thermodynamics). For open systems at constant pressure relevant to most biomolecular situations, this balance is formally represented by the Gibbs free energy change:

$$\Delta G = \Delta H - T\Delta S$$

where T is the absolute temperature (0 °C = 273.15 K). This quantity (ΔG) must be negative (i.e. go downhill in free energy) for any process to take place of its own accord ('spontaneously' in thermodynamics jargon), otherwise free energy or work has to be supplied to the system to make the process go. For reactions involving covalent bonds it is frequently possible to ignore entropy effects since the bond energy changes are so large. However, for non-covalent interactions this is definitely not the case. Typical thermal energies (kT per molecular degree of freedom) correspond to around 2.5 kJ/mol at room temperature, and are comparable to non-covalent bond enthalpies. Consequently, weak bonding interactions between molecules or groups, for example, are readily disrupted by thermal agitation (Brownian motion on the molecular scale), and the final outcome is a delicate balance between bonding (ΔH) and thermal disruption ($T\Delta S$) effects.

It is important to appreciate that the thermodynamic system comprises not only the (macro)molecules of interest, but also the solvent molecules and other components in the mixture. Indeed, the solvent can (and often does) play a dominant role in the thermodynamics of interaction, and apparently energetically unfavourable interactions (endothermic, ΔH positive) can be 'entropy-driven' ($T\Delta S$ positive and greater than ΔH) by solvation effects. The hydrophobic interaction is a classic example of this.

2.2 Equilibrium and 'standard' thermodynamic quantities

A system reaches thermodynamic equilibrium when $\Delta G = 0$, that is when the overall sum of interaction enthalpy (ΔH) terms is exactly balanced by the overall thermal disruptive entropy ($T\Delta S$) terms. This is true for all systems at equilibrium so does not convey any measure of the relative strengths of different kinds of interaction. Therefore, for comparative purposes, it is useful to define 'standard' thermodynamic quantities – designated by superscript o: ΔG^o ΔH^o ΔS^o – with reference to some arbitrary standard state. For example, for the sort of binding interactions of interest here, it is conventional and useful to describe the equilibrium binding situation thus:

$$P + Q \rightleftharpoons PQ \ ; \ K = [PQ]/[P][Q]$$

where K is the equilibrium constant, also variously referred to as the association or affinity constant, and its inverse ($1/K$) is the dissociation constant. This shows how the equilibrium binding situation may be shifted by changing the relative concentrations of the molecules (P and Q) involved. (Note: strictly speaking the molar concentrations $[P]$, $[Q]$, etc., should be thermodynamic 'activities', but the errors introduced by this approximation are usually assumed to be insignificant at the low concentrations used in most experiments.) K may be thought of as reflecting the relative probabilities of finding the molecules in the bound and free states at equilibrium, and also the dynamic equilibrium where the rate of formation of complex is exactly balanced by the rate of dissociation.

K can also be expressed as the 'standard' Gibbs free energy change for the process:

$$\Delta G^\circ = -RT \ln(K) = \Delta H^o - T\Delta S^o \qquad (R = \text{gas constant, } 8.314\ \text{J/K.mol}) \tag{1}$$

representing the free energy change ΔG^o, and the constituent enthalpy ΔH^o and entropy ΔS^o changes, that would take place in the (hypothetical) standard state in which reactants (P and Q) and products (PQ) were all present at 1 molar concentration (or activity). (The convention of 1 M concentration for standard states in solution, clearly unrealistic for biomolecular systems, is a historical consequence of our choice of standard units for measuring concentration. Nonetheless, this provides a convenient way of comparing interaction free energies, etc., on the same comparative scale.) A convenient way to view ΔG^o is simply as the equilibrium constant, K, expressed on a logarithmic energy scale. The standard enthalpy, ΔH^o (practically indistinguishable from the heat, ΔH, measured under non-standard conditions), is the molar heat of formation of the complex; and ΔS^o represents all the thermally disruptive effects including the entropy of mixing of P, Q, and PQ at molar concentrations.

For a calorimetric experiment in which P and Q are mixed, the overall heat effect measured (δH) will depend on the extent of reaction and the (molar) enthalpy of the process (expressed per mole of P):

$$\delta H = \Delta H^o[PQ]/([P] + [PQ])$$

and this forms the basis for ITC methods described below.

Alternatively, the enthalpy and entropy components of ΔG° may be obtained from measurements of the temperature dependence of K. Following on from *Equation (1)*

$$\ln(K) = -\Delta H^\circ/RT + \Delta S^\circ/R$$

or

$$\partial \ln K/\partial(1/T) = -\Delta H^\circ/R \text{ (van't Hoff equation)}$$

so that a plot of $\ln(K)$ versus $1/T$ is linear with slope $-\Delta H^\circ/R$ and intercept $\Delta S^\circ/R$. This is the familiar van't Hoff plot which, until the advent of sufficiently sensitive calorimetric methods, was the only method commonly available for estimating reaction enthalpies. It is an indirect method requiring prior knowledge (or assumption) about the nature and stoichiometry of the process under investigation. Its further drawbacks include inaccuracies arising from the limited (absolute) temperature range normally available for biomolecular experiments and non-linearities in cases where ΔH is itself temperature dependent (ΔC_p) effects (15).

2.3 Heat capacity

The heat capacity (or specific heat) of any substance (usually termed C_p at constant pressure) is the ability of the substance to absorb heat energy without an increase in temperature, and this is central to DSC measurements. Liquid water has a relatively high C_p by virtue of its extensive ice-like H-bond network that allows heat energy to be used up in breaking bonds between water molecules rather than increasing their kinetic energy (i.e. temperature). Proteins, nucleic acids and other organic materials have a lower specific heat than water, except possibly when undergoing some process such as unfolding or melting involving breaking of bonds. Consequently, the heat capacity of a dilute biomolecule solution is dominated by the water in the system and great care has to be taken to subtract this out in DSC measurements to give the excess differential heat capacity contribution arising from the process of interest.

Heat capacity is the fundamental quantity from which both absolute enthalpies and entropies may be derived:

$$H = \int C_p dT \text{ and } S = \int C_p/T dT$$

where both integrals are from absolute zero (0 K) to the temperature of interest. As a consequence, wherever a process (such as unfolding or binding) involves a change in heat capacity, ΔC_p (and they normally do), then both the enthalpy (ΔH) and entropy (ΔS) changes will be temperature dependent. For example, if ΔC_p is finite but constant, the enthalpy and entropy changes with respect to some standard reference temperature (e.g. 25 °C, 298 K) will be given by:

$$\Delta H(T) = \Delta H(298) + (T - 298)\Delta C_p$$

and

$$\Delta S(T) = \Delta S(298) + \Delta C_p \ln(T/298)$$

As a further consequence (16) these effects will largely cancel in the standard free energy expression (*Equation 1*) to give a ΔG that is relatively less affected by temperature change. This is one example of 'entropy–enthalpy compensation' or 'linear free energy' effects that are frequently found in systems involving multiple weak interactions (16,17). Finite ΔC_p effects are frequently interpreted in terms of changes in buried macromolecular surface area during the process (18–20) but this cannot always be the case (11,16).

2.4 Linked functions: ion binding and protonation changes

Whenever two (macro)molecules bind in solution there are inevitably other changes such as binding or release of ions, changes in hydration, and so forth, that are frequently invisible or disregarded, yet nonetheless affect the binding process. Consider, for example, a simple binding process involving release of a number (ν) of ions (Y) or other small species from the macromolecules P and Q. The true equilibrium expression here is:

$$P + Q \rightleftharpoons PQ + \nu Y \; ; \; K = [PQ][Y]^{\nu}/[P][Q]$$

which shows how the equilibrium situation will depend not only on macromolecule concentrations but also on the concentration of species Y. We might not realize this at first, and would simply determine an apparent or observed equilibrium constant in terms of just the macromolecular species of interest:

$$K_{obs} = [PQ]/[P][Q] = K/[Y]^{\nu} \; ; \text{or } \log(K_{obs}) = \log(K) - \nu\log[Y]$$

Variation of K_{obs} with changes in $[Y]$, under otherwise identical conditions, reveals the existence of such invisible processes. In the simplest cases a logarithmic plot of $\log(K_{obs})$ versus $\log[Y]$ is linear with slope ($-\nu$) giving the apparent number of Y species released in the process. Frequently, this is used to follow uptake or release of cations (Na^+, K^+, etc.) during DNA binding processes. If Y is the hydrogen ion (H^+) then the above can be transformed into a familiar expression for the variation in observed equilibrium (or binding) constant with pH:

$$\partial \log K_{obs}/\partial \text{pH} = \nu_{H+}$$

These considerations apply to any equilibrium process, no matter how measured, and are just one manifestation of the general theory of linked thermodynamic functions that have been described in considerable detail by Wyman (21, 22). For calorimetric measurements we must also consider the additional heat effects arising from these changes (23, 24). For example, in the particular case of H^+ release:

$$P + Q \rightleftharpoons PQ + \nu H^+$$

in a well-buffered system the hydrogen ions will be picked up by buffer ions (B) so that the overall reaction is:

$$P + Q \rightarrow PQ + \nu H^+ \; ; \Delta H$$

and:

$$\nu \times H^+ + B \rightarrow BH^+ \; ; \Delta H_i$$

where ΔH_i is the heat of proton ionization of the buffer species. Consequently, in a calorimetric binding experiment, the total heat change observed is:

$$\Delta H_{obs} = \Delta H + \nu \Delta H_i$$

The existence of such effects is shown by doing repeat experiments under identical pH and temperature conditions, but using buffers with different heats of ionization: ΔH_i can vary from around 0 kJ/mol (e.g. acetate, or phosphate at neutral pH) up to -50 kJ/mol for amine buffers such as Tris (1), so that markedly different heats of binding, for example, might be observed simply as a consequence of choice of buffer system. This can be disconcerting if one is unaware of the protonation changes involved, but can be turned to advantage for the additional information it yields—thus, a plot of ΔH_{obs} versus ΔH_i will be linear, with a slope giving the stoichiometry of proton release (ν) and the intercept giving the enthalpy of reaction (ΔH) independent of buffer ionization effects (23,24). [Note: It is not generally appreciated that, depending upon how one conducts the experiment, such effects can also occur in non-calorimetric methods to determine the reaction enthalpy from the temperature dependence of K using the van't Hoff equation (above). This arises because of the temperature variation of buffer pH, which is greater the larger the heat of buffer ionization. If, as is most commonly the case, one makes no adjustment to buffer/sample pH during the temperature variation, then any observed K will reflect not only the temperature dependence of the reaction itself (a manifestation of ΔH) but also the indirect effect of the pH change arising from the change in temperature of the buffer (reflecting ΔH_i). To avoid this effect one should choose buffers with low or insignificant heats of ionization or, alternatively and more tediously, adjust the buffer pH at each temperature back to its value at the reference temperature.] Similar arguments hold for other ions such as Ca^{2+}, Mg^{2+}, and even non-ionic species, that might be bound or released concomitant with the macromolecular process, provided that there is some adequate buffering of their concentrations in solution - for example in the presence of metal ion chelators such as EGTA, EDTA, etc.

3 Microcalorimetry instrumentation and methods

For protein–nucleic acid interactions studies, although a range of microcalorimetry techniques are available in principle, only two—DSC and ITC—are currently in common use. DSC is used to study heat energy changes in processes such as the thermal unfolding of proteins, or melting of nucleic acids, that are brought about by an increase (or decrease) in temperature. Its relevance to protein–DNA interactions lies in the way the thermal stabilities can be affected by complexation. ITC is used for more direct studies of binding brought about by mixing one component with the other, in solution, at constant temperature. Commercial DSC instruments suitable for work with biological macromolecules in solution are available from Microcal, CSC – this is the commercial successor to the original DASM-4 and earlier instruments from the Privalov group (5) – and Setaram. ITC and similar instruments are manufactured by Microcal, CSC, and Thermometric. Each has its own merits and idiosyncrasies, differing marginally in sensitivity and features available, but the majority of current work is based on Microcal instruments, and this will be assumed in subsequent descriptions here. [Note: It is important here to distinguish between DSC instruments designed specifically for use with dilute solutions of biomolecules, such as those referred to here, and the more general-purpose DSC or DTA instruments commonly used for less demanding studies of solids, liquids, pastes, etc. The latter typically take a sample size of only 50 μl or so (usually in sealed pans) and, although of comparable thermal sensitivity, require unacceptably high concentrations of protein or DNA.]

The basic design of both DSC and ITC instruments is very similar. Both comprise a pair of matched calorimetric cells—one for sample, the other as reference—usually made of inert metal, suspended in a temperature-controlled enclosure. The cells are of total-fill design, with active volume typically 1–2 ml, and are fitted with sensitive thermocouples that monitor temperature differences between each of the cells and between the cells and the surroundings. Electronic/computer-controlled feedback circuits drive heater coils attached to each cell to supply measured amounts of heat energy to sample or reference, as appropriate, to compensate for any temperature differences between them. It is this differential heat energy that is of interest here, since it directly reflects the energetics of the process(es) going on in the sample cell. In DSC, the temperature of the entire system—cells + enclosure—is raised at a constant rate, typically 60 °C/h, while monitoring the temperature difference between sample and reference cell. If both cells are identical then there will be no differential temperature lag between the two cells and no additional feedback heat will be required. However, if, for example, at some temperature the sample solution begins to undergo some endothermic process (e.g. protein unfolding or DNA melting), then the sample cell temperature will lag behind the reference and extra feedback heat will be required to maintain temperature equality. This extra heat energy is the differential excess heat capacity (C_p) of the sample solution with respect to the reference. Typical temperature ranges for DSC of biomolecules in aqueous solution are 0–110 °C, or higher, and inert gas pressure of around 1–2 atm (100–200 kPa) is applied to the cells to inhibit bubble formation or boiling at higher temperatures. (Downscan facilities for DSC are sometimes available for studying processes during cooling, but these are currently less satisfactory than the more conventional upscan procedure.)

For ITC, the sample cell is fitted with an injection syringe which also acts as a stirrer rotating at constant speed (typically 400 r.p.m.) during the thermal titration process. Injection of small aliquots (2–20 μl) of ligand solution into the sample solution gives rise to small temperature changes – increases or decreases, that again are compensated by feedback heaters. In this case the calorimeter (ITC) is maintained at constant temperature (or imperceptibly slowly scanning) so that the feedback heat is a direct measure of the thermal energy change induced by the injection. This will include heats of ligand binding to the macromolecule in the sample cell, if any, together with heats of dilution and mixing of the macromolecule, ligand, buffer and other components. A typical ITC experiment comprises a sequence of injections of ligand (which may itself be a macromolecule) into the sample, with concentrations spanning the anticipated binding range, giving a sequence of heat pulses which, when corrected for heats of dilution and mixing (obtained from separate control experiments) represent the differential thermal titration curve for the process, which may be analysed to give the K, ΔH, and stoichiometry (n) of the reaction.

Protocol 1

DSC of protein-DNA complex

Equipment and reagents

- DSC instrument (Microcal MC2, VP-DSC, or equivalent)
- Dialysis tubing or cassettes (e.g. Pierce Slide-A-Lyser)
- Degassing equipment (vacuum dessicator, magnetic stirrer)
- Protein + DNA: typically 2 ml at a (protein) concentration of 1 mg/ml.
- Buffer. (Most buffers, including organic solvent mixtures, are compatible with DSC, but mercaptoethanol is best avoided because of adverse thermal effects due to oxidation and thermal degradation. Other reducing agents such as DTT or DTE are usually satisfactory, if needed.)

A. Sample preparation

1 Prepare the protein + DNA mixture and dialyse several changes of appropriate buffer. Each DSC run will typically require 1–2 ml of protein solution at a concentration of around 1 mg/ml, together with similar stoichiometric amounts of DNA. Depending on conditions, it may be preferable to dialyse the protein and DNA in separate dialysis bags (although in the same pot) for subsequent mixing.

2 Retain the final dialysis buffer for DSC reference, equilibration and dilutions.

3 Determine the protein and DNA concentrations by 280/260 nm absorbance or other appropriate method.

3 Immediately prior to the experiment, degas portions of the sample mixture and buffer for 2–3 min under gentle vacuum with gentle stirring. Be careful to avoid excessive degassing or frothing of the mixture at this stage.

B. DSC procedure

1 Load DSC sample and reference cells with degassed buffer and collect baseline scan(s) using appropriate temperature range and scan rate (typically 20–100 °C, 60 °C/h).

2 Allow the DSC cells to cool and refill the sample cell with protein + DNA mixture.

3 Repeat the DSC scan(s) using the same parameters as in 1.[a]

4 After final cooling, remove the sample and examine for turbidity, aggregation or other visible changes.[b]

5 Process data using instrumental software. This normally involves subtraction of buffer baseline (from 1), concentration normalization, followed by deconvolution of the resultant thermogram using an appropriate model.

Protocol 1 continued

[a] Depending on circumstances, it may be useful to carry out repeat scans with the same sample to establish reversibility and reproducibility.

[b] Precaution: traces of aggregated protein or other contaminants in the DSC cell will cause erratic baseline behaviour. Routine vigorous cleaning of the DSC cells, using detergents or strong acids, as recommended by the manufacturer, is essential for reliable DSC operation, especially when working with readily aggregating systems.

Protocol 2

ITC of protein-DNA complex formation

Equipment and reagents

- ITC (Microcal Omega, MCS, or equivalent)
- Dialysis tubing or cassettes (e.g. Pierce Slide-A-Lyser)
- Buffer (including any necessary co-factors)
- Degassing equipment (vacuum desiccator, magnetic stirrer)
- Protein and DNA

The concentration of macromolecule in the ITC cell should be at least 10 μM and that in the injection syringe should be normally 15- to 20-fold greater. This will ensure satisfactory titration for a reasonably tight binding system using a 100 or 250 μl injection syringe. The choice of which component – protein or DNA – to put in the ITC cell or injection syringe will depend on operational matters such as solubility and sample availability. For nucleic acids, in particular, the higher concentrations required in the injection syringe may result in unwanted aggregates or unusual folds.

A. Sample preparation

1. Prepare the protein and nucleic acid samples by dialysis against an appropriate buffer and retain some of the final dialysis buffer for ITC equilibration and dilution controls. This step is crucial: any slight mismatch in buffer composition or pH between ITC cell and syringe solutions will give large heat of mixing/dilution artefacts that can mask the desired heat of binding. (Alternatively, if one or both components are available as suitably pure solids – e.g. freeze-dried protein and/or DNA – then they may be simply dissolved in dialysis buffer without further equilibration. However, be wary of possible salt and other contamination in lyophilized samples.)
2. Determine concentrations of protein and DNA solutions.
3. Immediately prior to the experiment, briefly degas all solutions (as in *Protocol 1*) to avoid bubble formation during the calorimetric titration.

B. ITC procedure

In what follows, 'ligand' refers to what is in the injection syringe and 'macromolecule' to what is in the ITC cell at the start of titration, even though both may be macromolecules.

Protocol 2 continued

Note that ITC is relatively non-destructive and samples can be recovered and re-cycled after use if desired, although some degradation can occur as a result of long periods of stirring in the ITC cell during equilibration and titration.

1. Rinse both the ITC cell and injection syringe with several changes of dialysis buffer.[a]
2. Select an appropriate injection programme, typically 25 × 10 μl injections, possibly including a small (1–2 μl) preinjection to allow for irregularities sometimes seen in the first injection because of dilution/mixing in the tip of the injection syringe during loading. A 3-min period between injections is usually adequate unless binding effects are slow.
3. Fill both cell and injection syringe with degassed dialysis buffer.
4. Perform a series of trial buffer-into-buffer injections to confirm instrumental stability.
5. Fill the injection syringe with ligand and perform the complete injection program to give the ligand-into-buffer dilution control. (Frequently, the heat of ligand dilution changes during an injection sequence as the ligand concentration builds up in the ITC cell, so complete data must be collected to allow proper correction of experimental data.)
6. Rinse the ITC cell with buffer and fill with degassed macromolecule solution. Re-fill the injection syringe with ligand solution and repeat the complete injection programme to give the ligand-into-macromolecule titration.
7. Leave the macromolecule–ligand solution in the cell, rinse and refill the syringe with buffer and repeat the injection sequence to give the buffer-into-macromolecule mix dilution control.[b]
8. Analyse the integrated heat of binding data (step 6) after correction for dilution heats from steps 5 and 7 (and step 4 if necessary, although this is usually insignificant) using an appropriate binding model (several standard models are provided with instrumental software).

[a] Thermal equilibration and calorimeter stability are better if the temperatures of rinsing solutions and samples inserted into the cell are close to the working temperature of the calorimeter – usually 25 °C, but temperatures in the range 10–50 °C are typically feasible.)

[b] Buffer-into-macromolecule dilution controls may be performed separately, if required, but the above sequence is most economical in use of both time and material.)

4 Examples

Some examples of DSC and ITC data are briefly given here to illustrate the sort of data that may be obtained and what sort of information they may contain. Consult references for more details.

4.1 DSC of DNA–repressor interactions

Figure 1 shows DSC data for thermal unfolding of mixtures of the methionine repressor protein (MetJ) and a 16-bp DNA consensus recognition sequence

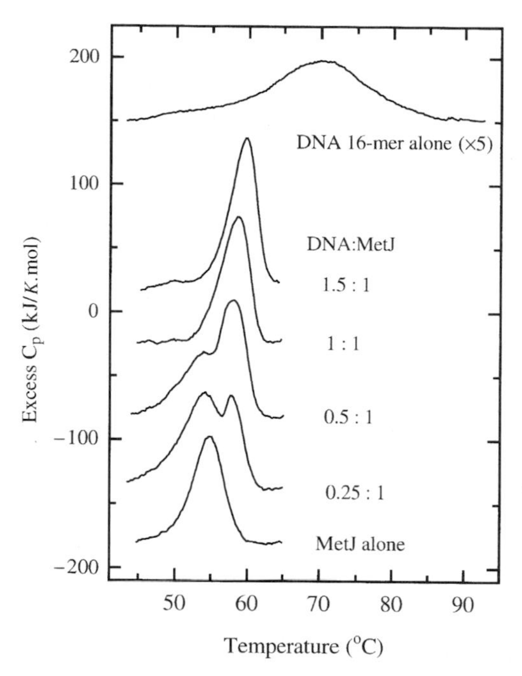

Figure 1 DSC thermograms (concentration normalized and corrected for buffer baselines) for MetJ protein, repressor DNA 16 bp consensus fragment, and protein–DNA mixtures at different stoichiometric ratios (25 mM phosphate, 0.1 M KCl, 1 mM DTT, pH 7; DSC scan rate 60 °C/h). The data for melting of the DNA alone (upper curve) have been expanded for clarity. (Adapted from ref. 6.)

(dAGACGTCTAGACGTCT, comprising two Met boxes) in the presence of SAM co-repressor (6).

In the absence of DNA, the protein undergoes a simple two-state unfolding transition with a mid-point transition temperature (T_m) around 54 °C (25). The addition of increasing stoichiometric amounts of DNA fragment results in a decrease in this peak and the appearance of a second peak at higher temperature ($T_m \approx 59$ °C) corresponding to the unfolding of the more stable repressor-DNA complex. (The thermal unfolding of the DNA duplex alone is quite broad and occurs at even higher temperatures under these conditions.) Bimodal DSC data have also been observed in other protein–DNA complexes under certain conditions (10), but in this case the second peak can be attributed to melting of the more stable DNA duplex after dissociation from the protein at high temperature.

4.2 ITC of DNA–repressor interactions

Direct ITC binding data for DNA-protein complex formation are illustrated in *Figure 2* for the case of the Trp repressor (A. Cooper and A. Lane, unpublished data; see also ref. 9). Similar data have been obtained for the MetJ–DNA system also studied by DSC as described above (6, 7) and for Cro protein-DNA complexes (8).

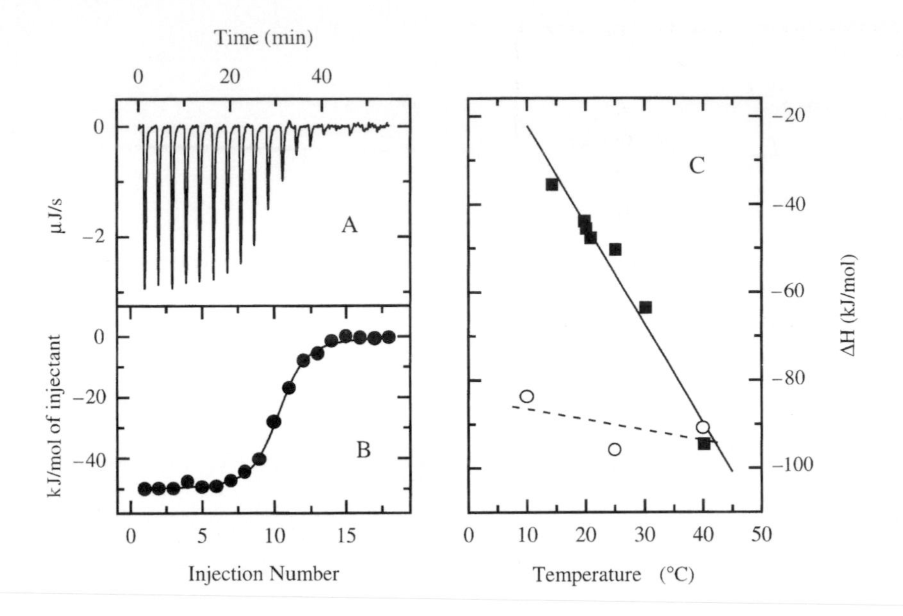

Figure 2 (A) ITC data for binding of Trp repressor to TrpO DNA (A. Cooper, A. Lane, unpublished data; and ref. 6). Raw data for sequential 10 μl injections of Trp repressor protein (0.165 mM) into TrpO DNA (23 μM, 20-mer): 25 °C, 10 mM phosphate pH 7.5, 0.25 M KCl. (B) Integrated heat data with theoretical fit to a stoichiometric binding model with $K \approx 10^7$/M and $\Delta H \approx -50$ kJ/mol. (C) Temperature dependence of the enthalpy of binding of Trp repressor to TrpO DNA (solid squares) and, for comparison, binding of MetJ to the consensus met box DNA (open circles).

Initial injections of Trp repressor into TrpO DNA solutions (*Figure 2A*) give heat pulses corresponding to the exothermic formation of complex. The magnitude of these pulses decreases with subsequent injections as the protein concentration reaches stoichiometric excess over DNA-binding sites. Later injections give only very small heat effects corresponding to dilution heats. The integrated heat data (*Figure 2B*) can be fitted to an appropriate model to give the stoichiometry, K and ΔH. As is frequently the case, enthalpies of complex formation are temperature-dependent (*Figure 2C*), implying a finite ΔC_p effect. Such effects are frequently attributed to changes in accessible molecular surface area during complexation (18–20), but some caution is advised in taking such interpretations too literally (11, 16).

4.3 DSC of protein–single-stranded DNA complexes

DSC of complexes between gene 5 protein and single-stranded DNA (13) at increasing protein–DNA ratios (*Figure 3*) show a gradual upward shift in T_m and shape of the protein unfolding endotherm characteristic of a rapid (on the DSC time-scale) equilibrium multimeric complex. Analysis of the DSC thermogram shapes in terms of a complex dissociation-equilibrium model (13) indicates that

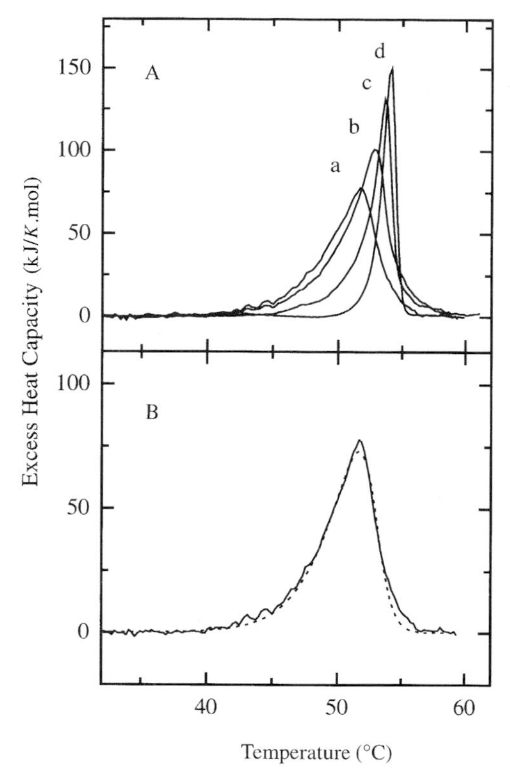

Figure 3 (A) DSC thermograms (concentration normalized and corrected for buffer baselines) of gene 5 protein–viral DNA complexes at different protein concentrations: (a) 18.9 μM, (b) 30.2 μM, (c) 61.0 μM and (d) 116 μM (10 mM Tris-HCl, pH 7.5, scan rate 60°C/h). (B) Theoretical fit (dotted line) of curve (a) to a model in which four protein monomers dissociate and unfold cooperatively from the single-stranded DNA. The size of the cooperative unit increases at higher concentrations. (Adapted from ref. 13.)

the cooperative unit for unfolding on the DNA corresponds to clusters of protein molecules ranging in size from about four protein monomers at low concentrations to 10–20, or more, at higher protein-DNA ratios.

4.4 ITC of protein–RNA interactions

To illustrate what are probably the current experimental limits on ITC of protein-nucleic acid complexes, *Figure 4* shows data (N. J. Stonehouse, P. G. Stockley and A. Cooper, unpublished data) for the heat of insertion of a specific RNA fragment into empty viral capsids. To my knowledge, this is the only example yet seen of calorimetry of a protein-RNA system. The relatively small absolute heat effects, with attendant thermal noise, reflect the difficulty in getting sufficient amounts of capsids into suspension. The relative scarcity, cost and instability of RNA tends to make such experiments challenging.

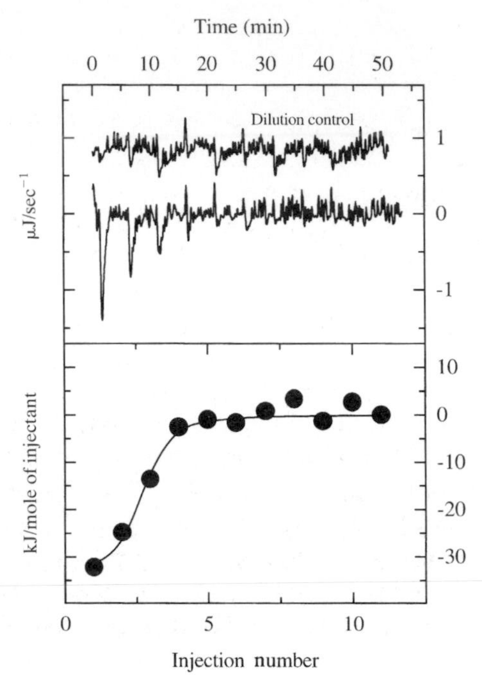

Figure 4 ITC data illustrating the interaction of a specific RNA fragment with empty MS2 viral capsids: sequential 10-μl injections of a specific RNA 19-mer hairpin (5′-ACAUGAGGAUUACCCAUGU-3′; 0.2 mM) into a suspension of empty viral capsids; 10 mM HEPES, 0.1 M NaCl, 1 mM EDTA, pH 7.2., 25 °C. (N. J. Stonehouse, P. G. Stockley and A. Cooper, unpublished data.)

References

1. Cooper, A. and Johnson, C. M. (1994). In *Methods in molecular biology: microscopy, optical spectroscopy, and macroscopic techniques*, (ed. C. Jones, B. Mulloy and A. H. Thomas), Vol. 22, pp. 109–150. Humana Press, Totowa.
2. Cooper, A. (1997). In *Methods in molecular biology: protein targeting protocols*, (ed. R. A. Clegg), Vol. 88, pp. 11–22. Humana Press, Totowa.
3. Wiseman, T., Williston, S., Brandts, J. F., and Lin, L. N. (1989). *Anal. Biochem.* **179**, 131.
4. Sturtevant, J. M. (1987). *Annu. Rev. Phys. Chem.* **38**, 463.
5. Privalov, P. L. and Potekhin, S. A. (1986). In *Methods in enzymology*, Vol. 131, pp. 4–51. Academic Press, London.
6. Cooper, A., McAlpine, A., and Stockley, P. G. (1994). *FEBS Lett.* **348**, 41.
7. Hyre, D. E. and Spicer, L. D. (1995). *Biochemistry* **34**, 3212.
8. Takeda, Y., Ross, P. D., and Mudd, C. P. (1992). *Proc. Natl. Acad. Sci. USA* **89**, 8180.
9. Ladbury, J. E., Wright, J. G., Sturtevant, J. M., and Sigler, P. B. (1994). *J. Mol. Biol.* **238**, 669.
10. Carra, J. H. and Privalov, P. H. (1997). *Biochemistry* **36**, 526.
11. Lundbäck, T., Hansson, H., Knapp, S., Ladenstein, R., and Härd, T. (1998). *J. Mol. Biol.* **276**,775.
12. Oda, M., Furukawa, K., Ogata, K., Sarai, A. and Nakamura, H. (1998). *J. Mol. Biol.* **276**, 571.
13. Davis, K. G., Plyte, S. E., Robertson, S. R., Cooper, A., and Kneale, G. G. (1995). *Biochemistry* **34**, 148.

14. Wittung, P., Ellouze, C., Maraboeuf, F., Takahashi, M. and Nordèn, B. (1997). *Eur. J. Biochem.* **245**, 715.
15. Naghibi, H., Tamura, A. and Sturtevant, J. M. (1995). *Proc. Natl. Acad. Sci. USA* **92**, 5597.
16. McPhail, D. and Cooper, A. (1997). *J. Chem. Soc. Faraday Trans.* **93**, 2283.
17. Dunitz, J. D (1995). *Chem. Biol.* **2**, 709.
18. Spolar, R. S. and Record, M. T. (1994). *Science* **263**, 777.
19. Janin, J. (1995). *Prot. Struct. Funct. Genet.* **21**, 30.
20. Freire, E. (1994). In *Methods Enzymol*, (ed. M. L. Johnson and L. Brand) **240**, 502.
21. Wyman, J. (1964). *Adv. Prot. Chem.* **19**, 223.
22. Wyman, J. and Gill, S. J. (1990). *Binding and linkage: functional chemistry of biological macromolecules*. University Science Books, Mid Valley.
23. Sturtevant, J. M. (1962). In *Experimental thermochemistry, vol. II* (ed. H. A. Skinner), p. 427. Interscience, New York.
24. Cooper, A. and Converse, C. A. (1976). *Biochemistry* **15**, 2970.
25. Johnson, C. M., Cooper, A. and Stockley, P. G. (1992). *Biochemistry* **31**, 9717.

Chapter 10
Protein–DNA crosslinking with formaldehyde *in vitro*

Konstantin Brodolin
Laboratory Molecular Genetics of Microorganisms, Institute of Molecular Genetics, Kurchatov Square 46 Moscow 123182, Russia

1 Introduction

Formaldehyde is a highly reactive reagent which produces protein-DNA and protein-protein crosslinks between macromolecules in close contact. The distance between the crosslinking groups of protein and DNA should be around 2 Å, suggesting an interaction at the range of van der Waals radii (1). Therefore, the formaldehyde crosslinking procedure provides a tool for identification of proteins and protein domains closely positioned to DNA. Formaldehyde cross-linking had been widely used for studies of protein-DNA interactions in chromatin *in vivo* and has been described in detail recently (2). In this chapter, the focus is on applying formaldehyde crosslinking to studies of specific protein-DNA interactions *in vitro*. This approach is principally different from the one mentioned above since relatively mild crosslinking conditions will be described that allow preferential formation of protein crosslinks of RNA polymerase from *Escherichia coli* with single-stranded (locally melted) regions of DNA in open complexes. In this context, crosslinking provides a simple way to identify domains on RNA polymerase bound to promoter DNA in open complexes as well as the identification of those subunit domains contacting the melted DNA. It can also be used for monitoring conformational changes in nucleoprotein complexes induced by different factors (temperature, salt conditions, pH) (3, 4). In general, this approach can be applied to the study of any protein contacting distorted or melted regions of the DNA double helix.

2 Chemistry of formaldehyde crosslinking

The exact mechanism of the crosslinking reaction between proteins and DNA is not known. Nevertheless, the reaction of formaldehyde with nucleotides and DNA is well studied. The crosslinking reactions between certain amino acids have been described (1, 5, 6). It has been shown that formaldehyde reacts with

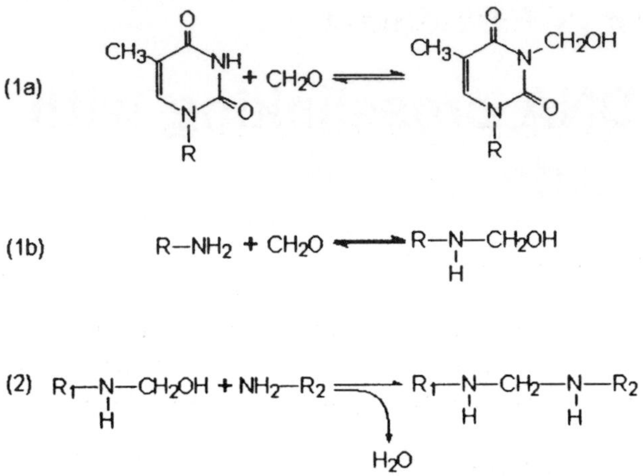

(1a) $\text{Thymine (N3-H)} + CH_2O \rightleftharpoons \text{Thymine (N3-}CH_2OH\text{)}$

(1b) $R{-}NH_2 + CH_2O \rightleftharpoons R{-}NH{-}CH_2OH$

(2) $R_1{-}NH{-}CH_2OH + NH_2{-}R_2 \rightleftharpoons R_1{-}NH{-}CH_2{-}NH{-}R_2 + H_2O$

Figure 1 Reactions mediated by formaldehyde. The first stage of crosslinking results in the formation of methylol derivatives of the thymine imino group (reaction 1a) or the amino group of other bases and amino acid side-chains (reaction 1b) The second stage results in a stable condensation product with methylene bridging groups: R_1 and R_2.

the amino groups of cytosines, guanines and adenines, and the imino groups of thymines and probably guanines. In proteins, potential candidates for crosslinking are Lys, Arg, Trp and His residues. Several possible mechanisms of crosslinking reactions are shown in *Figure 1*.

The first stage includes the reaction with amino or imino groups and results in the formation of methylol derivatives, which react with the adjacent second-reacting group. In conditions where short crosslinking times are used and crosslinks occur only in the melted DNA regions, the imino groups of thymine (or guanine) bases are the probable targets of the crosslinking reaction. This assumption is based on two known observations:

(1) Kinetic studies demonstrated slow formaldehyde reaction rates with amino groups while reaction rates of imino group were fast (6).

(2) Reactions with the amino groups of purines can take place with double-stranded DNA as well as with melted DNA regions while the reaction with imino groups of thymine occurs only in melted DNA regions (5).

The product of the first reaction step (methylol derivative) is very unstable while the final crosslinked derivatives are stable at 24 °C. Crosslinked complexes of RNA polymerase subunits can be incubated for several hours at room temperature without significant degradation. At 50 °C, slow degradation of protein–DNA crosslinks was detected after 20 min of incubation in Tris-HCl containing buffer ($t_{1/2}$ = 90 min); incubation for 2 min at 95 °C lead to full disruption of the crosslinks.

3 Practical application of formaldehyde crosslinking

3.1 Crosslinking of RNA polymerase–promoter complexes

3.1.1 Basic approach

The standard experimental scheme which can be applied for the studies of RNA polymerase–promoter interactions is shown in *Figure 2.*

Simply, it involves formation of the RNA polymerase-promoter complex with promoter DNA labelled at the 3′ end, followed by crosslinking. The crosslinking reaction is terminated by addition of SDS-containing buffer which denatures the nucleoprotein complex and therefore destroys any non-covalent contacts

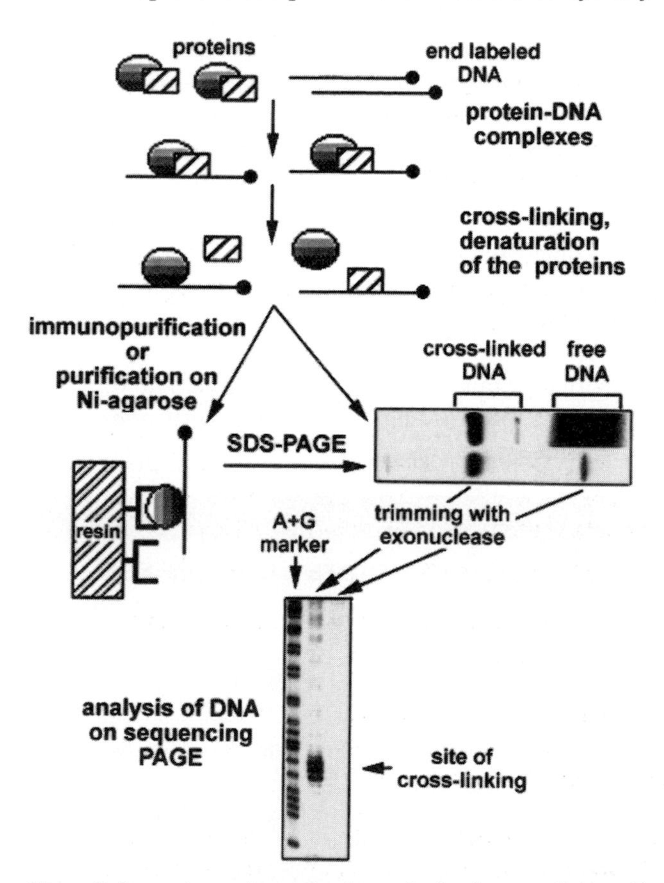

Figure 2 General experimental scheme for *in vitro* crosslinking. The protein–DNA complexes are treated with formaldehyde, denatured and analysed on SDS-PAGE. For identification of the protein composition, the crosslinked complexes are precipitated with antibodies and immunofractionated on protein A–Sepharose CL4B. If His-tagged RNA polymerase is used, crosslinked complexes are fractionated on Ni^{2+}-NTA agarose and analysed by SDS-PAGE. An autoradiogram of an SDS-PAGE separation of RNA polymerase–*lac*UV5-promoter crosslinked complexes is shown. Separated crosslinked complexes and free DNA are treated with λ exonuclease and the positions of the crosslinked complexes visualized on a sequencing gel.

between protein and the DNA. Crosslinked complexes can be easily resolved by SDS-PAGE using a standard Laemmli system (7) on 5% acrylamide gels, except that stacking gels (pH 6.8) must not be used since they provoke crosslink degradation. Crosslinked complexes retarded in the gel can be visualized by autoradiography or phosphorimagery. It should be noted that the electrophoretic mobility of subunits with a crosslinked DNA fragment is affected not only by the apparent relative mass (M_r) of the protein but also by the position of the crosslinked protein relative to the ends of the DNA fragment. It is therefore sometimes useful to use 5–15% gradient PAGE to achieve better separation of the crosslinked complexes.

Identification of the proteins involved in a crosslinked complex can be done by means of immunoaffinity or affinity chromatography (but see Pemberton chapter 22, this volume) and the crosslinks positions on DNA can be mapped by exonuclease footprinting (3, 4) (Chapter 14). In the example shown in *Figure 2*, two protein-DNA complexes containing β′ and σ subunits were visualized following crosslinking of the *lac*UV5-promoter RNA polymerase open complex.

Special care should be taken to ensure the purity of the chemicals used for the buffer composition, since certain impurities can induce spontaneous crosslinks between RNA polymerase and DNA incubated for more than 15 min at 37 °C even without formaldehyde being added.

Warming of samples should be avoided following crosslinking because of the reversibility of crosslinks. Ideally, samples containing crosslinked complexes should be analysed on the gel immediately, although they can be frozen and stored at −80 °C.

Long crosslinking times (more than 2 min) lead to the accumulation of protein–protein crosslinks and multiple DNA–protein crosslinking with non-melted regions of DNA, and should therefore be avoided.

Protocol 1

Crosslinking of RNA polymerase–promoter complexes

Equipment and reagents

- 0.2 M formaldehyde aqueous solution freshly made from 37% formaldehyde (Aldrich)
- 2× SB (125 mM Tris-HCl pH 6.8, 2% SDS, 10 mM DTT, 10% (v/v) glycerol, bromophenol blue)
- Separating gel acrylamide stock solution: 30 g acrylamide, 1 g *N,N′*-methylene-bis-acrylamide in 100 ml
- Separating gel buffer: 0.375 M Tris-HCl pH 8.8, 0.1% SDS
- Running buffer: 25 mM Tris base, 0.192 M glycine, 0.1% SDS
- 5× CLB (0.25 M HEPES-NaOH pH 8.0, 0.5 M NaCl, 25 mM $MgCl_2$, 25% glycerol)
- SDS-PAGE gel apparatus

Method

1 Mix RNA polymerase (100–200 nM final concentration) with 3′-end labelled promoter DNA fragment (5–10 nM final concentration) in 10 μl of 1× CLB and incubate at 37 °C for 10 min to form an open complex.

Protocol 1 continued

2 Add 1 μl of 200 mM formaldehyde to the sample and mix rapidly, incubate at 37 °C for 10–30 s.

3 Terminate the reaction by the addition of equal volume of SB.

4 Separate crosslinked complexes on 5% SDS-PAGE or freeze and store at −80 °C. During the electrophoresis set constant power output to 2 W for standard 18 × 18 cm gel plates, and 0.6 mm thick gel. The temperature of the gel should not exceed 25 °C.

High formaldehyde concentrations can affect the formation of specific protein–DNA complexes. For example, RNA polymerase cannot form complexes with promoter DNA fragments pretreated with 0.2% (v/v) formaldehyde for 10 min at 37 °C; treatment with 1% (v/v) formaldehyde of RNA polymerase-promoter complexes formed at 14°C leads to their dissociation (9). Therefore the concentration of reagent used must be kept within the range of 2–30 mM. Crosslinking time can be varied from 10 to 30 s to obtain detectable amounts of crosslinked material (the yield of crosslinked DNA is usually 0.5–4%) without disrupting the complex.

3.1.2 Applications

The crosslinking method is very sensitive to the conformation of the nucleoprotein complex and can be used for the comparison of different complexes formed under different conditions. For example, the different crosslinking patterns of open complexes formed on the *lac*UV5 promoter at different values of pH and temperature are shown in *Figure 3*.

Comparison of the kinetics of crosslinking at 37 °C vs. 22 °C shows that the

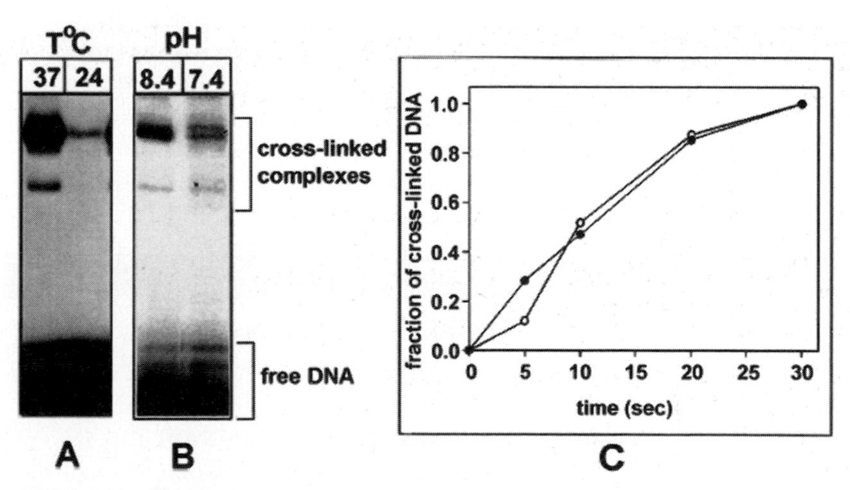

Figure 3 Visualization of conformational changes in nucleoprotein complexes. RNA polymerase is mixed with end-labelled *lac*UV5-promoter fragment in CLB buffer (see *Protocol 1*), incubated for 10 min at different temperatures (Panel A) or different pH values (Panel B) and crosslinked. Panel C: kinetics of crosslinking at 22 °C (open circles) and 37 °C (closed circles). The fraction of crosslinked DNA is plotted against the time of crosslinking for the β′-DNA complex.

reaction rate is not highly affected by temperature (*Figure 3*). (e.g. pseudo first-order rate constants calculated for one of the crosslinked complexes are 0.05/s (37 °C) vs. 0.04/s (22 °C)).

It is possible to use formaldehyde crosslinking for kinetic studies of RNA polymerase-promoter interactions (6). In this case, the samples crosslinked at different time points can be separated by SDS-PAGE and signals characterizing the crosslinks of individual subunits quantified. A disadvantage of formaldehyde crosslinking compared with faster techniques, such as UV laser crosslinking (see Chapter 14) is that the relatively slow crosslinking rate, restricts this method to slow processes generally taking place over the minute time range.

3.2 Identification of crosslinked species

3.2.1 Immunopurification

RNA polymerase interactions with promoter DNA can be used to illustrate this approach. The identification of those RNA polymerase subunits crosslinked under different experimental conditions, is crucial. Two methods described below in *Protocols 2* and *3* provide simple and direct ways to characterize these contacts. The first method requires rabbit polyclonal antibodies to individual subunits and is a modification of the method used for fractionation of crosslinked chromatin (2, 10). The method includes precipitation of the crosslinked complexes with subunit-specific antibodies and purification on protein A–Sepharose CL-4B. Analysis of bound and non-bound fractions on SDS-PAGE allows the identification of those subunits involved in any particular crosslinked complex (3). Before immunofractionation of crosslinked complexes the antibodies should be tested for binding of RNA polymerase subunits in high salt and detergent conditions to find the optimal proportion of antibodies to protein. Because the final elution volume is sometimes rather large it is helpful to precipitate the complexes by acetone and dissolve them in a small volume before analysis by SDS-PAGE (*Protocol 2*, steps 10–12). The electrophoretic pattern depends significantly on the removal from the eluted material of traces of detergents present in the IP buffer. Therefore, the washing step is important. Controls , where all steps are carried out but without added antibodies, should be made so that the non-specific absorption of crosslinking complexes can be taken into account.

Protocol 2

Immunofractionation of crosslinked complexes

Equipment and reagents

- Protein A–Sepharose CL-4B (Amersham-Pharmacia Biotech)
- Rabbit polyclonal antibodies to RNA polymerase subunits
- HEN: 20 mM HEPES-NaOH pH 7.4, 150 mM NaCl, 5 mM EDTA
- IP buffer: 20 mM HEPES-NaOH pH 7.4, 5 mM EDTA, 1 M NaCl, 1% (v/v) Triton-X 100, 0.1% SDS, 0.2% BSA
- SB buffer, see *Protocol 1*
- Vortex Genie-2 (Scientific Industries)

Protocol 2 continued

Method

1 Carry out procedures in 1.5 ml Eppendorf tubes. Prepare crosslinked complexes as described in *Protocol 1*, Step 1, but scale up fivefold the total reaction volume (e.g. 6 μg RNA polymerase for sample) and do not add SB to terminate crosslinking reaction.

2 Terminate crosslinking reaction by adding 10% SDS to 0.5% final concentration.

3 Add all components of IP buffer to adjust the sample volume to 300 μl.

4 Add 1–3.5 μg of antibodies for 1 μg of RNA polymerase and incubate for 2 h at 20 °C.

5 Add 20 μl of 50% (v/v) suspension of protein A-Sepharose CL-4B equilibrated with IP buffer and incubate the sample with gentle agitation using Vortex Genie-2 shaker for 3 h at 20 °C.

6 Pellet the resin by brief centrifugation in Eppendorf centrifuge and remove supernatant carefully.

7 Wash resin with 0.2 ml of IP buffer for 20 min, centrifuge and remove supernatant.

8 Wash the resin with 0.2 ml of HEN buffer containing 1% (v/v) Triton-X100 for 5 min, centrifuge and remove supernatant.

9 Wash resin with 0.2 ml of HEN for 5 min, centrifuge and remove supernatant. Repeat this step five times.

10 Add 20 μl of 1% SDS solution to the resin and shake for 5 min at room temperature, collect supernatant and repeat the procedure.

11 Add four volumes of cold acetone to the eluate and incubate for 30 min at −70 °C.

12 Precipitate the complexes by centrifugation for 10 min at 16 000 **g** in an Eppendorf centrifuge at 4 °C.

13 Wash pellet with 80% cold acetone, centrifuge 2 min at 16 000 **g** in Eppendorf centrifuge at 4 °C, dry pellet, dissolve in SB and load on SDS-PAGE (see *Protocol 1*).[a]

3.2.2 Affinity chromatography

The second method (see *Protocol 3*) uses genetically engineered subunits containing His-tags at the *C*- or *N*-terminal end. The RNA polymerase containing such subunits can be assembled by standard methods (11) and used for crosslinking studies. As was shown elsewhere, these RNA polymerases possess the properties of the wild-type enzyme. In this case, the crosslinked complexes must be denatured by urea and subjected to fractionation on Ni^{2+}-NTA agarose. Special care should be taken with the controls because the Ni^{2+}-NTA agarose also absorbs denatured RNA polymerase subunits without the His-tag. To reduce the non-specific binding, additional washing steps must be carried out. The procedure should also all be repeated with RNA polymerase which does not contain a His-tag, so as to adjust the conditions of fractionation and to obtain a low level of non-specific binding.

Protocol 3

Purification of crosslinked complexes on Ni^{2+}-NTA agarose

Equipment and reagents

- Ni^{2+}-NTA agarose (Qiagen)
- RNA polymerase containing His-tag in one of the subunits
- 5× CLB, see *Protocol 1*
- 2× SB, see *Protocol 1*
- Vortex Genie-2

Method

1. Prepare crosslinked complexes as described in *Protocol 1*, step 1.
2. Terminate crosslinking by addition of 2.5 volumes of 10 M urea.
3. Add 10 μl of 50% (v/v) suspension of Ni^{2+}-NTA agarose equilibrated with 0.5× CLB containing 7 M urea to the Eppendorf tube and incubate for 40 min at 20 °C with gentle agitation using Vortex Genie-2 shaker.
4. Centrifuge on Eppendorf centrifuge 30 s and discard supernatant.
5. Wash resin with 150 μl of 0.5× CLB containing 7 M urea by shaking for 5 min at 20°C. Centrifuge resin and remove supernatant.
6. Repeat step 5 three times.
7. Elute bound fraction by shaking the resin with 20 μl SB containing 20 mM EDTA 5 min at room temperature, add 0.2 volumes of SB and load on SDS-PAGE (see *Protocol 1*).

3.3 Mapping of crosslinking sites on DNA

Thermolability of formaldehyde crosslinks and the fact that formaldehyde does not induce DNA cleavage at the site of crosslinking puts some constraints on crosslink mapping. The primer extension technique, widely used for localization of UV-induced crosslinks (see Chapter 14), cannot be applied here. The phage λ exonuclease utilizing double-stranded DNA is a simple way of mapping the linear arrangement of crosslinks on DNA (3). The exonuclease cannot hydrolyse DNA when it encounters a crosslinked base. It does not stop exactly at the base which is involved in crosslinking but more probably one or two bases before. Thus, the positions of crosslinks relative to the labelled 5′-end of the DNA are obtained. An advantage of using phage λ exonuclease is that it digests DNA from the 5′ end and therefore DNA fragments labelled at the 3′ end by Klenow fragment could be used. To localize a crosslinking site for any particular subunit the crosslinked complexes should be separated by SDS-PAGE. The complexes are then visualized by autoradiography and treated with exonuclease in gel (*Protocol 4*). As a control, non-crosslinked DNA must be excised from the same gel and treated with exonuclease. This is important, since exonuclease stops may appear at positions on the DNA that are not real sites of crosslinking.

The intensity of these artificial bands can be reduced by variation of exonuclease concentration and by adjusting the conditions of digestion (see *Figure 2*).

Protocol 4

Exonuclease footprinting in gel

Equipment and reagents

- Phage λ exonuclease (Boehringer)
- 5× exonuclease buffer 0.33 M glycine-KOH pH 9.4, 0.25 mg/ml BSA
- 50 mM $MgCl_2$
- Sequencing gel apparatus

Method

1. Prepare crosslinked complexes as described in *Protocol 1*, step 1, but scale up fivefold the total reaction volume (standard reaction contains 6 μg RNA polymerase in 50 μl volume) add 0.2 volumes of SB to terminate reaction.
2. Resolve crosslinked complexes on preparative 5% SDS-PAGE. Standard 0.6 mm thick gel with 15 × 10 mm comb wells.
3. Transfer the wet gel to any solid film (e.g. X-ray film), cover with laboratory film and attach fluorescent marker (Stratagene).
4. Autoradiograph the gel at room temperature (it usually takes 0.5 h) and excise slices containing radioactive bands.
5. Wash the gel slices with water three times for 5 min.
6. Crush the gel slices well and place in Eppendorf tubes, adjust total volume to 100 μl with exonuclease buffer and add 0.5–2 units of λ exonuclease. Incubate for 20 min at room temperature. The amount of exonuclease should be adjusted for every experiment.
7. Add $MgCl_2$ to 2.5 mM final concentration and incubate for 20 min at 30 °C.
8. Stop the reaction by addition of 0.01 volume 0.5 M EDTA and 0.01 volume 10% SDS.
9. Add Tris-HCl pH 8.0 to 0.3 M final concentration and elute DNA for 4 h at 65 °C.
10. Precipitate DNA, dissolve in formamide and load on sequencing gel.

References

1. Kunkel, G. R., Menrabian, M., and Martinson, H. G. (1981). *Mol. Cell. Biochem.* **34,** 3.
2. Orlando, V., Strutt, H., and Paro, R. (1997). *Methods* **11,** 205.
3. Brodolin, K. L., Studitsky, V. M., and Mirzabekov A. D. (1993). *Nucl. Acids Res.* **21,** 5748 .
4. McGhee, J. D. and von Hippel, P. H. (1975). *Biochemistry* **14,** 1281.
5. Feldman, M. Y. (1973). *Prog. Nucl. Acids Res. Molec. Biol.* **13,** 1.
6. Laemmli, U. K. (1970). *Nature* **227,** 680
7. Buckle M., Pemberton, I. K., Jacquet, M-A., and Buc, H. (1999). *J. Mol. Biol.* **285**, 955.

8. Brodolin, K. L., Studitskii, V. M., and Mirzabekov A. D. (1993). *Mol. Biol.* **27,** 671 (in Russian).
9. Postnikov, Y. V., Shick, V. V., Belyavsky, A. V., Khrapko, K. R., Brodolin, K. L., Nikolskaya, T. A., and Mirzabekov, A. D. (1991). *Nucl. Acids Res.* **19**, 717.
10. Kashlev, M., Nudler, E., Severinov, K., Borukhov, S., Komissarova, N., and Goldfarb, A. (1996). *Methods Enzymol.* **274**, 326

Chapter 11
Solid-phase DNAse I footprinting

Raphael Sandaltzopoulos
Laboratory of Molecular and Cell Biology, National Cancer Institute, National Institutes of Health, Blg. 37, Rm 5E-24, Bethesda, MD 20892–4255, USA

Peter B. Becker
Adolf-Butenandt Institut,Ludwig-Maximilians Universität München, Schillerstrasse 44, D–80336 Munich, Germany

1 Introduction

Classical DNase I footprinting is used to map the DNA targets of sequence-specific binding proteins (1). An end-labelled DNA fragment encompassing a specific target sequence is incubated with the DNA-binding protein to form a protein–DNA complex. Mild digestion of the fragment with DNase I gives rise to a distinct fragmentation pattern which can be visualized owing to the end-labelling. Tightly bound protein will protect its target sequences from DNase I cleavage. The direct comparison of the cleavage pattern derived from the DNA/protein complex with that derived from free DNA reveals the sequences that are protected by the bound protein. Ordinary footprinting protocols are laborious, with most of the time needed being spent on the preparation of the labelled template and purification of DNA fragments prior to gel electrophoresis.

Solid-phase DNase I footprinting emerged from the combination of traditional footprinting with the analysis of DNA coupled to a solid support. The labelled DNA fragments to be analysed for protein binding are immobilized on paramagnetic beads, enabling instant and quantitative concentration of the DNA in a magnetic field and facilitating the purification of the DNA from complex reaction mixtures. The ease of purification accounts for faster protocols, higher quality of results, flexibility of experimental design and reduced exposure of the researcher to radiation (2). In this chapter we describe solid-phase footprinting protocols and illustrate the advantages of this novel technique. The outline of the experimental design is presented in *Figure 1*.

2 Solid-phase footprinting protocol

2.1 Using a radioactive probe

A PCR fragment harbouring the putative target sequences is generated using a pair of 5′ end-labelled primers. One oligonucleotide is biotinylated during synthesis

1. Generation of a PCR fragment

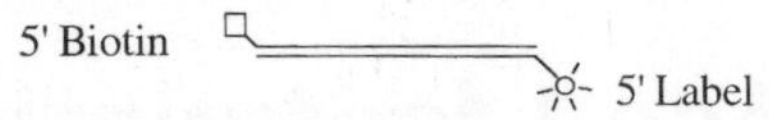

2. Immobilisation of streptavidin-coated paramagnetic beads

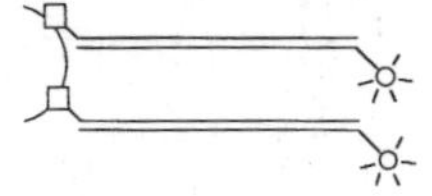

3. Incubation with sequence-specific DNA-binding protein

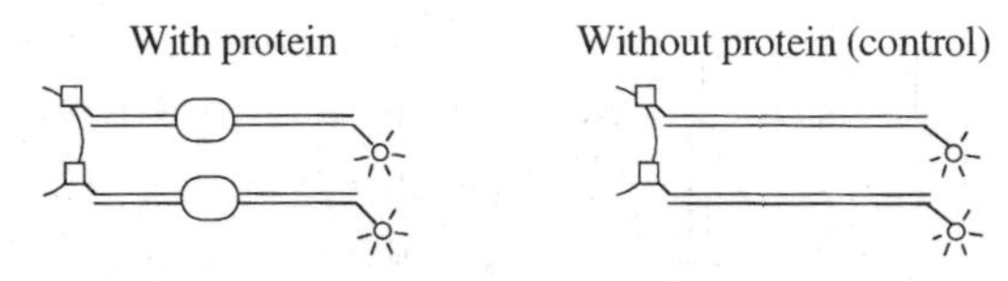

4. Limited DNAse I digestion

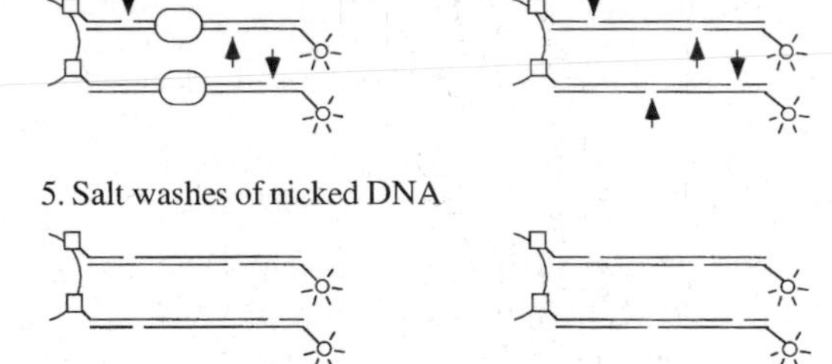

6. Denaturation and gel electrophoresis

Figure 1 Schematic outline of the solid-phase footprinting procedure.

and serves as an anchor for the immobilization of the fragment to streptavidin-coated paramagnetic beads. The other primer is radioactively labelled to allow detection of the DNase I cleavage products. If the protein sample is contaminated with phosphatase activity, the target DNA can be prepared by filling-in the 5′ overhangs of suitable restriction fragments with appropriate biotinylated dNTPs, [α-^{32}P]dNTPs and Klenow exo$^-$ polymerase (3).

Protocol 1

Preparation of labelled template

Equipment and reagents

- T4 PNK 10 units/ml and 10x PNK buffer (Biolabs)
- dNTPs and [γ-^{32}P]ATP, 7000 Ci/mmol, 167 mCi/ml (ICN)
- PCR cycler
- Sephadex G-25 TE spin column (Boehringer)
- Taq DNA polymerase 5 units/ml and 10 x buffer (Promega)

Protocol 1 continued

Method

1. Kinase the non-biotinylated oligo A. Mix 5 μl of 10 pmol/ml oligo, 3 μl 10× PNK buffer, 3 μl [γ-^{32}P]ATP, 18 μl H_2O and 1 μl PNK 10 units/ml.
2. Incubate at 37°C for 45 min to 1 h.
3. Resuspend the matrix of a Sephadex G-25 TE spin column. Remove the top lid first, then the bottom of the column.
4. Place in a reaction tube provided (without lid) and let it drain in a vertical position at room temperature (≈ 5 min).
5. Empty the reaction tube and put the column (together with the reaction tube) in a 15 ml Falcon tube.
6. Spin for 1 min at 1100 **g**.
7. Discard flow-through and spin at 1100 **g** for 2 min.
8. Replace the collection tube by a fresh one. Apply the kinase reaction slowly at the centre of the resin without touching the resin.
9. Spin at 1100 **g** and collect flow-through. The volume of your sample should be ≈ 30 μl.
10. Set up a PCR reaction. Mix 30 μλ purified radiolabelled primer A (from step 9), 5 μλ 10× Taq polymerase buffer, 5 μλ biotinylated primer B (10 pmol/ml), 2 μl 10 mM mix of each dNTP, 2 μl template plasmid 10 ng/ml, 5 μl H_2O and 1 μl Taq DNA polymerase. Perform 30 cycles of PCR. The annealing temperature should be determined taking into account the melting temperature of the primers. The probe is then immobilized on streptavidin-coated paramagnetic beads. This is achieved by simply incubating the biotinylated DNA with the beads. The high ionic strength of the coupling buffer reduces electrostatic repulsion of DNA molecules and increases immobilization efficiency. The amount of biotinylated primer used in the PCR reaction and the capacity of the beads to bind biotin must be considered. In general, a substantial excess of beads must be used. During the incubation of the streptavidin-coated beads in the crude PCR reaction all biotinylated molecules will be immobilized, which includes, in addition to the desired PCR fragment, non-specific or incomplete byproducts of the reaction as well as unincorporated primer. The efficiency of the coupling reaction may be evaluated by the 'missing band' assay (see Protocol 2). The extraordinary high affinity of biotin-streptavidin bond ($K_D = 10^{-15}$ M) guarantees that the probe is stable during storage or sequential washings involved in the footprinting procedure.

A typical result of solid-phase footprinting is presented in *Figure 2* showing the protection of five tandem Gal4 binding sites by a recombinant Gal4-HSF fusion protein.

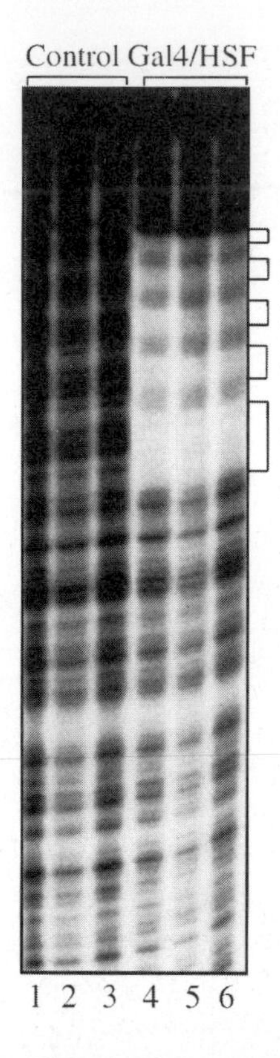

Figure 2 An example of solid-phase footprinting. A Gal4–HSF fusion protein comprising the DNA-binding domain of GAl4 (aa 1–147) and Drosophila HSF (aa 321–691) (11) protects five regions on the adenovirus E4 minimal core promoter engineered to contain five consecutive GAl4-binding sites (UAS_G–E4; see ref. 12). The 337-bp long PCR fragment was generated using as template the pGIE-0 plasmid (12). The biotinylated primer (upper strand) annealed at positions –238 to –218 while the radiolabelled primer (lower strand) at +99 to +79 relative to the transcription start site. For each reaction, 0.3 μl of bead suspension (prepared as explained in Protocols 1 and 2) were used. Samples 1–3 received no protein. Samples 1 and 4, 2 and 5, and 3 and 6 were treated with 0.001, 0.003 and 0.01 units of DNAse I, respectively, for exactly 1 min, as explained in Protocol 3 and analysed by 6% Urea-PAGE; bands were visualized after exposure of the dried gel for 3 h.

Protocol 2

Immobilization of labelled template

Equipment and reagents

- Dynabeads M-280 (Dynal) or equivalent from other suppliers
- Rotating wheel
- Magnetic particle concentrator rack (MPC-6, Dynal)
- Wash buffer (WB: PBS pH 7.5/0.01% BSA/0.05% Nonidet P40)
- Coupling buffer (CB : 2 M NaCl/10 mM Tris-HCl pH 7.5/1 mM EDTA)

Method

1. Pipette 300 μl of bead suspension into a 1.5 ml reaction tube. Place the tube for 30 s in the MPC. Discard supernatant and resuspend the beads in 300 μl of WB. Repeat this step and resuspend in 300 μl CB.
2. If you used mineral oil for PCR, remove excess oil. Dilute the PCR reaction without purification of the fragment 1 : 1 in 2× CB (100 μl final volume) in a 500 μl reaction tube.
3. Keep 1 μl as a reference for monitoring coupling efficiency.

Protocol 2 continued

4 Add the bead suspension to the PCR products (final volume is 400 μl).

5 Place the tube on a tilted rotating wheel at room temperature. Rotate for at least 30 min (1 h for fragments longer than 400 bp).

6 Transfer the coupling reaction into a 1.5 ml reaction tube. Concentrate beads on the MPC-6 for 1 min. Keep 10 μl of the supernatant and discard the rest.

7 Wash beads twice with 300 μl of WB. Resuspend in 100 μl CB. The template is ready for footprinting analysis (protocol 3). 0.3–1 μl should be enough for one footprinting reaction, depending on the efficiency of the radiolabelling reaction.

8 To monitor coupling efficiency, analyse 1 μl of the original PCR reaction and 8 μl of supernatant after the immobilization by electrophoresis on a 1% agarose gel. Dry the gel after electrophoresis onto DE81 paper (Whatman) and detect by autoradiography. The absence of radiolabelled full-length PCR fragment from the supernatant indicates complete immobilization ('missing band test').

9 The immobilized DNA is then incubated with the protein of interest to allow formation of protein–DNA complexes. When an equilibrium has been reached, limited treatment of the complexes with DNase I (or other DNA nicking agents) creates single-strand nicks. The reaction is tuned such that, on average, only one nick per fragment occurs. (i.e. there are still many fragments that remain intact). Although nicked, the fragment remains on the beads unless the DNA is denatured. The reaction is terminated by addition of EDTA that chelates divalent cations necessary for the enzymatic activity of DNase I. At the same time, the high ionic strength of termination buffer efficiently strips all proteins from DNA. Hence, there is no need for organic extractions or precipitation of DNA. The labelled probe is then washed again to remove excess salt that might interfere with band resolution during gel electrophoresis. Finally, the fragments are released from the beads by a brief denaturation step in presence of formamide.

Protocol 3

Solid-phase DNase I footprinting

Equipment and reagents

- Binding buffer (e.g. 12.5 mM HEPES-KOH pH 7.9, 2% polyvinylalcohol, 10% glycerol/1 mM DTT, 1 mM EDTA pH 8.0, 100 μg/ml BSA, 50 mM KCl)
- Stock solution of DNAse I, 10 units/ml (Boehringer) in 5 mM $CaCl_2$/10 mM $MgCl_2$
- Reaction stop buffer (S4E100: 4 M NaCl, 100 mM EDTA)
- High salt washing buffer (S2E20 : 2 M NaCl/20 mM EDTA)
- Loading buffer (LB: 96% formamide, 0.05% xylene cyanol, 0.05% bromophenol blue, 10 mM EDTA, freshly mixed 3 : 1 with 100 mM NaOH)
- Waterbath or Eppendorf thermomixer

Protocol 3 continued

Method

1. Resuspend bead-template well. Pipette out an appropriate amount of immobilized template sufficient for N + 1 reactions (N = desired number of footprinting reactions).
2. Concentrate beads on the magnet and discard supernatant.
3. Resuspend beads in (N + 1) × 25 μl binding buffer of choice.
4. Add 1 μl of protein solution to be assayed.
5. Incubate for desired time (usually between 5 and 20 min) at binding temperature (frequently room temperature, = 20°C). Occasional agitation of the tube by gentle tapping is required to keep the beads well dispersed during extended incubations. An Eppendorf thermomixer (shaker setting 10) can be used, but avoid harsh vortex mixing.
6. Meanwhile prepare DNase I dilutions in 5 mM $CaCl_2$/10 mM $MgCl_2$. It is advisable to digest each sample with at least two DNase I concentrations. The DNase concentrations need to be determined empirically, but 0.001–0.01 units may serve as a good starting point.
7. Prepare two fresh tubes (for two DNase I concentrations) for each reaction containing 20 μl of the an appropriate DNase I dilution. For convenience, DNase digestions are frequently done at room temperature.
8. Transfer 20 μl of binding reaction to each tube of DNase I.
9. Terminate DNase I digestion after exactly 1 min by addition of an equal volume (40 μl) of S4E100.
10. After having processed all samples to this point, concentrate beads on the MPC, discard supernatant and wash once with 100 μl of S2E20.
11. Wash beads with 50 μl 10 mM Tris-HCl, pH 8.0, 1 mM EDTA.
12. Concentrate beads again, remove supernatant completely (if necessary spin down briefly) and mix the beads with 8 μl loading buffer.
13. Denature samples for 5 min at 76 °C and load on a pre-run 6% denaturing polyacrylamide gel (standard sequencing gel). Beads do not interfere with electrophoresis, so their removal before loading is not necessary.

In order to achieve an average of one nick per DNA strand, a considerable fraction of labelled DNA should not be cleaved. Since the radiolabelled strand is not biotinylated, it will be released from the beads during denaturation even if it is still intact. The proportion of the full-length fragment indicates the extent of digestion. As an empirical guideline, there should be hardly any counts released during the wash steps after DNase I digestion (as monitored by a Geiger counter) if the samples were not over-digested.

2.2 Using a non-radioactive labelled probe

Solid-phase footprinting is also compatible with the non-radioactive analysis of protein–DNA interactions (4). The PCR product may be labelled using a primer modified by a fluorescent moiety (e.g. fluorescein, Texas Red, Cy-5, etc.) during synthesis. This eliminates the exposure to radiation and has the additional advantage that the labelled fragment does not decay. A large amount of fragment may be prepared once and used with comparable results repeatedly in the future. Essentially the same protocol described above can be followed, with the only exception that the generated DNA fragments are analysed on an ALF sequencer (Pharmacia). The sensitivity of this approach is at least as high as the radioactive method (4). Recently, Machida et al. (5) showed that the sensitivity can be increased further by at least tenfold using an infrared automated DNA sequencer. Remarkably, they were able to detect protein–DNA interactions on both strands of a 895-bp long DNA fragment with extremely high sensitivity (using only 3.1 fmol template). Thus, this method has the potential to be applied to the analysis of protein–DNA interactions in the context of large-scale genome projects.

As automated sequencing instruments become more popular, non-radioactive footprinting may be the approach of choice since it is faster (does not require drying and exposure of the gel), data can be stored electronically and results quantified with appropriate image-processing software. Importantly, since fluorescent moieties are not substrates for phosphatases, even crude extracts rich in phosphatase activity may be used for the detection of sequence-specific DNA binding activities.

3 Applications of solid-phase footprinting

Solid-phase footprinting is the method of choice for mapping binding sites of sequence-specific DNA binding proteins because it is the most time-efficient footprinting method, yet provides superior quality with high consistency. Among the proteins that have been analysed by solid phase footprinting so far are the *Drosophila* heat shock factor (2, 6, 7) and GAGA factor (7), human HNF-1 and HNF-1/DCoH (8), TATA binding protein (TBP) and hormone receptors, Sp1 (5) and Gal4 fusion hybrid proteins (*Figure 2* and unpublished results). Since the radiolabelled strand is not biotinylated, incomplete non-biotinylated PCR products are not immobilized and therefore do not contribute to the analysis. This built-in 'template-selection' process during the preparation of the immobilized probe improves the signal-to-noise ratio and allows even suboptimal PCR reactions to be used as a substrate for the analysis (9).

More importantly, the method can be applied to study the stability of protein–DNA complexes. The option for instant purification of the protein–DNA complexes from the binding reaction enables the determination of off-rates. For this purpose, protein–DNA complexes are formed under optimal conditions (e.g. high protein concentrations and/or presence of chemicals that promote molecular crowding) and then purified. The nature and speed of the concentration in the magnetic field minimizes the risk of complex disruption during the purification procedure. Excess unbound protein is removed by subsequent washes with buffer and the isolated complexes are incubated during a time course before DNase I digest. Quantification of the footprinting profile at each time-point of the experiment allows visualization of the dissociation of the protein from the DNA. This strategy has been applied to analyse the *Drosophila* heat shock transcription factor–DNA complexes (6) and to characterize the effect of the dimerization cofactor of HNF-1 (DCoH) on the DNA-binding properties of HNF-1 (8).

Instead of simply diluting the reaction (10), the separation of the complex from unbound factor allows one to analyse specifically the parameters that influence the stability of complexes after formation. The stability of the purified complexes can be tested in competition with other transcription factors (e.g. DNA-binding proteins recognizing overlapping cis-elements) or modifying enzymes (e.g. kinases). Challenging the protein–DNA complexes with buffers of increasing ionic strength, Rhee et al. (8) showed that DCoH affects the mode of interactions of HNF-1 with its cognate DNA sequence.

Since solid-phase footprinting is compatible with crude extracts (2), the option of washes with low and increasing stringency before DNAse I treatment may facilitate the analysis of factors in crude protein mixtures by discriminating between specific and non-specific protein–DNA complexes; non-specific interactions of lower affinity should be less resistant to stringent washes and hence selectively destabilized.

Solid-phase footprinting can be conveniently integrated into sophisticated experimental designs. For example, a protein of interest may be purified from a crude nuclear extract by DNA affinity. While most of the purified protein/DNA complexes may be subjected to *in vitro* transcription assay, an aliquot may be tested by solid-phase footprinting in order to determine the percentage of the template that was bound by the protein.

In summary, the ease of manipulations, the speed of analysis and the high quality of results, as well as the increased freedom for experimental design, make solid-phase footprinting the method of choice for the analysis of sequence-specific protein/DNA interactions.

Acknowledgements

The authors thank Drs V. Ossipow and G. Mizuguchi for providing highly purified recombinant Gal4-HSF.

References

1. Galas, D. and Schmitz, A. (1978). *Nucl. Acids Res.* **5**, 3157–3170.
2. Sandaltzopoulos, R. and Becker, P. B. (1994). *Nucl. Acids Res.* **22**, 1511.
3. Sambrook, J., Fritsch, E. F., and Maniatis, T. (1989). *Molecular cloning. A laboratory manual, 2nd edn*, p. 100. Cold Spring Harbor Laboratory Press, Cold Spring Harbor.
4. Sandaltzopoulos, R., Ansorge, W., Becker, P. B., and Voss, H. (1994). *BioTechniques* **17**, 474.
5. Machida, M., Kamio, H., and Sorensen, D. (1997). *Biotechniques* **23**, 300.
6. Sandaltzopoulos, R., Quivy, J. P., and Becker, P. B. (1995). *Methods Molec. Cell. Biol.* **5**,176.
7. Sandaltzopoulos, R., Mitchelmore, C., Bonte, E., Wall, G., and Becker, P. B. (1995). *Nucl. Acids Res.* **23**, 2479.

8. Rhee, K., Stier, G., Becker, P. B., Suck, D., and Sandaltzopoulos, R. (1997). *J. Mol. Biol.* **265**, 20.
9. Sandaltzopoulos, R., and Bonte, E. (1995). *BioTechniques* **5**, 77.
10. Lieberman, P. M., and Berk, A. J. (1991). *Genes Dev.* **5**, 2441.
11. Wisniewski, J., Orosz, A., Allada, R., and Wu, C. (1996). *Nucl. Acids Res.* **24**, 367.
12. Pazin, M. J., Kamakaka, R. T., and Kadonaga, J. T. (1994). *Science* **266**, 2007.

Chapter 12
Hydroxyl radical footprinting

Annie Kolb
Unité de Physicochimie des Macromolecules Biologiques, Institut Pasteur, F–75015 Paris, France

Tamara Belyaeva and Nigel Savery
School of Biosciences, The University of Birmingham, Birmingham B15 2TT, UK

1 Introduction

Hydroxyl radicals have now been used for more than 15 years to probe the structure of various macromolecules. They have also been widely used to footprint DNA–protein and RNA–protein complexes (1, 2). Hydroxyl radicals are very small, comparable in size to a water molecule, are most commonly generated from Fe-EDTA in solution by an innocuous and inexpensive procedure (see *Protocols 2* and *3*) and are capable of cutting the DNA or RNA backbone at every accessible nucleotide without sequence specificity (3–5). A footprinting reagent in which the EDTA moiety is coupled to an intercalator can also be used to generate hydroxyl radicals (6). This reactant, methidium-propyl Fe-EDTA, binds to some preferred positions along the DNA, and consequently the pattern of DNA backbone cleavage is much less random than with Fe-EDTA alone.

Recently, it was discovered that the small Fe-chelate could also cleave accessible peptide bonds, apparently independently of the amino acids involved (7). Despite some uncertainty in the chemistry, this method can now be used in combination with the new technology of end-labelled proteins to study the solvent accessible surface of a protein. Furthermore it can detect a conformational change of a protein induced by a ligand, as for the cAMP receptor protein (CRP) after addition of cAMP (8), and allows interactions between two proteins or a protein and a nucleic acid to be mapped by protein footprinting (9, 10). The Fe-EDTA chelate may also be tethered to a single amino acid (e.g. a cysteine side chain) and used to map its proximity to an individual peptide or nucleotide bond (11, 12). Recent applications of the Fe-BABE DNA-cleaving reagent illustrates the powerful methodology now associated with hydroxyl radical cleavage (13–15).

In this chapter, we focus on the use of hydroxyl radicals to make footprints of protein–DNA complexes. A great number of excellent reviews have already appeared on the subject (1, 16). Our aim is to provide short protocols to enable the reader to obtain clean and reproducible patterns of hydroxyl radical cleavage of DNA, and to apply this method to obtain high-resolution protection and interference footprints of protein–DNA complexes.

2 Principle of the procedure

2.1 Chemistry of the reaction

Hydroxyl radicals are most commonly generated according to the Fenton reaction in which Fe(II) reduces hydrogen peroxide to give hydroxyl radicals:

$$Fe(EDTA)^{2-} + H_2O_2 \rightarrow Fe(EDTA)^{1-} + OH^- + OH^\bullet$$

The Fe(III) generated in this reaction is subsequently reduced by ascorbate back to Fe(II), thus allowing the reaction to continue and decreasing the amount of $Fe(EDTA)^{-2}$ that needs to be present in the reaction mixture. Since the metal chelate is negatively charged it is not thought to bind DNA, and hydroxyl radicals can diffuse in the solution up to distances of 10 Å before interacting with their target.

The cleavage of DNA by hydroxyl radicals occurs primarily through attack of the deoxyribose, removing a hydrogen atom located in the minor groove of DNA. Secondary reactions cause the breakage of the sugar–phosphate backbone leaving a gapped duplex with 3′ and 5′-phosphate termini, and also some proportion of 3′-phosphoglycolate derivatives which migrate faster on high-percentage polyacrylamide sequencing gels (17).

The hydroxyl radical cleavage procedure offers several advantages over other nucleases used to footprint protein–DNA complexes:

(1) Hydroxyl radicals cut DNA in a sequence-independent manner [with the exception of bent DNA fragments which show reduced reactivity in the regions with decreased accessibility of the minor groove, as pointed out by Price and Tullius (18)].

(2) Because of their small size and lack of specificity, they provide the highest resolution cleavage pattern.

(3) They can be used in a variety of conditions with different buffers, ionic strength, temperature and pH (although free radical scavengers such as glycerol can greatly decrease the intensity of the cleavage).

2.2 Principle of hydroxyl radical footprinting

As with other footprinting techniques, hydroxyl radical footprinting relies on the ability of a ligand to protect the DNA to which it is bound from nuclease attack. A DNA fragment is radioactively end-labelled with ^{32}P at one end (*Protocol 1*). A portion of this sample is incubated alone or with a ligand (e.g. a small molecule, a protein or even crude extracts) in the appropriate conditions to obtain optimal binding and subjected to hydroxyl radical cleavage under 'single hit' attack conditions such that each molecule of DNA is cleaved once at most. After quenching of the reaction and DNA purification, both samples are electrophoresed on a sequencing polyacrylamide gel (*Protocol 2*). Then, after autoradiography, the patterns of the cleavage for the free DNA and the ligand bound DNA are compared. A regular ladder corresponding to cleavage of the DNA at every nucleotide is observed for the free DNA. Any difference in the intensity of the cleavage between the free and the bound DNA at a particular nucleotide is due to the protection of this nucleotide by the ligand, or possibly to a conformational change in the DNA induced by binding to the ligand (for example see *Figure 1A*, which shows the hydroxyl radical footprint of a bacterial transcription factor, CRP, in the presence and absence of RNA polymerase).

2.3 Interference method: the missing nucleoside assay

The missing nucleoside assay complements the protection assay described above and provides information about the function of specific nucleotides in complex formation rather than simply indicating proximity to the bound ligand. In this assay the population of DNA fragments is first modified by hydroxyl radical cleavage to introduce, at most, one gap per duplex DNA and then assayed for ligand binding. The free and bound DNA are separated on a native polyacrylamide gel and each fraction is analysed on a sequencing polyacrylamide gel. The molecules that contain a nucleoside important for DNA binding by the ligand will not be present in the bound DNA sample and lead to a 'missing' or weak band in the lane of bound DNA (*Figure 2*). In contrast, these positions are enriched in the sample of free DNA. The opposite pattern can be observed for the nucleosides that do not make energetically important contacts for binding in the location of the binding sites: they are more abundant in the lane corresponding to the bound DNA than in the lane corresponding to the free DNA.

As with any interference method, this approach will detect not only the contacts that are important for the stability of the bound complex, but also any intermediate contacts, which are essential for the formation of the complex. However, it is not likely to be convenient for the study of a multi-protein complex where a missing nucleoside can affect differently the binding of the various components. In addition, the flexibility of the DNA is enhanced by the presence of gaps and missing nucleotides can therefore affect the binding of proteins, which induce conformational changes in DNA in an unpredictable manner (19).

2.4 Quantification and interpretation of hydroxyl radical footprinting data

Since hydroxyl radical cleavage provides high-resolution patterns with some subtle differences in the presence of the ligand, it is recommended that the intensities of the bands in the different lanes are quantified in order to extract maximum information from each experiment. This can be achieved either by scanning the film using a densitometer or by measuring the ^{32}P disintegrations directly from the gel by using a phosphorimager. *Figure 1B* shows information from a phosphorimager scan of lanes from the gel shown in *Figure 1A*, in which the regions of decreased band intensity caused by binding of the CRP protein

A.

CRP
holo RNAP

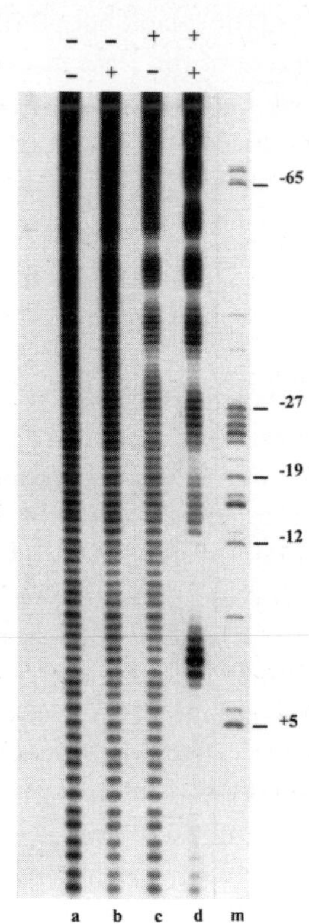
– – + +
– + – +
-65
-27
-19
-12
+5
a b c d m

B.

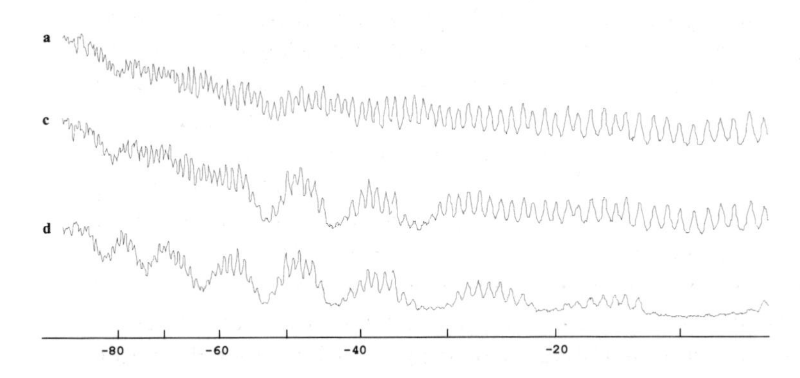
a
c
d
-80
-60
-40
-20

C.

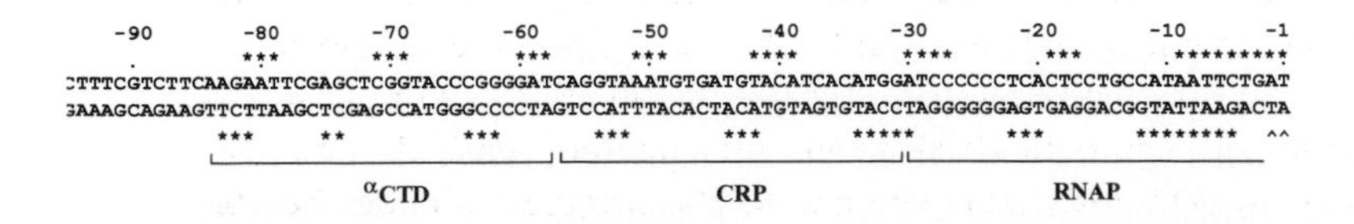
-90 -80 -70 -60 -50 -40 -30 -20 -10 -1
CTTTCGTCTTCAAGAATTCGAGCTCGGTACCCGGGGATCAGGTAAATGTGATGTACATCACATGGATCCCCCTCACTCCTGCCATAATTCTGAT
GAAAGCAGAAGTTCTTAAGCTCGAGCCATGGGCCCCTAGTCCATTTACACTACATGTAGTGTACCTAGGGGGAGTGAGGACGGTATTAAGACTA
αCTD
CRP
RNAP

Figure 1 Hydroxyl radical footprint of binary and ternary complexes containing RNA polymerase, the transcription factor CRP and promoter DNA. End-labelled fragments containing a promoter with a CRP-binding site centred at −41. 5 – *CC(-41. 5)* – were incubated with combinations of CRP (100 nM) and holo-RNA polymerase (100 nM), as indicated in the figure, and subjected to hydroxyl radical attack as described in *Protocol 2*. (A) Autoradiogram of an acrylamide gel showing hydroxyl radical cleavage pattern of the lower strand of the promoter in the absence (a) or presence (b–d) of proteins. Lane m contains a Maxam–Gilbert 'G' reaction, numbered with respect to the transcription start site (+1). (B) Profile of lanes a, c and d generated from a phosphorimager data file using Imagequant software supplied by Amersham Pharmacia BioTech: CRP protects three regions close to −41.5 and RNA polymerase protects DNA on either side of the CRP-binding site. (C) Summary of the protection data obtained from hydroxyl radical footprints of both strands of the *CC(−41. 5)* promoter. Asterisks denote protected positions. A model of the promoter showing this protection is shown in Plate 4

and RNA polymerase are readily visible. Calibration of such gels allows the position of nucleosides protected in the protein–DNA complexes to be plotted on the sequence of the target DNA (*Figure 1C, Plate 4*). Note that binding of CRP causes three regions of decreased reactivity centred 10 bp from each other. A similar pattern is obtained on the non-template strand with an offset of 2–3 bp in the 3′ direction (*Figure 1C*). This offset indicates equivalent protections of the sugars across the minor groove (20) and demonstrates that the protein binds on a single face of the DNA helix. In contrast, if both strands are protected in the same region at the base pair level, it indicates that the protein wraps around DNA when bound.

3 Experimental procedures

3.1 Preparation of the end-labelled DNA

The DNA fragment to be footprinted should be purified from an acrylamide gel and labelled specifically at one end of one strand with ^{32}P (using either T4 polynucleotide kinase to label a dephosphorylated 5′ end or the Klenow fragment of DNA polymerase to label the 3′ end of an appropriate restriction fragment). In order to achieve clear footprints the DNA should be of as high a quality as possible and great care should be taken not to introduce nicks into the DNA during the fragment preparation procedures. Suitable fragments can be prepared by restriction endonuclease digestion of CsCl-purified or comparable quality plasmid DNA. If possible, the restriction sites should be chosen such that the labelled end of the DNA is 30–40 bp away from the area in which the proteins are expected to bind. To decrease the risk of introducing nicks into the DNA during fragment preparation, restriction enzyme concentrations should be carefully matched to DNA concentration according to the manufacturers instructions and incubation times should be kept short (i.e. overnight incubation should be avoided). If the fragment is to be labelled at the 5′ end the plasmid should be cut with restriction enzyme I, treated with calf intestinal alkaline phosphatase, extracted thoroughly with phenol–chloroform, precipitated, washed with 70% ethanol and then cut with restriction enzyme II. The fragment should then be

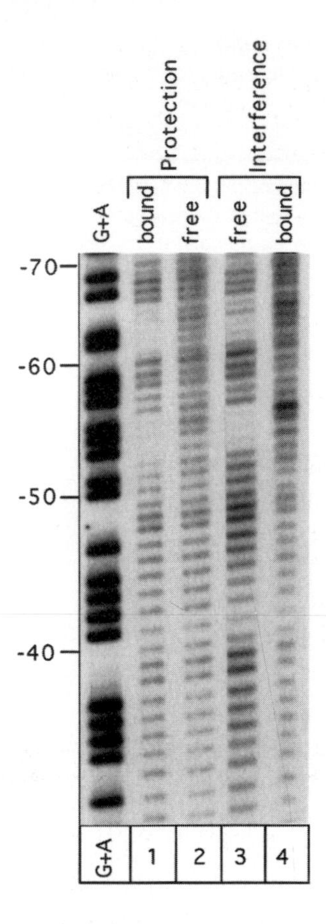

Figure 2 Comparison between CRP protection and interference data at the *lac* CRP-binding site. The pattern of the bound sample in the protection experiment (lane1) is similar to the pattern of the free sample in the interference assay (lane3) showing that the residues which are protected by CRP are also needed for optimal binding of the protein. The 139-bp *lac* DNA was synthesized by PCR as described in *Protocol 1* using the following primers 5′ ^{32}P 5′GCTCACAATTCCACACATTATACGAG (from +10 to −16) and 5′GCTGGCACGACAGGTTTCCCGA (from −124 to −103), purified on a glass fibre column and directly used for hydroxyl radical cleavage as in *Protocol 3* for interference studies. The fragment was shown to be at least 96% pure on a sequencing polyacrylamide gel. For protection experiments, labelled DNA (3 nM) was incubated with 10 nM CRP in the presence of 200 μM cAMP at room temperature, in a buffer containing 40 mM HEPES pH 8. 0, 10 mM magnesium chloride, 100 mM potassium glutamate and 200 μg/ml BSA and incubated with 50 μM ferrous ammonium sulphate–100 μM EDTA, 200 μM sodium ascorbate and 0. 03% hydrogen peroxide for 1 min, mixed with loading buffer and immediately loaded onto the running electrophoretic mobility gel in TBE buffer containing 200 μM cAMP (*Protocol 2*^{c}). Free and bound DNAs were purified from the gel. For interference studies, labelled DNA was first treated with hydroxyl radicals, as described above, purified and then incubated with 50 nM CRP and 200 μM cAMP in the same binding buffer before separation of free and bound DNA on the electrophoretic mobility gel.

purified from an acrylamide gel (agarose should not be used as it contains impurities which can interfere with subsequent reactions) and labelled with T4 polynucleotide kinase and [γ-^{32}P]ATP. If the Klenow fragment of DNA polymerase is used to label the 3′ end of one strand, the labelling reaction must be done before cutting with the second restriction enzyme (and the subsequent steps must be undertaken with a radioactive DNA) or restriction enzyme II should be chosen to give a non-recessed 3′ end.

Fragments for footprinting reactions can also be prepared using carefully optimized PCR reactions. The PCR procedure (*Protocol 1*) offers a number of potential advantages over the conventional procedure outlined above. Firstly, it is quick: the complete procedure takes only 2 days and there is no need to produce a high-concentration, high-purity plasmid preparation. Furthermore, if the PCR conditions can be optimized to produce a 98% pure fragment, as determined by electrophoresis on a sequencing polyacrylamide gel, it is possible to save time by purifying the fragment away from DNA polymerase and nucleotides on a glass fibre column. Secondly, PCR allows easy 5′ end-labelling of either strand of the DNA fragment and, since the location of the labelled end depends only on the location of the synthetic primer, the labelled end can always be at the optimum distance from the area to be footprinted, without having to rely on the positioning of restriction sites.

Protocol 1

Isolation of end labelled DNA using PCR

Equipment and reagents

- Perspex screens and boxes to protect from radioactivity
- Thermal cycler for PCR
- A pair of highly purified oligonucleotides A and B (18–25 nucleotides), designed to PCR amplify the fragment and which contain no heterogeneity at their 5′ ends
- T4 polynucleotide kinase and 10× kinase buffer (usually supplied with enzyme: typical composition 50 mM DTT, 100 mM $MgCl_2$, 700 mM Tris HCl, pH 7.6)
- [γ-^{32}P]ATP (3000 Ci/mmole)
- *Pfu* DNA polymerase and 10× *Pfu* buffer (usually supplied with enzyme: typical composition 100 mM KCl, 100 mM $(NH_4)_2SO_4$, 1 mg/ml BSA, 1% Triton X-100, 200 mM Tris HCl, pH 8.8), deoxyribonucleotide stock containing each dNTP at a final concentration of 2.5 mM, and mineral oil
- Purified plasmid DNA containing the insert to be amplified
- Acrylamide mix (6% acrylamide (acrylamide: bisacrylamide ratio 29:1), 1× TBE buffer)
- TBE buffer (89 mM Tris-borate, 89 mM boric acid, 2 mM EDTA)
- Gel loading buffer (50% sucrose, 1 mM EDTA, 0.025% bromophenol blue, 0.025% xylene cyanol)
- 10% Ammonium persulphate (freshly made)
- TEMED
- Elution buffer (0.5 M ammonium acetate, 1 mM EDTA)
- TE buffer (10 mM Tris-HCl, 0.1 mM EDTA pH 8.0)

Protocol 1 continued

Method

1. Mix the following in a 0.5 ml microfuge tube: 5 μl water, 1.5 μl 10 × kinase buffer, 5 μl 1 μM oligonucleotide A, 3 μl [γ-^{32}P]ATP (3000 Ci/mmol), and 2–5 units of T4 polynucleotide kinase (final volume 15 μl).
2. Incubate at 37 °C for 30 min.[a]
3. Prepare the following in a PCR tube 13 μl water, 5 μl 10× *Pfu* buffer, 5 μl 2.5 mM deoxyribonucleotide stock, 5 μl 1 μM oligonucleotide B, 2 μl of 2–5 nM plasmid DNA (about 5 ng of a 5000 bp plasmid), and the kinased reaction from step 1. Add 1 μl of *Pfu* DNA polymerase (2.5 Units), mix, centrifuge, cover with mineral oil and keep on ice until the thermocycler has reached 94 °C.
4. Subject the sample to 30 cycles of denaturation, annealing and extension using a thermal cycler.[b] Remove the mineral oil from the sample.[c]
5. Add 10 μl of loading buffer and separate the labelled fragment from impurities by electrophoresis through a polyacrylamide gel (typically 6% acrylamide made up in TBE buffer).[d] Let the non-incorporated [γ-^{32}P]ATP run out of the gel.
6. Remove the upper electrophoresis plate, cover the gel with cling film transparent cellophane, tape a plastic rule with fluorescent markings (Stratagene) onto the border of the gel and visualize the DNA by autoradiography.
7. Excise the band of the gel containing the DNA and let it stand overnight at 37 °C in 0.3 ml of elution buffer.
8. Remove the supernatant, transfer it to a 1.5 ml microfuge tube, rinse the polyacrylamide gel piece with 0.15 ml of elution buffer and add to the supernatant.
9. Precipitate the DNA by adding 2.5 volumes of ethanol. Incubate for 30 min at −70 °C. Centrifuge at 12 000 **g** for 15 min at −20 °C. Check that most of the radioactivity is in the pellet with a Geiger counter. Wash with 70% ethanol, centrifuge briefly. Remove and discard the ethanol.
10. Dry the pellet under vacuum. Resuspend in 40 μl of 10 mM TE buffer[e]. Estimate the concentration of the labelled fragment by measuring 1–3 μl by Cerenkov counting.

[a] Estimate the specific radioactivity of the labelled oligonucleotide by spotting 0.5 μl of the kinased reaction on the bottom of a PEI cellulose paper. Run the chromatography with 1 N HCl: the oligonucleotide remains near the bottom and the [γ-^{32}P]ATP migrates with the eluent. Divide the chromatograph into two parts and count them in a β-counter using the Cerenkov effect. Some oligonucleotides are only partly deprotected and cannot be labelled with high efficiency.

[b] The PCR conditions have to be carefully optimized with unlabelled primers. The annealing temperature is ≈ 5 °C below the T_m of the primer. For a primer with a T_m of 55°C a suitable cycle profile for generating a 200 bp fragment is: 3 × (94 °C 2 min, 50 °C 1 min, 72 °C 1 min), 27 × (94 °C 1 min, 50 °C 1 min, 72 °C 1 min), 1 × (94 °C 1 min, 50 °C 2 min, 72 °C 10 min).

[c] If the PCR fragment does not require purification on a polyacrylamide gel, use a PCR high-purification kit (Boehringer) to remove oil and non-DNA contaminants, as indicated by the manufacturer. Use 70 μl water to elute the sample at the last step.

Protocol 1 continued

[d] The recovery of the labelled fragment is better if the electrophoresis is performed on a 0.4 mm thin polyacrylamide gel; in this case, siliconize one glass plate before pouring the gel.

[e] It is important to ensure that the samples are fully resuspended. This often requires that the samples are incubated in buffer for 10–15 min at 37 °C after being vortexed vigorously. Labelled fragments should be stored at −20 °C and used as soon after labelling as possible.

3.2 Formation of the DNA–protein complexes and attack by the hydroxyl radicals.

The basic protocol for footprinting in solution described below assumes full occupancy of the DNA site by the protein being analysed. Even after optimization of the binding conditions, however, some fraction of the DNA can remain unbound. To increase the intensity and the resolution of the footprints, immediately after hydroxyl radical attack (*Protocol 1*, step 5), free and bound DNA can be separated on a native polyacrylamide gel (see *Protocol 3* for EMSA) (22, 23). The free and bound DNA are eluted from the gel and analysed on a sequencing polyacrylamide gel. However, this technique requires about three to five times more labelled DNA than is needed in solution because of the inefficient recovery of the materials from the EMSA gel. In addition, it is essential that the equilibrium between free and bound DNA should not be modified during the footprinting reaction and loading on the EMSA gel.

Protocol 2

Hydroxyl radical cleavage

Equipment and reagents

- DNA fragment labelled specifically at one end with ^{32}P as prepared above (*Protocol 1*)
- Protein samples - some proteins are kept in a buffer containing 50% glycerol and should be diluted appropriately to perform the reaction at a concentration of glycerol lower than 0.5% (v/v)
- 10× binding buffer (50 mM $MgCl_2$, 500 mM potassium glutamate, 10 mM DTT, 5 mg/ml BSA, 200 mM HEPES pH 8.0)
- 10 mM sodium ascorbate (freshly made)
- Fe/EDTA solution (freshly made: mix an equal volume of 0.2 mM ferrous ammonium sulphate (stored in aliquots at −20 °C) and 0.4 mM EDTA)
- Stop solution (freshly made: mix 16 μl of 0.2 M EDTA (pH 8.0) and 5 μl of 0.1 M thiourea (freshly made) per reaction)
- 0.3% H_2O_2 (freshly diluted from 30% stock)
- 20 mg/ml glycogen
- Phenol/chloroform/isoamyl alcohol (25 : 24 : 1 by volume saturated with 0.1 M Tris HCl pH 8.0)
- Sequencing gel loading buffer (40% deionized formamide, 5 M urea, 5 mM NaOH, 1 mM EDTA, 0.025% bromophenol blue, 0.025% xylene cyanol)
- Denaturing acrylamide mix (6% acrylamide (acrylamide-bisacrylamide, ratio 19 : 1), 7 M urea, 1× TBE buffer)
- 10% Ammonium persulphate (freshly made)
- TEMED
- 10% acetic acid/20% ethanol

Protocol 2 continued

Method

1. Mix end-labelled DNA fragment (1–4 nM) containing about 50–100 000 d.p.m. with the proteins to be footprinted in a total volume of 35 μl of 1× binding buffer in a 1.5 ml microfuge tube.[a]
2. Incubate at 37 °C for 20 min, or at least for a sufficient amount of time to allow the complex to be formed (the temperature may also be altered if necessary).
3. Place 5 μl of Fe/EDTA solution, 5 μl of 0.3% H_2O_2 and 5 μl of sodium ascorbate solution as separate droplets onto the wall of the tube.[b]
4. Tip the tube to allow the droplets to fall into the solution.
5. Incubate the reaction mixture at 37 °C for 1 min.[c]
6. Add 21 μl of stop solution and mix well.
7. Extract with 70 μl of phenol/chloroform/isoamyl alcohol. Transfer the aqueous phase to a fresh microfuge tube.
8. Add 150 μl water and precipitate the DNA by adding 1 μl of 20 mg/ml glycogen and 500 μl of cold 100% ethanol.
9. Centrifuge for 15 min at 12 000 **g** at 4 °C, and discard the supernatant.
10. Add 600 μl of cold 70% ethanol.
11. Centrifuge for 15 min at 12 000 **g** at 4 °C, and discard the supernatant.
12. Dry the DNA pellet under vacuum.
13. Resuspend the pellet in 8 μl of sequencing gel loading buffer.
14. Incubate the samples at 90 °C for 2 min, then load 3 μl of each sample onto a denaturing polyacrylamide gel.[d]
15. When electrophoresis is complete the gel should be fixed in acetic acid/ethanol mix and dried onto 3MM paper. The results can be visualized using autoradiography or a phosphorimager system.

[a] Different binding buffers can be used, and some experimentation may be necessary in order to find the optimal conditions for studying any particular system. In particular, it is essential to ensure that any cofactor required for activity of the proteins under study is included in the binding reaction (for example, the *Escherichia coli* cAMP receptor protein requires the presence of cAMP in order to bind its specific target on the DNA). The water used in the preparation of all buffers should be of the highest quality available. Some unlabelled non-specific competitor DNA can also be added to reduce non-specific binding to the fragment of interest.

[b] The cleavage conditions have to be optimized whenever a new binding buffer or DNA fragment is used for the first time. The most appropriate conditions are those that leave ≈ 70% of the fragment uncut. Perform the cleavage reaction on naked DNA (i.e. no added proteins) and alter the concentration of the Fe/EDTA and sodium ascorbate solutions used in this step until an appropriate level of DNA cleavage is observed (sodium ascorbate should always be present at several times the concentration of the Fe/EDTA).

[c] Alternatively, the reacted complex can be immediately loaded on a running electrophoretic mobility shift gel after addition of 10 μl of loading buffer (50% sucrose in 1× binding buffer,

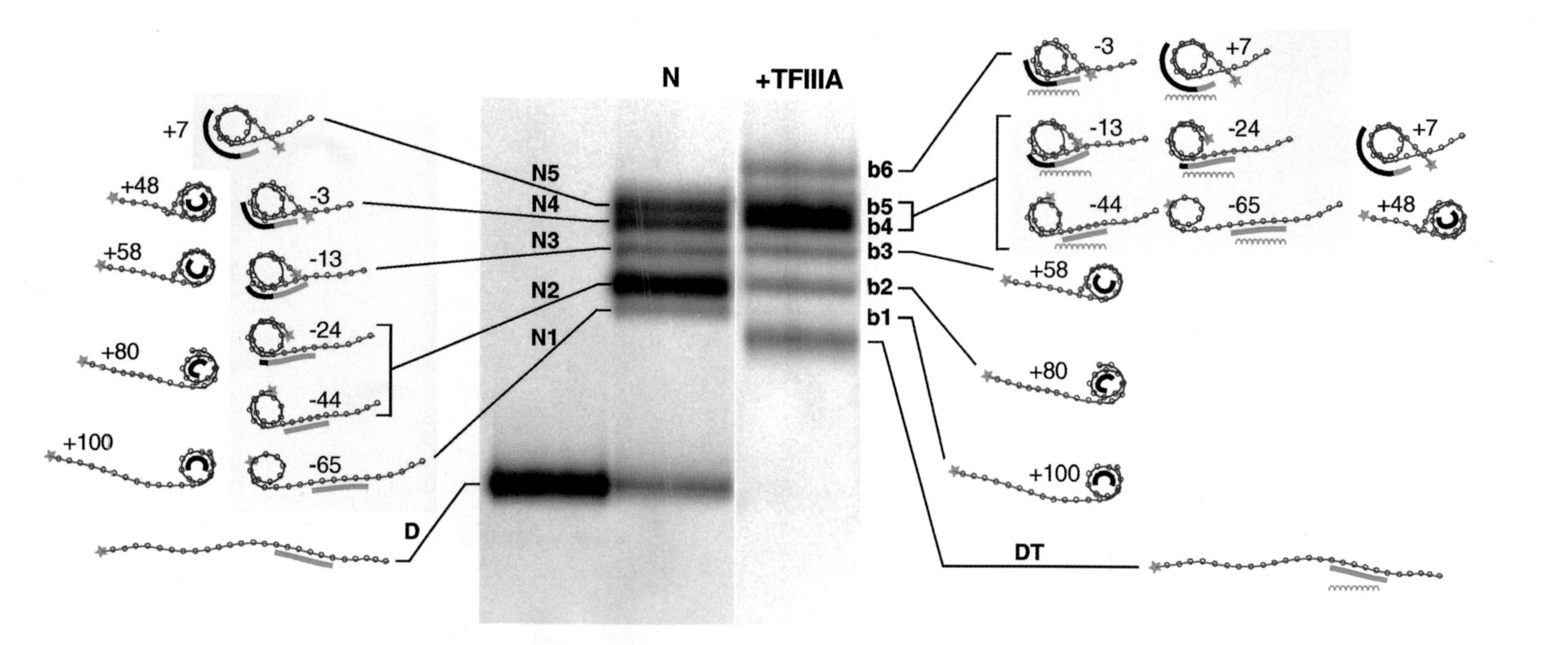

Plate 1. Fractionation of transcription factor IIIA binding to nucleosomal complexes in a nucleoprotein polyacrylamide gel. Lane N shows nucleosomes reconstituted on the somatic 5S rRNA gene. Bands N1-N5 represent nucleosome particle which only differ by the histone octamer translational positions along the DNA. The intensities of the bands vary, indicating that some positions are more abundant than others. The lane marked +TFIIIA shows nucleosome complexes in the presence of TFIIIA. Bands b1-b6 represent nucleosome particles as in lane N, but after incubation with TFIIIA. D and DT indicate free DNA and TFIIIA-bound DNA, respectively. The nucleosome positions were deduced from a nucleosome dyad mapping experiment which involves using site-directed hydroxyl radical cleavage reaction prior to fractionation on a nucleoprotein polyacrylamide gel followed by analysis in a denaturing polyacrylamide gel (9). The nucleosome positions present in each gel band are represented on either side of the gel (drawn using a prediction program for the DNA curvature which takes into account the position of the nucleosome and the DNA sequence (24)). Only the DNA helix axis is shown. In the the yellow panel, nucleosome positions that permit TFIIIA binding are indicated. The TFIIIA-binding site is indicated in green where exposed and in black where covered by a positioned nucleosome. The figure is modified from (9).

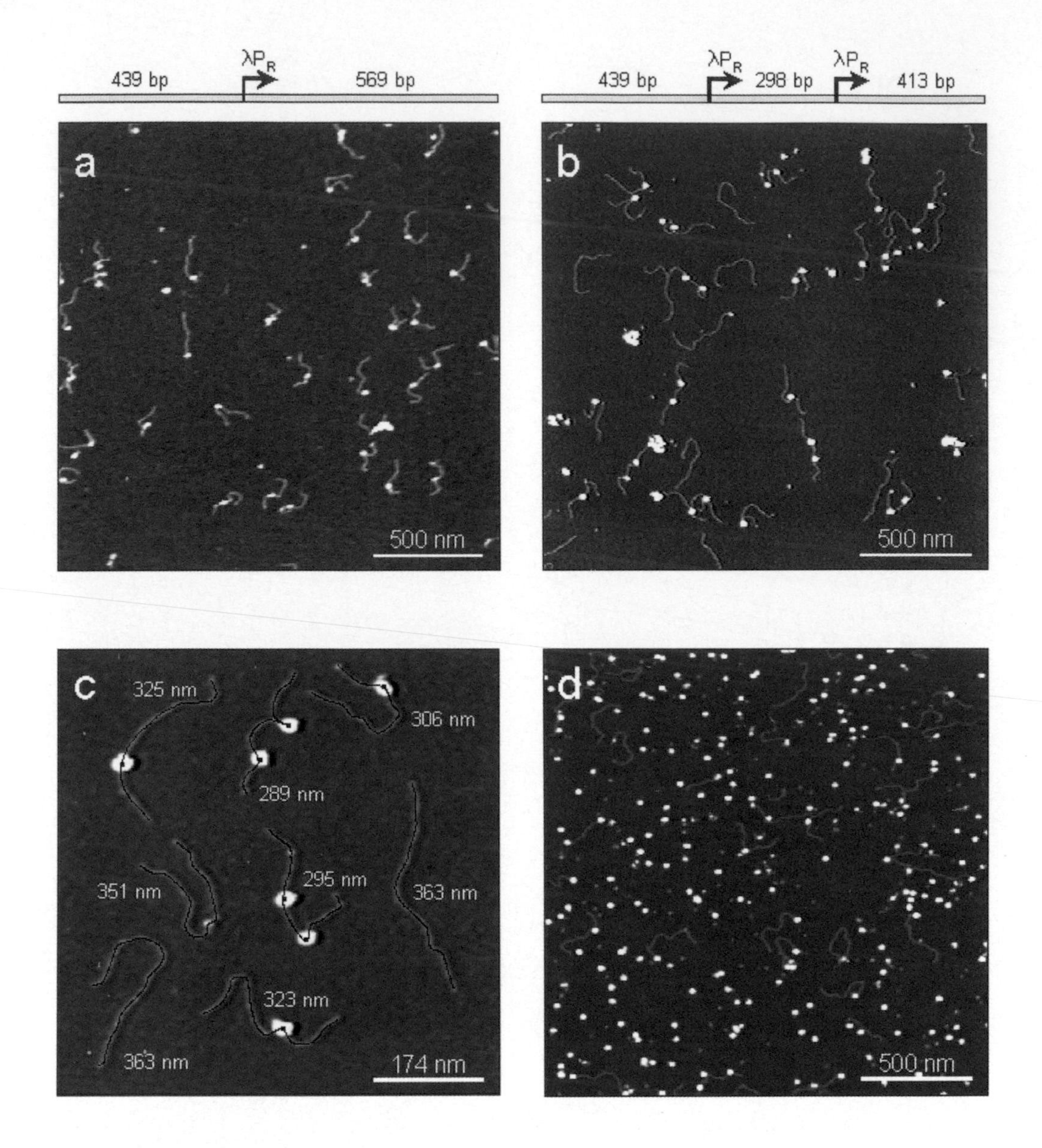

Plate 2. (a) Scheme of the 1008 bp DNA template (template A) showing the position of the λ_{PR} promoter and the direction of the transcription together with an SFM images of OPCs obtained with this DNA template. (b) Scheme of the 1150 bp DNA template (template C) showing the position of the two λ_{PR} promoters and the direction of transcription; also shown is an SFM images of OPCs assembled on template C in which DNA molecules alone, DNA molecules with one OPC and DNA molecules with two OPCs are clearly distinguishable. (c) Close-up of an SFM image of OPCs on template C: the lines represent the DNA contours traced as described in the text, while the small dots represent the centre of the RNAP. The contour length of each molecule is indicated on the image. (d) SFM Image of heparin-resistant RNAP-DNA complexes obtained with template B using the procedure given in *Protocol 2*. All the images were recorded in air with the tapping mode. The colour code corresponds to a height range of 5 nm from dark to clear

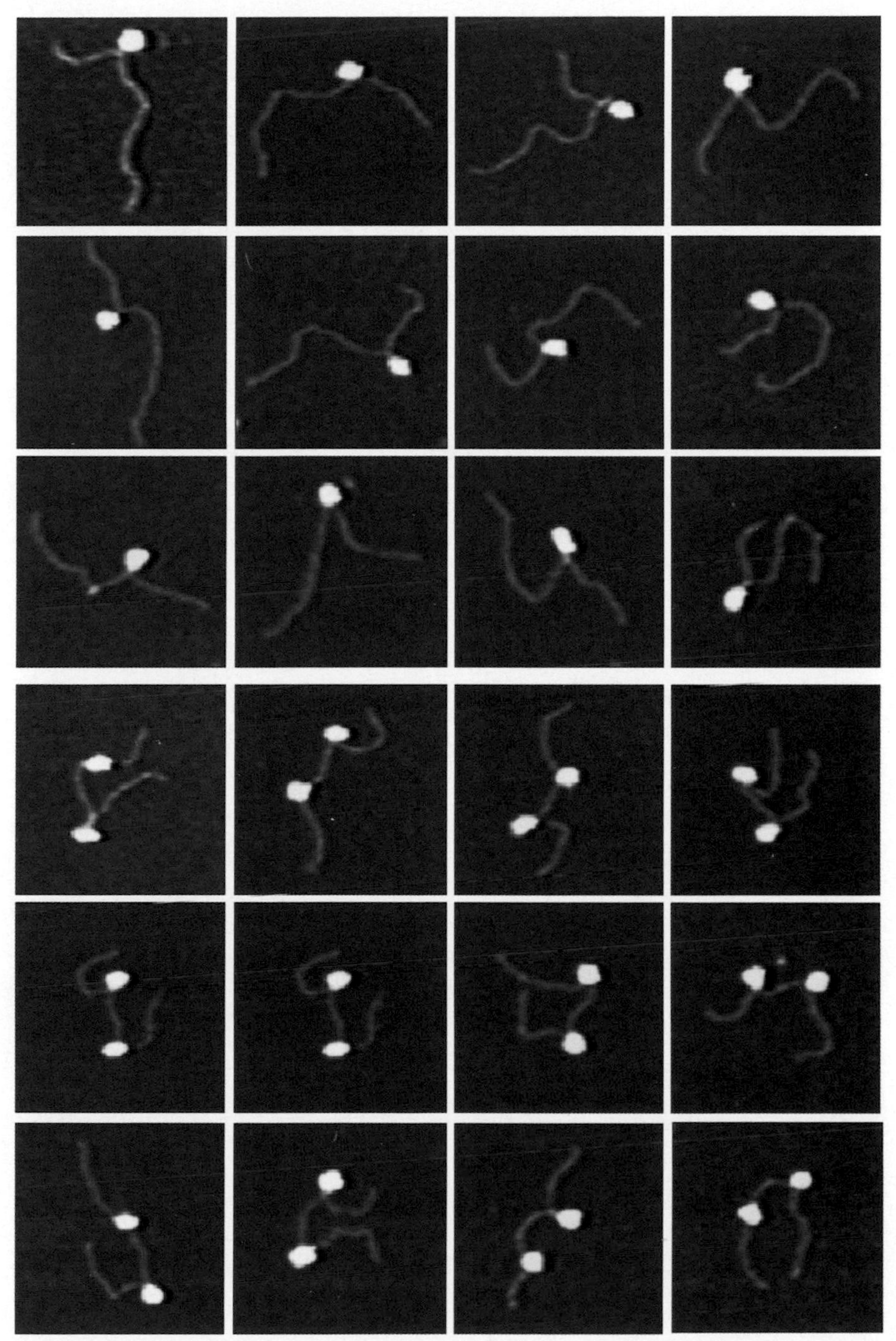

Plate 3. Montage of SFM images showing OPCs on template C suggesting a structure in which the DNA is wrapping around the RNAP. Image size: 250 nm. The images were recorded in air with the tapping mode. The colour code corresponds to a height range of 5 nm from dark to clear.

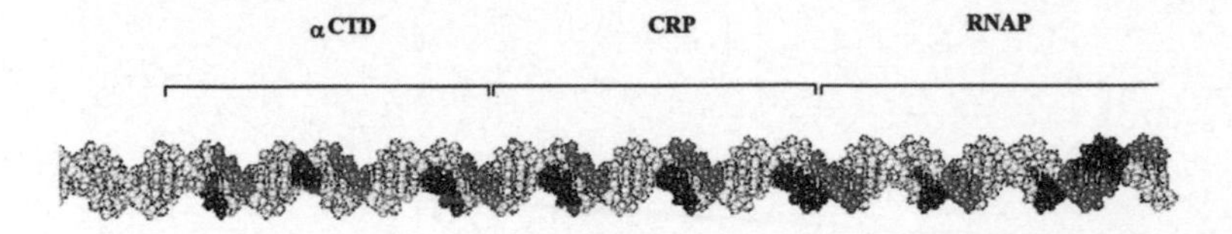

Plate 4. End-labelled fragments containing a promoter with a CRP-binding site centred at –41.5 were incubated with combinations of CRP (100 nm) and holo-RNA polymerase (100 nm), as indicated in *Figure 1*, and subjected to hydroxyl radical attack as described in *Protocol 2*. The model of the promoter shows the protection pattern of the CRP-RNAP-promoter ternary complex. Protections on the upper strand are marked in blue, protections on the lower strand are marked in red.

Protocol 2 continued

0.025% xylene cyanol), as described in *Protocol 3* (steps 12–16). Use a comb with wide wells to allow the complex to immediately enter the gel.

[d] The samples are analysed on a denaturing gel of the type typically used as sequencing gels. Mix 60 ml of denaturing acrylamide mix with 300 μl of 10% Ammonium persulphate and 60 μl of TEMED. Pour between 35 × 42 cm glass plates held apart by 0.4 mm spacers, and insert a suitable comb. Allow the gel to polymerize for at least 1 h before use. The gels should be pre-run at 60 W constant power for at least 30 min before the samples are loaded (the electrophoresis buffer is 1× TBE). The exact percentage of the gel and the duration of electrophoresis varies according to the distance of the footprinted area from the labelled end of the DNA fragment: for a footprint beginning 40 bp away from the labelled end of the fragment a 6% acrylamide gel is suitable and electrophoresis should be stopped when the dark blue dye in the loading buffer is 3–5 cm from the bottom of the gel. A 'G + A' or 'G' ladder should be run in parallel with the footprinting samples to act as a marker. The ladders are prepared from the same labelled DNA fragment stock that is used for the footprinting experiments using standard Maxam–Gilbert sequencing techniques (24).

3.3 Interference studies: the missing nucleoside assay

This method complements the footprinting method and is able to provide further information on the DNA conformational changes induced by ligand binding and, possibly, on the intermediates involved in the formation of the complex. It is therefore recommended that the samples from footprinting and interference studies are run on the same sequencing gel for direct comparison.

Protocol 3

The missing nucleoside assay

Equipment and reagents

Equipment and reagents are as for *Protocol 2*, plus:

- Acrylamide mix (6% acrylamide (acrylamide: bisacrylamide ratio 29 : 1), 1 × TBE buffer)
- Gel loading buffer (50% sucrose, 1 mM EDTA, 0.025% bromophenol blue, 0.025% xylene cyanol)
- Elution buffer (0.5 M ammonium acetate, 1 mM EDTA, 0.01% SDS)
- TE buffer (10 mM Tris-HCl, 0.1 mM EDTA, pH 8.0)

Method

1. Treat 0.3–0.6 pmol of ^{32}P end-labelled DNA (about 1–2 000 000 d.p.m.) as described in *Protocol 2*, steps 1–7 (without the addition of proteins).
2. Precipitate the DNA by adding 7 μl of 3 M sodium acetate and 170 μl of cold 100% ethanol.
3. Centrifuge for 30 min at 12 000 *g* at −20 °C and discard the supernatant. There should be no radioactivity remaining in this solution.

Protocol 3 continued

4. Add 700 μl of cold 70% ethanol.
5. Centrifuge for 15 min at 12 000 **g** at 4 °C and discard the supernatant.
6. Wash again with 700 μl of cold 70% ethanol.
7. Centrifuge again for 15 min at 12 000 **g** at 4 °C and discard the supernatant.
8. Dry the DNA pellet under vacuum.
9. Resuspend the pellet in 20 μl of TE buffer.
10. Mix 9 μl of gapped DNA, 1 μl of 10× binding buffer and the required amount of protein diluted in 1 × binding buffer in a total volume of 15 μl.
11. Incubate at 37 °C for 20 min (or in the appropriate conditions to allow the complex to form).
12. Add 3 μl of loading buffer, mix carefully, and load immediately onto a running native polyacrylamide gel (typically 6% acrylamide made up in TBE buffer pre-run for 10 min at 10–15 V/cm).
13. When the separation of free and bound DNA is achieved, stop the current and proceed as described in *Protocol 1*, steps 7–9, except that the addition of 1 μl of 20 mg/ml glycogen and 0.01% SDS in the elution buffer is recommended.
14. Wash again with 70% ethanol.
15. Dry the pellet, count the samples, and resuspend them in the appropriate volume of sequencing gel loading buffer in order to load the same amount of d.p.m. for the free and bound DNA.
16. Proceed exactly as described in *Protocol 2*, steps 14 and 15.

3.4 Data collection and analysis

Quantifiable data can be obtained from hydroxyl radical footprinting gels either by densitometry of an autoradiograph or by use of a phosphorimager. Once the data is in electronic form there are a number of commercially and freely available analytical software programs which allow individual bands to be quantified, or 'profiles' of entire lanes to be generated (e.g. see *Figure 1B*). The information can then be transferred to spreadsheet software with which, for example, subtraction plots can be generated in order to see more clearly the differences between the ladders generated in the presence and absence of protein.

However, there are a number of precautions that should be taken before attempting to interpret or quantify data from any hydroxyl radical cleavage experiment:

(1) The footprints should be performed on both strands with a sequencing ladder performed on each labelled DNA.

(2) The native labelled DNA should migrate as a single band with a background less than 2% on a sequencing polyacrylamide gel, indicating the integrity of the DNA fragment (it is, however, possible that a lower band migrating very

close to the full-length DNA fragment may appear owing to incomplete denaturation of the duplex DNA).

(3) 'Single hit' conditions must have been achieved during the cleavage reaction (an indication that this condition has been met is that $> 70\%$ of the fragment should remain uncut in the cleavage reaction lanes).

(4) The ligand should not cleave the DNA *per se* (i.e. in the case of crude extracts, avoid the presence of magnesium in the binding buffer). Some proteins bind the Fe(II) ion or the Fe(II) chelate which, after oxidation, generate a high population of cleavages in their vicinity (25). This is the case with RNA polymerase where the Fe(II) chelate binds close to the catalytic centre inducing hyper-reactivity to hydroxyl radical attack at positions around -1 on the template strand (see *Figure 1A*, lane d). Interestingly these reactivities are not decreased by the addition of glycerol, a radical scavenger which can absorb diffusing hydroxyl radicals (26).

Quantification of hydroxyl radical cleavage ladders is subject to the well-known limitations of the various procedures used for collecting the data from radioactive gels (in particular, care must be taken that the intensity of the bands being analysed falls within the linear response range of the phosphor screen or autoradiograph used: in the case of autoradiographs this often means that several different exposures must be taken in order to quantify accurately bands which differ significantly in intensity). In addition, as the hydroxyl radical cleavage reaction generates such a high-resolution picture (cleaving at every position within the DNA) individual bands become difficult to resolve towards the top of the gel (see *Figure 1B*). Although gel analysis software which aims to resolve this problem of overlapping bands is now becoming available (27), it is still often necessary to run the samples on a number of different gels to bring different areas of the footprint into clearer resolution when footprinting complexes that extend over large regions of DNA.

References

1. Dixon, W. J., Hayes, J. J., Levin, J. R., Weidner, M. F., Dombroski, B. A., and Tullius, T. D. (1991). In *Methods Enzymol.*, 208, 376.
2. Noller, H. F., Green, R., Heilek, G., Hoffarth, V., Huttenhofer, A., Joseph, S., *et al.* (1995). *Biochem. Cell Biol.* **73**, 997.
3. Tullius, T. D. and Dombroski, B. A. (1986). *Proc. Natl. Acad. Sci. USA.* **83**, 5469.
4. Tullius, T. D. (1987). *TIBS* **12**, 297.
5. Latham, J. A. and Cech, T. R. (1989). *Science* **245**, 276.
6. Hertzberg, R. P. and Dervan, P. B. (1984). *Biochemistry* **23**, 3934.
7. Greiner, D. P., Hughes, K. A., Gunasekera, A. H., and Meares, C. F. (1996). *Proc. Natl. Acad. Sci. USA* **93**, 71.
8. Baichoo, N. and Heyduk, T. (1997). *Biochemistry* **36**, 10830.
9. Nagai, H. and Shimamoto, N. (1997). *Genes Cells* **2**, 725.
10. Colland, F., Orsini, G., Brody, E., Buc, H., and Kolb, A. (1998). *Mol. Microbiol.* **27**, 819.
11. Dervan, P. B. (1991). In *Methods Enzymol.*, 208, 497.

12. Dumoulin, P., Ebright, R. H., Knegtel, R., Kaptein, R., Granger-Schnarr, M., and Schnarr, M. (1996). *Biochemistry* **35**, 4279.
13. Rana, T. M. and Meares, C. F. (1991). *Proc. Natl. Acad. Sci. USA* **88**, 10578.
14. Miyake, R., Murakami, K., Owens, J. T., Greiner, D. P., Ozoline, O. N., Ishihama, A., and Meares, C. F. (1997). *Biochemistry* **37**, 1344.
15. Murakami, K., Owens, J. T., Belyaeva, T. A., Meares, C. F., Busby, S. J., and Ishihama, A. (1997). *Proc. Natl. Acad. Sci. USA* **94**, 11274.
16. Schickor, P. and Heumann, H. (1994). In *Methods in molecular biology*, Vol. 30 (ed. Kneale, G. G.), p. 21. Humana Press Inc., Totowa.
17. Knapp-Pogozolski, W. and Tullius, T. D. (1993). *J. Biomol. Struct. Dynam.* **10**, 153.
18. Price, M. A. and Tullius, T. D. (1992). In *Methods Enzymol.*, 212, 194.
19. Werel, W., Schickor, P., and Heumann, H. (1991). *EMBO J.* **10**, 2589.
20. Oakley, M. G. and Dervan, P. B. (1990). *Science* **248**, 847.
22. Garner, M. and Revzin, A. (1981). *Nucl. Acids Res.* **9**, 6505.
23. Fried, M. (1989). *Electrophoresis* **10**, 366.
24. Maxam, M. A. and Gilbert, W. (1980). In *Methods Enzymol.*, 65, 499.
25. Mustaev, A., Kozlov, M., Markovtsov, V., Zaychikov, E., Denissova, L., and Goldfarb, A. (1997). *Proc. Natl. Acad. Sci. USA* **94**, 6641.
26. Zaychikov, E., Denissova, L., Meier, T., Gotte, M., Heumann, H. (1997). *J. Biol. Chem.* **272**, 2259.
27. Shadle, S. E., Allen, D. F., Guo, H., Pogozelski, W. K., Bashkin, J. S., and Tullius, T. D. (1997). *Nucl. Acids Res.* **25**, 850.

Chapter 13
Radiolytic cleavage of DNA. Mapping of the protein interaction sites

Michel Charlier and Mélanie Spotheim-Maurizot
Centre de Biophysique Moléculaire–CNRS, rue Charles-Sadron, F–45071 Orléans Cedex 2, France

1 Introduction

In aerobic conditions, DNA lesions induced by ionizing radiations such as γ-, β- and X-rays result mainly from attack by the oxidative hydroxyl radicals ($OH^•$) arising from the radiolysis of water (see *Figure 1*). The $OH^•$ radicals either abstract H atoms from the deoxyribose moiety, or add to the bases (see *Figure 2A*). The final damage (see *Figure 2B*) can be breaks of the phosphodiester backbone, directly

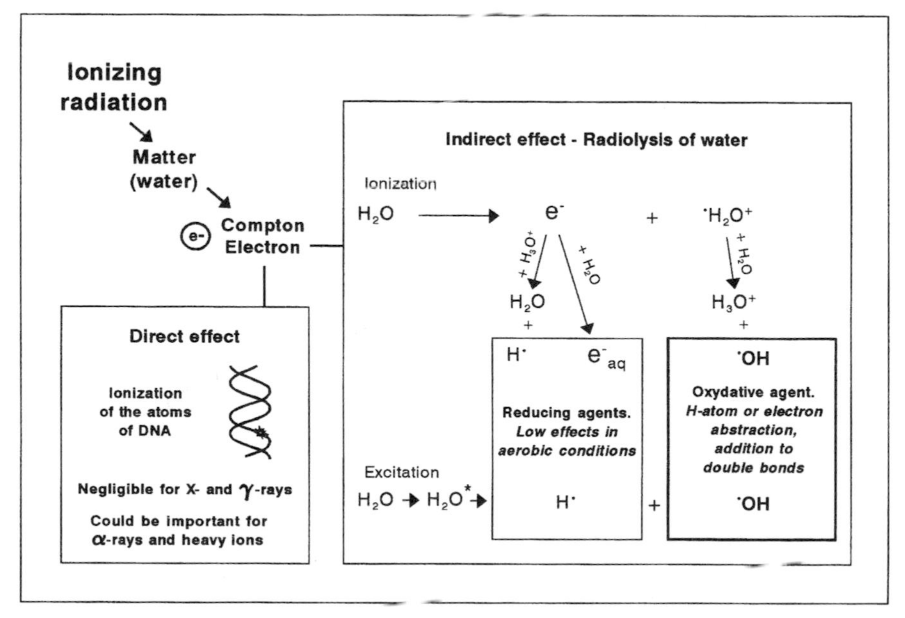

Figure 1 Effects of ionizing radiations such as X-, β- and γ- rays, on matter: direct and indirect effects.

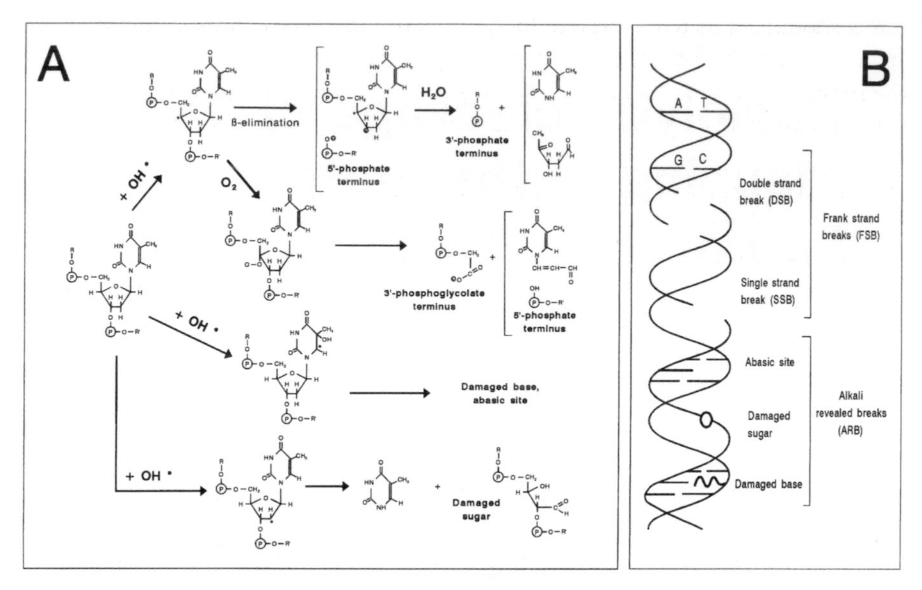

Figure 2 Mechanisms of DNA attack by the OH$^•$ radicals and the resulting damage. (A) Reactions leading to strand breaks with 3′-phosphate and 3′-phosphoglycolate termini, to damaged sugars or bases, and to abasic sites. (B) Classification of damage into FSB and ARB.

observable at neutral pH and called frank strand breaks (FSB), or abasic sites, modified sugars and modified bases, that can be partially revealed as strand breaks after alkaline treatment, and called alkali-revealed breaks (ARB). FSB and ARB occur at each nucleotide site along a linear canonical B-DNA with a strong modulation by the local sequence-dependent structure (1).

We have observed variations of the probability of damage induction at nucleotides belonging to single-stranded DNA (2), to regions with narrow minor groove (1), to Z-DNA (3) and to quadruplexes (4). Moreover, ligands such as polyamines or proteins modify radiolysis of DNA. Sequence-specific binding proteins protect their binding site by local scavenging of OH$^•$ radicals (screening effect) and local conformational changes of DNA (e.g. bending) (5, 6). Non-specific binding proteins (7) or polyamines (8) protect by global modification of DNA conformation (compaction) and by radical scavenging.

Therefore, radiolysis can be used to detect sequence-dependent variations of DNA structure, to identify protein-binding sites and to reveal protein-induced conformational changes in DNA. This constitutes the basis of the radiolytic footprinting method.

2 Irradiation of samples

For this type of experiments, ionizing radiations such as γ-, β- and X-rays are more suitable than α-rays and heavy ions, since they act almost exclusively via the radicals induced by water radiolysis, i.e. by indirect effect (see *Figure 1*). The direct effect should not lead to a protein footprint, since this type of attack should be influenced neither by the structure nor by any type of ligand.

An appropriate irradiation source could be a ^{60}Co or ^{137}Cs irradiator for medical use, blood products or biological samples, or a X-ray generator for medical use.

For an accurate determination of the dose (in Gy = J/kg) delivered to the irradiated samples, the irradiation must be performed in the same conditions as for dosimetric calibration. In our case (irradiation on the treatment table in the Radiotherapy Department of the Centre Hospitalier Régional d'Orléans, France), the dose calibration is performed every day using an ionization chamber included in a phantom of Plexiglas. To get the same conditions, samples had to be immersed under 1–1.5 cm of water. Moreover, we used polypropylene tubes with almost the same electronic density as water (see *Figure 3*). If the dosimetric calibration is performed using the Fricke's dosimeter (9) in air, irradiation has to be performed in the same type of tubes, preferably in the same conditions of filling, in air.

For experiments on purified DNA and proteins, the dose rate is not a determining parameter, as long as radical recombinations are rare (no influence was observed in our experiments from 0.1–50 Gy/min). At very high dose rate (using a 1 ns pulsed electron beam for example), recombinations strongly decrease the amount of radicals available for DNA attack. For living matter such as cells, the dose rate, even for low values, obviously becomes an important parameter (cell cycle, repair processes, etc.).

The sample temperature can be controlled during irradiation using a thermostatically controlled bath. For irradiation at 0 °C, it is convenient to simply use an ice-water bath.

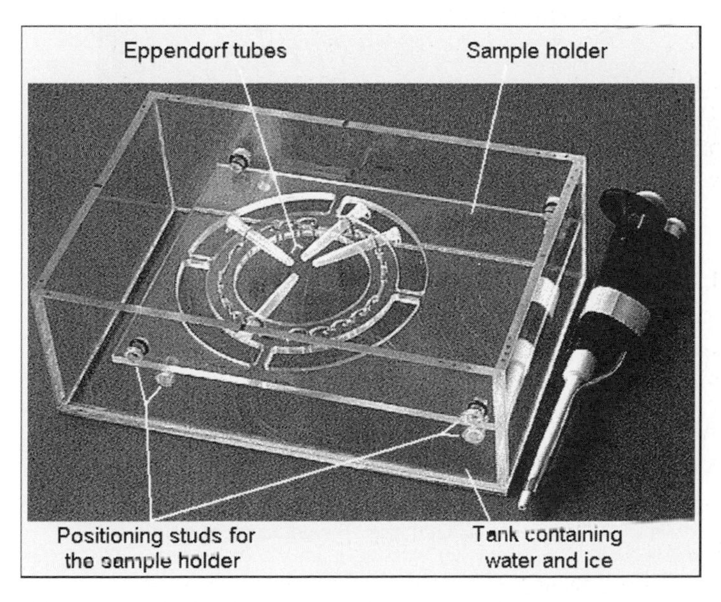

Figure 3 Tank for irradiation in water or water–ice mixture.

3 Determination of FSB yield using plasmids. Protection factor

Two types of FSB can be induced: single-strand breaks (SSB) and double-strand breaks (DSB). DSB are formed by two SSB, one on each strand and separated by few base pairs (less than the length of stability of the double helix). The true DSB result from two SSB induced by the same ionizing particle. The true DSBnumber linearly depends on the dose. At high doses (higher than those used in the experiments described here), DSB can also result from the random localization of non-correlated SSB; their number depends on the square of the dose.

The yields G_{ssb} and G_{dsb} (expressed as breaks/Da.Gy) of single- and double-strand breaks are determined from the relaxation/linearization of a supercoiled plasmid DNA. The presence of a radioprotector such as a protein decreases the yield. The protection factor, PF, is the ratio between the yield G_0 in the absence and G_R in the presence of radioprotector:

$$PF = G_0/G_R.$$

3.1 Principle of the method

The first SSB converts a supercoiled plasmid (form I) into a circular relaxed molecule (form II). The first DSB converts a circular plasmid (form I or II) into a linear molecule (form III). An additional SSB does not change the topological form (II or III), except when located close to a SSB on the opposite strand. In this last case, a DSB occurs. An additional DSB results in two shorter linear molecules. Assuming that the DSB occur randomly along the plasmid, the sizes of these two pieces are randomly distributed (short form III).

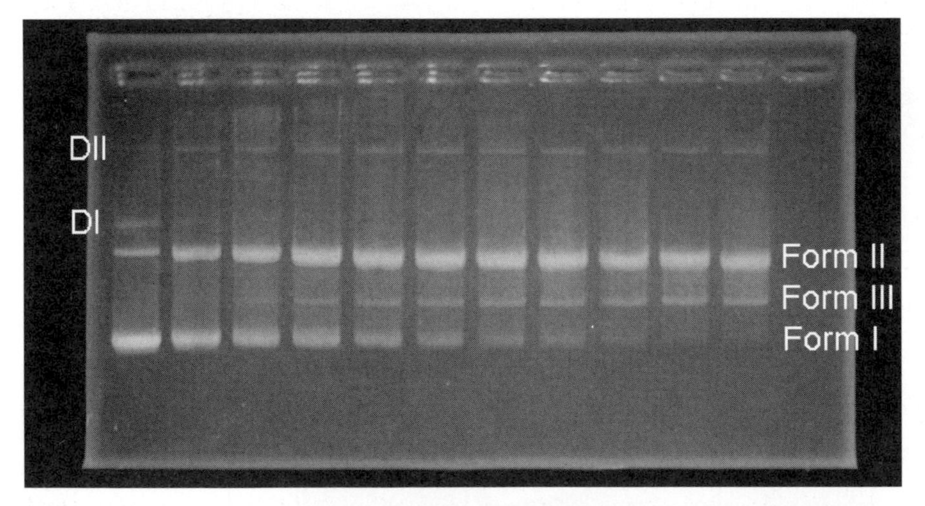

Figure 4 Agarose gel electrophoresis of irradiated plasmid DNA. From left to right: non-irradiated plasmid and plasmids irradiated with increasing doses. The gel separates the supercoiled plasmid (form I), the circular relaxed plasmid (form II), the linear plasmid (form III). Plasmid dimers always contaminate the preparation. The weak bands corresponding to the forms I (DI) and II (DII) of the dimer are visible in the upper part of the gel.

The different forms can be separated as distinct bands by submarine agarose gel electrophoresis (*Figure* 4), except for the short form III molecules which smear. After staining by ethidium bromide (BET), the fractions of forms I, II and III were determined by direct fluorescence measurements using a video camera and image analysis software. The relaxed DNA (forms II and III) binds 1.5 more BET than the supercoiled one (form I). Statistical treatment allows the determination of average number of SSB and DSB per plasmid as a function of the irradiation dose (10).

Protocol 1

Determination of FSB yield in a supercoiled naked plasmid

Equipment and reagents

- Plasmid (pBR322) of the best quality available (more than 90% supercoiled form, see ref. 11 for preparation)
- Irradiation buffer (as far as possible, should not contain hydroxyl radical scavengers): Phosphate (Na or K) buffer (1–100 mM) is the most suitable
- Polypropylene tubes (Eppendorf 400 μl).
- Submarine agarose gel electrophoresis apparatus
- BET stock solution (10 mg/ml in water)
- Gel buffer: Tris 40 mM, sodium acetate 20 mM, EDTA 2 mM, pH 8 (adjusted with glacial acetic acid)
- Video camera and image analysis software (Bioprobe)

Determination of the SSB yield

1. Dialyse extensively the plasmids (pBR322) solution (50–200 μg/ml) against the irradiation buffer to remove traces of radical scavengers (e.g. ethanol or Tris used during plasmid purification).
2. Immerse tubes (400 μl polypropylene tubes) containing 10 μl of solution in a thermostatically controlled water bath.
3. Irradiate the samples at increasing doses. At least 3-5% of the plasmid should remain supercoiled (form I) at the highest dose.
4. Load 1–4 μl of irradiated solution (containing around 150–200 ng of DNA) onto the agarose gel (1.2% in gel buffer, 6 cm long) and run at 800 V/m (reservoir buffer: gel buffer).
5. Stain the gel for 30 min in a BET solution (0.5 mg/l) in water.
6. Under UV light, the bands corresponding to forms I, II and III may be observed in each lane. The image of the gel is recorded using a video camera and analysed using an image-processing software.
7. The area of a peak in the densitogram is proportional to the amount of BET bound to DNA. Taking into account that form I binds less BET than forms II and III (1.5 times in our experimental conditions), determine the fractions of the three forms (f_I, f_{II} and f_{III}) in each lane.

Protocol 1 continued

8 Calculate the average number of breaks *<ssb>* and *<dsb>* according to (10):

$$<ssb> = \ln[(1 - f_{III})/f_I] \qquad <dsb> = f_{III}/(1 - f_{III}).$$

9 Plot *<ssb>* and *<dsb>* as the function of the dose to obtain straight lines whose slopes are the number of breaks per plasmid and per unit dose. G_{ssb} and G_{dsb} are calculated by dividing the slope by the molecular mass (in Da) of the plasmid.

3.2 Protection of plasmids by a DNA-binding protein

In this example, we have studied the protection by the chromosomal protein MC1 extracted from an archæbacterium, *Methanosarcina* sp. CHTI55 (7). This protein contains 93 amino acids and is naturally bound to DNA (one protein for 150 bp). No sequence-specific binding on a bacterial DNA has been observed. Consequently, the binding along DNA can be assumed as random. In our experimental conditions, the affinity constant is large enough to ensure that all proteins are bound as long as the DNA is not fully covered (r = protein/bp < 0.1), i.e. there is no free protein in the solution.

For different values of r (from 0 to 0.08), the protection factor, *PF*, is determined by the measurement of the breakage yields in the absence (see *Protocol 1*) and in the presence (see *Protocol 2*) of protein (see *Figure 5*).

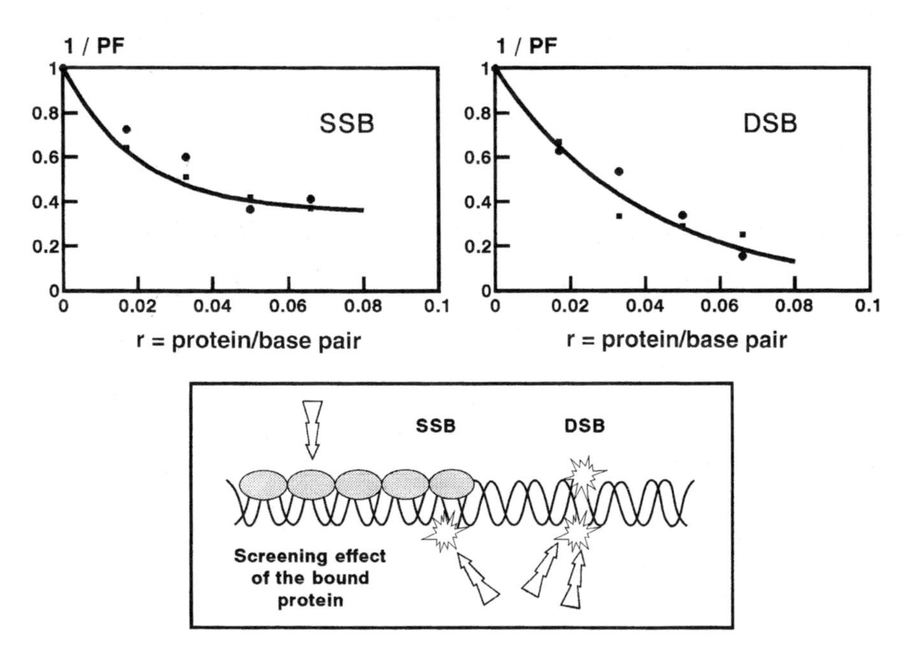

Figure 5 DNA radioprotection by the protein MC1. Upper part: increase of the protection factors PF (shown as a decrease of 1/PF) with increasing covering of DNA by the protein. Lower part: Scheme of the model accounting for the complete protection of DNA against DSB and of the limited protection against SSB.

Protocol 2

Determination of FSB yield in a supercoiled plasmid bearing proteins.

Protocol 2 is similar to *Protocol 1*. The following items should be added or modified.

Equipment and reagents (additional)

- MC1 protein stock solution in irradiation buffer
- SDS buffer (0.1% w/v SDS in TAE buffer)
- Sodium phosphate buffer 50 mM pH 7.25.

Determination of the SSB yield (modifications)

1 Plasmids (pBR322) are extensively dialysed against the irradiation buffer to remove traces of radical scavengers (e.g. ethanol or Tris used during plasmid purification), at a concentration of 50–200 μg/ml. Add the required amounts of protein.

4 1–4 μl of irradiated solution (containing around 150–200 ng of DNA) are layered onto a dissociating agarose gel (1.2% in SDS buffer, 6 cm long) and run at 800 V/m (reservoir buffer: SDS buffer). The gel is then incubated overnight in 50 mM sodium phosphate buffer to remove SDS.

The protection factor increases with the covering of DNA by the protein: the protection becomes total only for DSB and reaches a plateau at around 50% protection for SSB (see *Figure 5*). This can be explained by the screening of DNA by the protein. At complete coverage of DNA with MC1 protein, one face of the double helix remains accessible, whereas the other is protected from $OH^{\bullet}$ radical attack.

4 Determination of breakage sites using DNA restriction fragments. Protected sites: footprints

Two sets of information can be drawn from the analysis of the irradiated fragments: the average number of breaks (SSB + DSB) leading to evaluation of $G_{ssb+dsb}$, and the relative probability of breakage at each nucleotide site. The pattern of this relative probability along the fragment may exhibit local variations, i.e. a footprint. These variations reflect either a decrease or an increase of the radiolytic attack of that regions. They originate in the structural modifications (sequence-dependent or ligand-induced), and in the protection by the bound ligands. Thus, comparison of the radiolysis patterns of naked and protein-bound DNA can reveal the sites of interaction.

4.1 Principle of the method

One strand of the restriction fragment is labelled at the 5′-terminus, using the phosphatase/kinase procedure and [γ-^{32}P]ATP (12). For an accurate determination of the relative probabilities of breakage, the number of breaks per strand considered must not exceed one. Keeping the mean value of breaks per strand smaller than 0.1 limits to 1% the fraction of strands bearing more than one break. The order of magnitude of the dose $D_{0.1}$ (in Gy) corresponding to this yield for a strand of m Da is determined from the yield $G_{ssb+dsb}$ ($= G_{ssb} + G_{dsb}$) obtained, in the same buffer conditions for a supercoiled plasmid:

$$D_{0.1} = 0.1/(m \times G_{ssb+dsb}).$$

The irradiated samples (naked DNA or DNA-protein complexes) are analysed by denaturing sequencing gel electrophoresis after removal of proteins by phenol treatment. For the FSB determination, the samples are layered onto the gel without further additional treatment. For the ARB determination, the samples are submitted to alkali treatment using piperidine. In this case, the sum FSB + ALS is detected.

Labelled oligonucleotides of different lengths generated by the radiation-induced breaks are separated as discrete bands. The attack at a nucleotide site N generates a labelled oligonucleotide ending at the $(N-1)$th nucleotide. Two types of 3′-termini are obtained: 3′-phosphate and 3′-phosphoglycolate, leading to a splitting of the bands, visible in the bottom of the gel (13). A Maxam–Gilbert sequencing of the same strand (14) is run on the same gel. This procedure leads also to a 3′-phosphate ended fragment with $N - 1$ nucleotides. Therefore, the breakage site can be identified.

The radioactivity in each band is quantified either by autoradiography and densitometry, or using a PhosphorImager (Molecular Dynamics) and image analysis. It is proportional to the amount of fragments of the length $N - 1$, i.e. to the probability of breakage at the Nth nucleotide.

For a given lane, the upper band is the most intense (radioactivity $= U_b$), since it corresponds to the intact strand of L nucleotides. The rest of the radioactivity is distributed in the discrete bands corresponding to the generated fragments (radioactivity $= B_i$ in the ith band).

The fraction of intact fragments is

$$F_{int} = U_b/(U_b + \sum_0^L B_i)$$

and the relative probability of breakage at ith site is

$$p_i = B_i/(U_b + \sum_0^L B_i).$$

The variations of p_i along a lane reveal local modulations of breakage. The discrepancy between F_{int} for the naked DNA and for the DNA–protein complex reveals the global radioprotection or radiosensitization of the fragment by the protein.

Labelling the DNA strand at the 3′-terminus (using Klenow enzyme and [α-^{32}P]ATP) allows the same type of study, except that breakage (radiolysis or Maxam–Gilbert procedure) at nucleotide N leads to a labelled fragment beginning to the $(N + 1)$th nucleotide with a 5′-phosphate end. Note that in this case only 5′-phosphate ended oligonucleotides are produced (no splitting of the bands).

Protocol 3

Radiolytic footprinting of a protein-binding site

Equipment and reagents

- 1-cm thick Plexiglas or acrylic screens (for body protection) and tanks (for storage) for safe use of ^{32}P labelled material
- β-ray counter
- Labelled DNA fragments (bearing one or several protein-binding sites)
- DNA-binding protein (stock solution)
- Polypropylene tubes (Eppendorf 400 μl)
- Binding/irradiation buffer - depends on the DNA–protein system studied (as far as possible, should not contain hydroxyl radical scavenger, see *Protocol 1*)
- Phenol–chloroform solution for protein extraction (11)
- Cold (−20 °C) ethanol and sodium acetate (3 M) stock solution
- Piperidine solution in water (1 M)
- Loading buffer: 98% deionized formamide, 10 mM sodium EDTAcetate, 0.025% xylene cyanol, 0.025% bromophenol blue
- TBE buffer : 89 mM Tris, 89 mM boric acid, 2 mM sodium EDTAcetate, pH 8.3
- Gel buffer: TBE buffer supplemented with 7 M urea
- Sodium EDTAcetate stock solution 0.5 M (pH 8)
- Acrylamide and bis-acrylamide of the best grade available. Sodium persulfate and TEMED
- Sequencing gel electrophoresis apparatus (Base Runner, IBI), with plates of 45–60 cm high and wedged spacers (0.4 mm at the top, 0.8 mm at the bottom)
- Ethanol (10%)–acetic acid (10%) aqueous solution
- Gel dryer
- Films for autoradiography (3 M type R) and densitometer, or PhosphorImager and ImageQuant (Molecular Dynamics) software

Determination of the breakage pattern for naked or protein-complexed DNA

1. Prepare solutions of naked or of protein-complexed labelled DNA in the binding/ irradiation buffer. The amount of DNA per sample must be sufficient for loading 200–500 Bq in each electrophoresis lane (see below). This amount depends on the labelling, and on the yield of the subsequent treatments.
2. Irradiate the samples (Eppendorf 400 μl polypropylene tubes containing ≈10 μl solution) at the required doses (see *Protocol 1* and Section 2).
3. Remove proteins by phenol–chloroform treatment if proteins are present.
4. After adding sodium acetate to 0.3 M, precipitate DNA with two volumes of cold ethanol and dry. For FSB analysis, go directly to step 6.
5. For FSB + ARB analysis, a piperidine treatment must be done at this step. Add 100 μl of piperidine solution, heat at 90 °C for 30 min, chill at 77 K in liquid nitrogen, and lyophilize. Redissolve in 100 μl of pure water, chill and lyophilize. Repeat this last step until the piperidine is completely eliminated (generally three times).
6. Dissolve DNA in 3–6 μl of loading buffer and heat for 3 min in boiling water.

Protocol 3 continued

7 Prepare the electrophoresis gel as described in ref. 14. The length of the gel, the percentage acrylamide and the ratio acrylamide/bis-acrylamide (reticulation) are, in the case of a 80-bp fragment equal to 45 cm, 8% and 19 : 1, respectively.

8 Load the DNA sample (200–500 Bq) onto the gel, and run the gel (TBE buffer as reservoir buffer) using the dyes as migration references (15). In the case of an 80-bp fragment, a run of 2 h at 35 W on a 45-cm long gel allows the accurate analysis of at least 70 nucleotides.

9 Turn out the gel and rinse in the ethanol–acetic acid solution, then in pure water, to remove salts and urea, and dry the gel onto a Whatman paper sheet.

10 Expose for autoradiography or to PhosphorImager. Determine the radioactivity in each band by measuring the area of the corresponding peak on the densitogram. Owing to the low dynamic response of the films, the determination of the radioactivity in the upper band (U_b) is generally not possible by argentic autoradiography, since its activity (more than 90% input) is generally three orders of magnitude greater than that of the lower bands ($< B_i > \approx U_b/1000$) for a fragment of 100 bp.

4.2 Radiolysis of naked DNA

In a naked DNA, breakage can occur at each nucleotide sites. Nevertheless, p_i varies along the sequence. Since the radiolytic $OH^•$ radicals are small diffusing species produced in the bulk, this effect reveals local variations of accessibility of the attack sites due to the sequence-dependence of the DNA structure (1). The pattern of p_i for FSB along a fragment shows several regions of low radiosensitivity, especially 5′-AATT and 5′-AATTT sequences (see *Figure* 6).

Crystallographic determinations (16) and molecular modelling calculations show that these regions present a narrow minor groove (see *Figure* 6, upper part), and therefore a low accessibility of hydrogen atoms of deoxyribose H4′ and H5′2 (see *Figure* 6, lower part). Abstraction of H4′ and H5′ atoms by the $OH^•$ radicals leads to C4′ or C5′ centred radicals evolving to sugar degradation (17) and thus to strand breakage. The radiolytic pattern of naked DNA appears very suitable for investigating the structural microheterogeneity and the conformational changes in DNA.

4.3 Footprint of a protein on DNA

When DNA is complexed to a protein, some nucleotides can become less sensitive to the radiolytic attack. The corresponding bands are bleached (p_i decreases) revealing the DNA–protein contacts (footprint). This region can be delimited with a high precision because of the very small size of the $OH^•$ probe.

For example, bleaching of three zones along a fragment bearing the CRP-specific binding site is observed upon CRP binding (6). They correspond to the

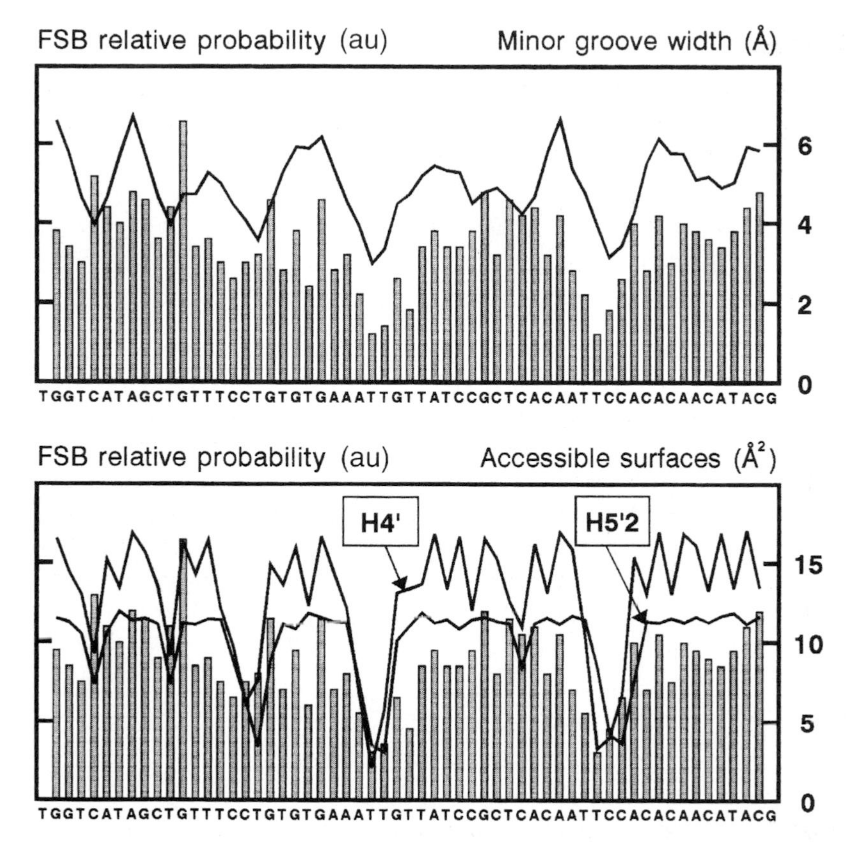

Figure 6 Sequence-modulated radiation-induced FSB along a DNA fragment. Correlation between experimental breakage probabilities in arbitrary units (histogram) and (upper part) the calculated minor groove width (solid line), and (lower part) the accessible surfaces of H4′ and H5′2 atoms of the sugars (solid lines).

three contact regions between the protein and its specific site on DNA (see *Figure 7*). This result is in good agreement with the crystallographic structure of the complex (18).

The same procedure was applied to the *lac* repressor–operator system (6) to reveal the contact area between the protein and its specific binding site. The results agree with the structure deduced from crystallographic studies (19).

The MC1 protein, whose binding is not sequence-specific, leaves a regular footprint along the DNA. This allows the determination of the length of the excluded binding site (7).

For the core nucleosome studies, we performed a β-irradiation by the [γ-^{32}P]ATP used for DNA labelling. The excess of [γ-^{32}P]ATP was kept for 1 day in the solution before separation from the labelled material (naked DNA or core nucleosome) (6). Perfectly regular footprints were obtained along the 146-bp fragments. The periodic protection/attack pattern with a period of 10.4 bp is in good agreement with the accepted model of core nucleosome (20).

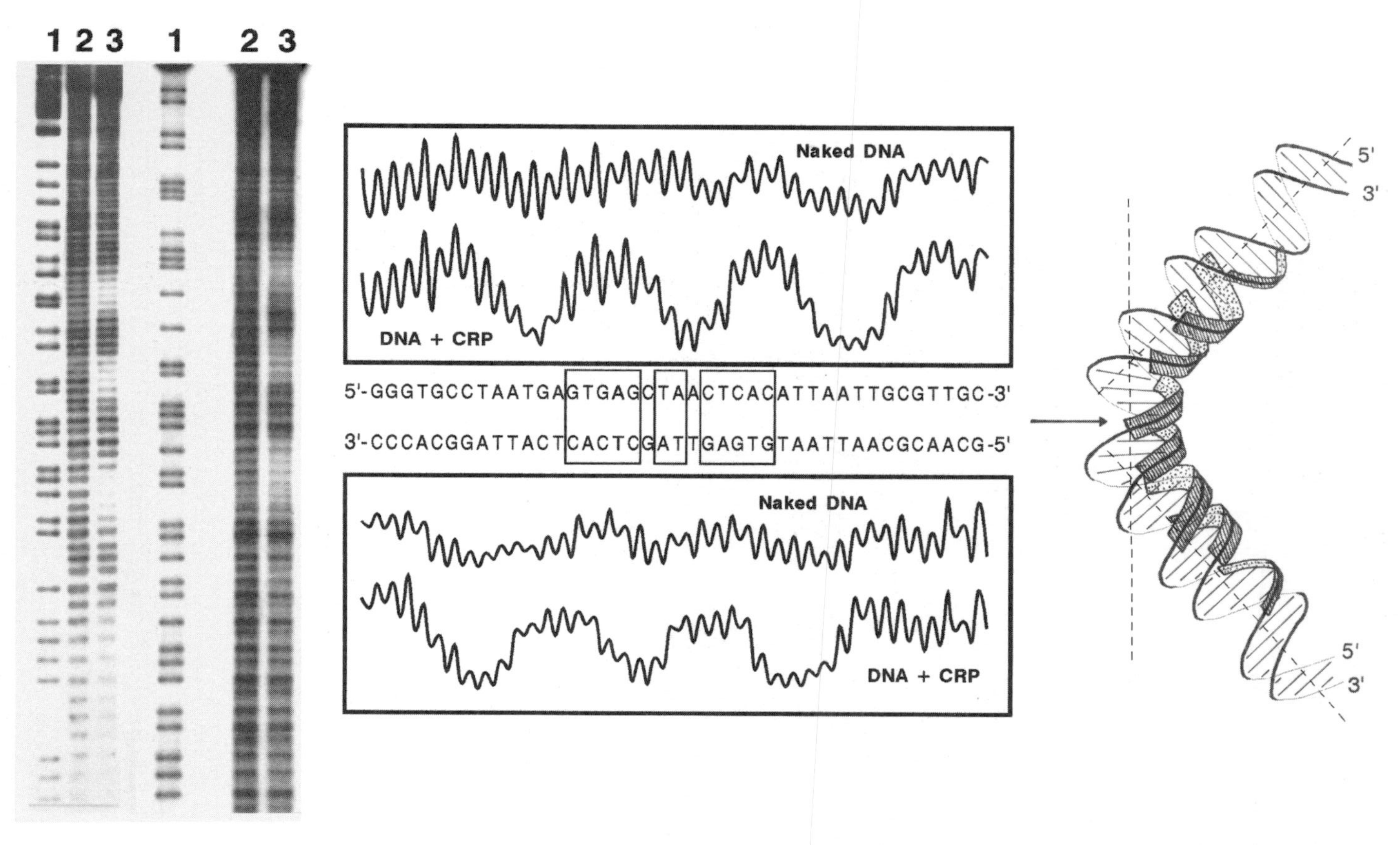

Figure 7 Radiolytic footprinting of the CRP on a 120-bp restriction fragment bearing the specific binding site. Left: autoradiography of the sequencing gel of the two strands (left part, upper strand; right part, lower strand); lanes 1, Maxam–Gilbert sequencing of purines; lanes 2, irradiated naked DNA; lanes 3, DNA–CRP complex. Centre: Densitograms of the lanes containing the naked DNA and the complexes, showing three bleached regions (upper part : upper strand, lower part : lower strand). Right: a scheme of DNA in the complex showing the protected regions on the two strands. The DNA has been bent as described in ref. 18. Part of this figure was reprinted from Franchet-Beuzit *et al.*, *Biochemistry*, 1993, **32**, 2104, with permission of the American Chemical Society (6).

Warning

For hot labelled fragments, radiolysis due to β-rays emitted by ^{32}P can be significant if DNA is stocked for long periods (several days) in a non-scavenging buffer (for example, K or Na phosphate buffer). The dose rate, D_s (Gy/s), due to self -irradiation can be estimated using the formula (21):

$$D_s = 1.6 \times 10^{-19} E_{eff} \times A/m$$

E_{eff} (in eV) is the effective energy of the β-ray, roughly equal to 1/3 of the maximum energy ($E_{eff} \approx 0.57 \times 10^6$ for ^{32}P), and A/m is the activity of the radionuclide per mass unit of the solution (Bq/kg).

If labelled fragments have to be stored before or after irradiation in a non-scavenging medium, a radical scavenger can be added to prevent self-radiolysis (10 mm Tris–EDTA buffer, and a few per cent ethanol, glycerol or DTT, for example). Another possibility is to keep the fragments precipitated in ethanol. The use of ^{33}P instead of ^{32}P (lower energy) strongly decreases this effect.

This holds good for all footprinting studies, regardless of the chemical, physical or enzymatic probe.

5 Conclusion

The radiolytic footprinting method has been perfected and tested using several known interaction systems: the *lac* repressor–operator complex, the CRP–DNA-specific complex and the core nucleosome. In all cases, the obtained information are in good agreement with the results from other techniques. Moreover, the sequence-dependent variations of radiosensitivity agreed with the known structural microheterogeneities of B-DNA. This method was then applied to the study of other systems: MC1–DNA complexes, Z-DNA, single-stranded DNA and DNA quadruple helices.

The advantages of this method over other footprinting methods are:

(1) The probe ($OH^\bullet$ radical) is produced *in situ*, without addition of chemical. This suggests its use for *in vivo* footprinting studies, since no toxicity or penetration into the cells nuclei should be encountered.

(2) The probe has no binding specificity for either DNA or for proteins. This avoids artefacts caused by the preferential cleavage of DNA around the probe binding sites.

(3) It is possible to use pulsed radiation beams to develop time-resolved footprinting studies, that could gives information about the kinetics of trans-conformations or interactions (22).

References

1. Sy, D., Savoye, C., Begusova, M., Michalik, V., Charlier, M., and Spotheim-Maurizot, M. (1997). *Int. J. Radiat. Biol.* **72**, 147.

2. Isabelle, V., Prevost, C., Spotheim-Maurizot, M., Sabattier, R., and Charlier, M. (1995). *Int. J. Radiat. Biol.* **67**, 169.
3. Tartier, L., Michalik, V., Spotheim-Maurizot, M., Rahmouni, A. R., Sabattier, R., and Charlier, M. (1994). *Nucl. Acids Res.* **22**, 5565.
4. Tartier, L., Spotheim-Maurizot, M., and Charlier, M. (1998). *Int. J. Radiat. Biol.* **73**, 45.
5. Tullius, T. D., Dombroski, B. A., Churchill, M. E. A., and Kam, L. (1987). In *Methods in enzymology* (ed. Wu, R.), Vol. 155, p. 537. Academic Press, London.
6. Franchet-Beuzit, J., Spotheim-Maurizot, M., Sabattier, R., Blazy-Baudras, B., and Charlier, M. (1993). *Biochemistry* **32**, 2104.
7. Isabelle, V., Franchet-Beuzit, J., Sabattier, R., Laine, B., Spotheim-Maurizot, M., and Charlier, M. (1993). *Int. J. Radiat. Biol.*, **63**, 749.
8. Spotheim-Maurizot, M., Ruiz, S., Sabattier, R., and Charlier, M. (1995). *Int. J. Radiat. Biol.* **68**, 571.
9. Fricke,H., and Hart,E.J. (1966) In *Radiation dosimetry* (ed. Attix, F. H. and Roesch, W. C.), 2nd edn, Vol. 2, p. 266, Academic Press, New York.
10. Spotheim-Maurizot, M., Charlier, M. and Sabattier, R. (1990). In *Frontiers in radiation biology* (ed. Riklis, E.), p. 493. VCH, Weinheim.
11. Sambrook, J., Fritsch, E. F., and Maniatis, T. (ed.) (1989). *Molecular cloning. A laboratory manual*, 2nd edn, Book 1, Chap. 1, p. 21. Cold Spring Harbor Laboratory Press, Cold Spring Harbor.
12. Sambrook, J., Fritsch, E. F., and Maniatis, T. (ed.) (1989). *Molecular cloning. A laboratory manual*, 2nd edn, Book 2, Chap. 10, p. 59 Cold Spring Harbor Laboratory Press, Cold Spring Harbor.
13. Henner, W. D., Rodriguez, L. O., Hecht, S. M., and Haseltine, W. A. (1983). *J. Biol. Chem.* **258**, 711.
14. Sambrook, J., Fritsch, E. F., and Maniatis, T. (ed.) (1989). *Molecular cloning. A laboratory manual*, 2nd edn, Book 2, Chap. 13, p. 11 Cold Spring Harbor Laboratory Press, Cold Spring Harbor.
15. Sambrook, J., Fritsch, E. F., and Maniatis, T. (ed.) (1989). *Molecular cloning. A laboratory manual*, 2nd edn, Book 1, Chap. 6, p. 36. Cold Spring Harbor Laboratory Press, Cold Spring Harbor.
16. Fratini, A. V., Kopka, M. L., Drew, H. R., and Dickerson, R. E. (1982). *J. Biol. Chem.* **257**, 14686.
17. von Sonntag, C. (ed.) (1987). *The Chemical basis of radiation biology*, p. 221. Taylor and Francis, London.
18. Schultz, S. C., Shields, G. C., and Steitz, T. S. (1991). *Science* **253**, 1001.
19. Lewis, M., Chang, G., Horton, N. C., Kercher, M. A., Pace, H. C., Schumacher, M. A., Brennan, R. G., and Lu, P. (1996). *Science* **271**, 1247.
20. Arents, G., Burlingame, R. W., Wang, B. C., Love, W. E., and Moudrianakis, E. N. (1991). *Proc. Natl. Acad. Sci. USA* **88**, 10148.
21. Kiefer, J. (ed.) (1990) *Biological radiation effects*, p. 83. Springer-Verlag, Berlin.
22. Sclavi, B., Woodson, S., Sullivan, M., Chance, M. R., and Brenowitz, M. (1997) *J. Mol. Biol.* **266**, 144.

Chapter 14
Laser UV-laser photoreactivity of nucleoprotein complexes *in vitro*

Malcolm Buckle, Christophe Place and
Iain K. Pemberton
Unité de Physicochimie des Macromolécules Biologiques (URA 1773 du CNRS),
Institut Pasteur, 25 Rue du Dr Roux, F–75724 Paris Cedex, France.

1 Introduction

In order to define a kinetically competent nucleoprotein complex, a complete description of all the contacts and conformational changes characterizing both its formation and subsequent function is required. Chemical, physical and biological probes are all used to generate this information. Such probes must, however, report a specific structure that is both structurally unique and also singular with respect to time. The two overall stringent requirements for such probes are that they are first non-perturbing and second, highly specific (1). In this chapter we describe the use of high-energy photons to excite and study nucleoprotein interactions *in vitro*. The use of a similar system *in vivo* is described in Chapter 15.

2 UV-laser configuration

2.1 Hardware set-up

The intrinsic photo-reactivity of DNA and RNA is a useful parameter to measure local DNA structure and identify regions in contact with proteins in given nucleoprotein complexes. In order to satisfy two main criteria of non-perturbation and selectivity, an optimum source of photons must be rapidly delivered to the sample. This is possible by using a UV-laser providing high energy photons (266 nm) in a brief pulse (5 ns) on a small volume of material (10–20 μl) (2–7). The minimal experimental set up that is required is shown in *Figure 1*. The laser source is a neodymium laser (Spectra Physics) consisting of Nd^{3+} ions at low concentration in yttrium–aluminium–garnet (YAG = $Y_3Al_5O_{12}$) that produces a continuous monochromatic beam at 1064 nm. In Q-switching mode, in conjunction with non-linear crystals this beam is quadrupled in frequency to give a homogenous

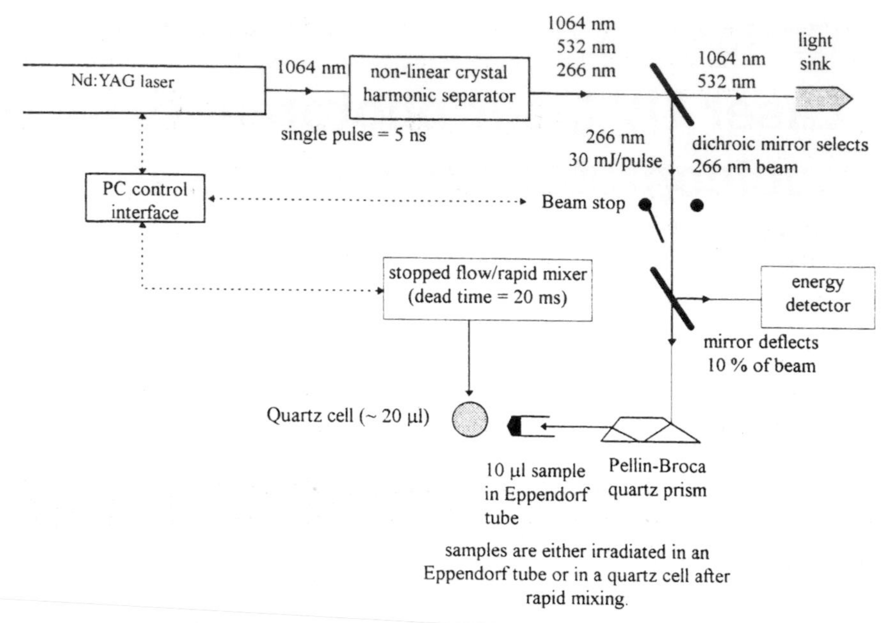

Figure 1 Typical UV-laser set up for irradiation of nucleoprotein complexes by single pulses at 266 nm.

polarized source of photons at 266 nm ($\approx 10^{17}$ photons/pulse, proving a dose of around 20–30 mJ/pulse). Since the beam has a diameter of $\approx$ 6 mm this gives complete irradiation of a solution of 10–20 μl at the bottom of an Eppendorf tube. The optimal beam form is a Gaussian rather than 'doughnut' configuration. The final 266 nm beam is selected using an array of dichroic mirrors and directed onto the sample using a Pellin–Brocca prism (Spectra-Physics) (*Figure 1*).

In setting up this system a number of operating and safety points should be observed, these are listed in *Table 1*.

Table 1 Safety operating guidelines

(1) Safety glasses (which must meet DIN standards, i.e withstand a direct hit of 10 s from a continuous wave laser of power $>$ 10 W, filter OD $>$ 3–4 at l $<$ 532 nm, and have $>$ 40°vision) should be worn at all times. Glasses rather than goggles are recommended since the latter tend to favour condensation and consequently reduced vision and the unfortunate tendency to remove them for cleaning.
(2) When adjusting the alignment of the beam the lamp energy should be maintained at a minimum.
(3) The beam should be visualized using thermal-sensitive paper or exposed film rather than developed film since the latter tends to disintegrate and the resulting particles may coat sensitive dichroic surfaces leading to reduced efficiency.
(4) The room housing the laser should be kept at a constant temperature ($<$ 22°C) and access to the room should be limited either to a single entrance equipped with a safety trip switch that stops the laser if the door is accidentally opened during use, or to a double security system involving an intermediate antechamber.
(5) The optical system should be designed so as to reduce all extraneous reflections. Walls should be dark and there should be no windows.

This configuration is by no means the only one available and can be adjusted according to space and experimental requirements. The laser works at optimal 10 Hz repetition. Although many lasers possess a single-pulse switch, this may not be used under frequency-doubling conditions since the efficiency of the non-linear doubling crystals is extremely temperature sensitive. Consequently, once the orientations of the doubling crystals have been optimized a beam-stop linked to the Q-switch should be used to interrupt the beam which then functions in continuous mode. This beam-stop should be set up such that during use one or, at most, two pulses are allowed to pass.

2.1.1 Setting up the laser for operation at 266 nm.

The following instructions refer to the Nd : YAG Spectra Physics DCR 11 model which is typical of the series of discontinuous lasers adequate for this technique

Protocol 1

Laser set up

Power up

1. Protective glasses on.
2. Switch on power on main power supply.
3. On console next to laser, make sure of the following
   (a) Computer switch in INT position
   (b) Single shot in up and NOT in REP position.
   (c) Lamp energy at start position (turned fully left). Attention ! do not touch internal Q-switch knob.
   (d) Q-SW MODE in position EXT.
4. Press ON: there should be a noise then a click and the SIMMER switch will glow orange, the CHANNEL 1 and CHANNEL 2 lights on the temperature control will glow orange - wait until these lights start to flicker. The temperature of the doubling crystals is now constant.
5. Turn on power meter, it should be on position Watts on the range 0.03.

Maximizing laser power

1. Turn single shot to REP position Turn LAMP ENERGY fully clockwise. Wait 15 min.
2. Turn Q-SW MODE to Q-SW position. The laser is now operational and firing at 10 Hz; it should be hitting the stopper.
3. Activate laser stop to allow beam to pass. The light will now hit the Pellin–Broca prism and be deflected onto the block. Use an exposed film to align the beam with a small (0.5 ml) Eppendorf tube placed in the block.
4. A fraction (10%) of the beam is deflected into the power meter detector; the meter should show in excess of 0.02 W.

Protocol 1 continued

5 When the power meter shows a constant value **> 0.02 W** (= 10% of total) place beam stop in the beam path

Irradiation

1 Place the Eppendorf tube containing sample in the block.

2 Release beam stop so that only one pulse hits the sample.

3 When all tubes are finished turn single shot to UP position; turn lamp energy to start; turn Q-SW MODE to EXT Wait 10 min for cooling, then turn console OFF.

2.2 Conditions for photo-irradiation.

Complexes may be preformed at the required temperature in a volume of 10–20 μl in a suitable buffer. The composition of the buffer is essentially limited by the requirements for the formation of a nucleoprotein complex. However, care should be taken to limit material that could attenuate the beam such as free nucleotides or high non-specific DNA concentrations. Samples should be pipetted into small polypropylene Eppendorf tubes held in a rigid support. These tubes are aligned with the incident laser beam such that the light descends completely to the bottom of the tube. Given that the beam has a diameter of around 6 mm, this ensures that all of a 10- or 20-μl sample is effectively irradiated.

Protocol 2

Sample irradiation conditions

Equipment and reagents

- Eppendorf tubes (0.5 ml)
- DNA fragments: it is essential that the DNA is as pure as possible and contains few if any modified bases or nicks. Commercially available oligonucleotides are unsuitable for crosslinking experiments in which the primer extension technique is to be used. This is because of the high degree of non-deprotected bases (generally > 10%) present, and which cause premature termination of extension by the Klenow fragment, and the presence of *n*–1, etc., oligonucleotides that can complicate binding studies on oligonucleotides containing multiple crosslinking sites with a given protein.
- Heating block recipient for the Eppendorf tubes.
- Ice in polystyrene container.
- *X–Y* plotter connected to energy-measuring device.

Method

Following incubation, a single (5 ns duration), high-intensity pulse of radiation at 266 nm (0.5×10^{11} W/m^2 intensity providing a dose of $\approx$ 100 J/m^2) is applied to the sample. The amount of energy applied to each sample is recorded using the *X–Y* plotter. A single pulse, which should be in excess of 20 mJ, is applied.

2.3 Identification of photoreactive species

2.3.1 Photochemistry of the reaction

To all intents and purposes only nucleotide bases are excited at 266 nm. At the photon fluxes used in the set-up described in *Figure 1*, each base in a solution of DNA, for example, will see at least one photon carrying $\approx$ 4 eV. Under the conditions used here, the ensuing photochemistry is monophotonic and the DNA is not ionized. This is generally confirmed by the absence of nicks on DNA strands following irradiation. It should be noted that the ionization energy for thymines is 7.6 eV in water. Furthermore, we can assume that, in simple terms, the excited state for the base follows the profile shown in *Figure 2* (8).

Irradiation thus produces excited singlet states of very short (picosecond) duration and a very small proportion of triplet states with longer (microsecond) lifetimes. In consequence, the photoreactive species are relatively short-lived and 98–99% are consumed within the 5 ns duration of the laser pulse. What then is the fate of the excited species? There are a number of potential photoreactive products but three in particular are of interest here. The first is the formation of pyrimidine dimers. Adjacent non-saturated conjugated 5–6 positions in the pyrimidine ring can, following excitation, form covalent cyclic products, notably 4–6 and 5–6 cyclobutanes. The quantum yield for the 4–6 cyclobutanes is around 4% (5) which essentially means that for a 100-bp DNA there will be one thymine dimer per molecule following a single pulse at 266 nm. The second event concerns modification of guanine residues (9). The nature of this reaction is somewhat obscure. The quantum yield may be relatively high and the overall chemistry may require a biphotonic event. The final photochemical event is perhaps the most interesting and involves the formation of covalent links between excited bases and amino acids. The conditions for this crosslinking are that the reactive species are within van der Waals radii distance of each other. These

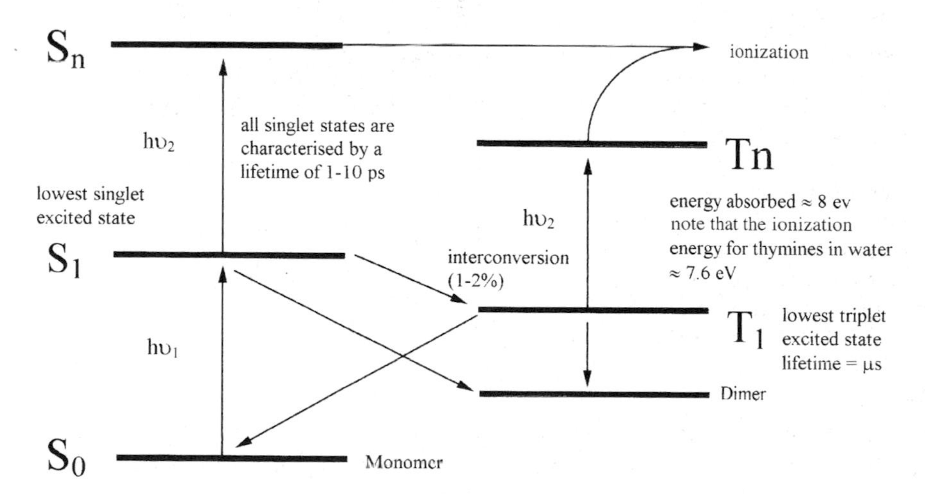

Figure 2 Jablowsky diagram illustrating the excitation states of bases following photo-irradiation.

limitations arise essentially because of the very short lifetimes of the reactive species (picoseconds). This limitation is of course present for pyrimidine dimer formation and represents one of the major advantages of the technique, namely that it is extremely rapid; thus it reports configurations that exist before perturbations occur.

2.3.2 Analysis of covalent crosslinking by SDS-PAGE

A simple approach to determine the extent of UV photocrosslinking to a DNA substrate is to separate the protein–DNA adduct(s) from non-crosslinked DNA by electrophoresis on SDS-PAGE. Quantification of the separated radiolabelled species is then performed by densitometry, e.g. using a PhosphorImager (Molecular Dynamics, Amersham Pharmacia BioTech). For most purposes, complexes are formed using 5′-end labelled oligonucleotides (i.e., labelled with [γ-^{32}P]ATP and polynucleotide kinase). After single-pulse UV irradiation, the samples are diluted 1:1 with SDS-PAGE loading buffer (final concentration 3% SDS, 5% glycerol, 31 mM Tris HCl pH 6.8, 100 mM DTT) and heated at 90 °C for 3 min before electrophoresis. The gel matrix (% acrylamide, linear versus gradient) should be optimized to allow a clear separation between the protein-DNA adduct and the DNA substrate. The electrophoretic migration of the UV-induced adduct will be dependent on its apparent mass, which in most cases equates directly to the sum of the covalently-linked components (i.e., the apparent mass of the crosslinked DNA strand plus that of the attached polypeptide chain). Assume $\approx$ 325 kDa for each nucleotide of the cross-linked strand. If the DNA fragment is too large to be separated clearly, then an alternative methodology must be considered, e.g. by first fragmenting the DNA or by degrading it with a nuclease (see Chapter 22).

When analysing crosslinking data, it is important to perform measurements over a range of protein concentrations. This is essential when comparing the relative crosslinking efficiencies of different DNA substrates, since it is the saturation isotherm and not the absolute level of DNA crosslinking that contains the information on binding affinity. Generally, titrations of fixed concentrations of the oligonucleotide are performed over a gradient of increasing protein concentration. A typical SDS-PAGE gel (visualized by autoradiography) is shown in *Figure 3A*.

The progressive fractional saturation of potential DNA-interacting sites is signalled by the increase in adduct formation, measured in terms of the percentage of the total DNA cross-linked (S_{obs}). Maximal cross-linking (S_{max}) is obtained under saturating concentrations of protein where all the DNA-binding sites are occupied. Adduct formation may thus be represented simply as:

$$S_{obs} = \theta S_{\max} = \theta \lambda_c$$

where θ ($= S_{obs}/S_{max}$) denotes the fractional saturation of the DNA lattice and λ_c represents the quantum yield of adduct obtained at saturation of the lattice with ligand (a constant factor dictated by the photochemical reactivity of the oligonucleotide sequence occluded and the number and nature of protein-DNA contacts formed). At 266 nm, λ_c is highly dependent on the base composition of the

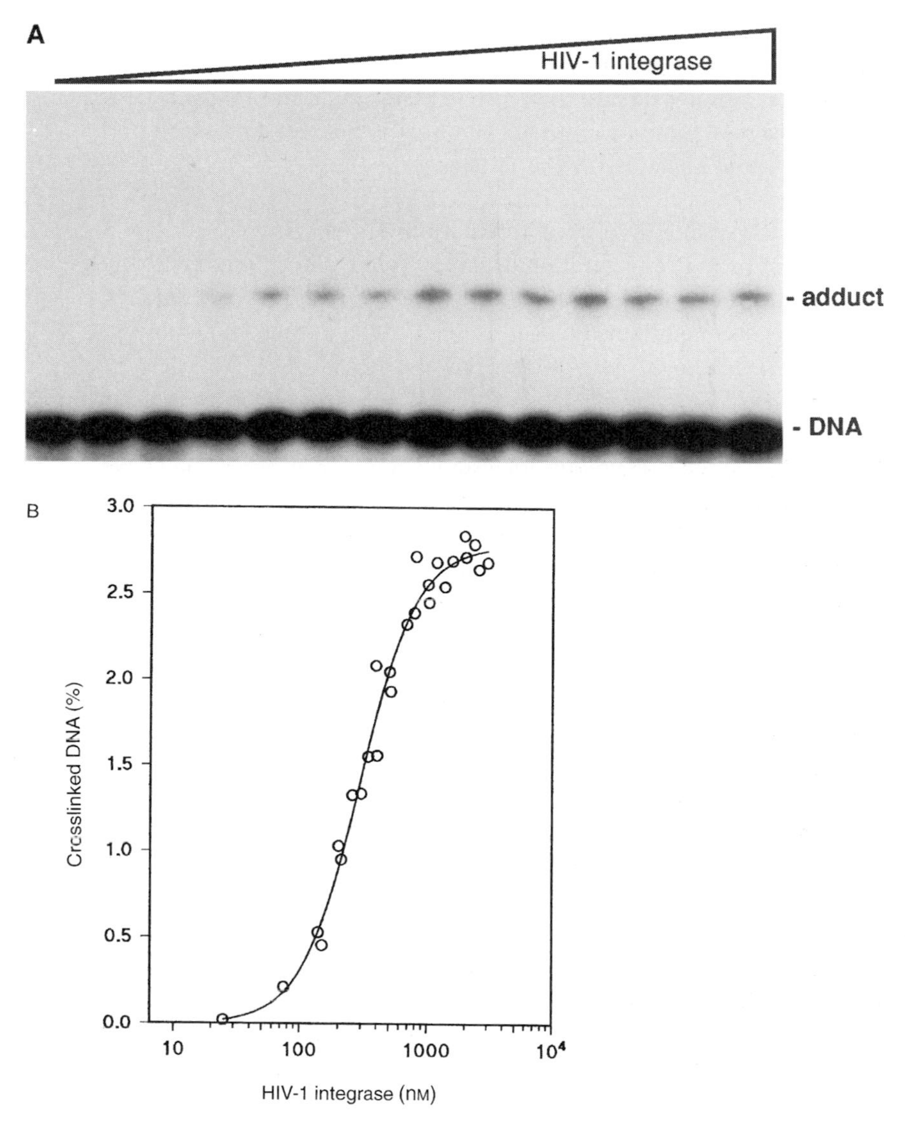

Figure 3 SDS-PAGE separation of crosslinked nucleoprotein complexes following UV-laser irradiation. Complexes between a protein and end-labelled DNA were irradiated with a single 5 ns UV-laser pulse at 266 nm. (A) Samples were denatured and separated on 10–15% polyacrylamide gels covalently crosslinked DNA migrated slower than the free DNA, and the amount of retarded material is a function of the affinity of the protein for the DNA (see text). (B) Binding isotherm showing the percentage retarded crosslinked material (quantified by phosphorimager densitometry), as a function of total protein (HIV integrase) concentration.

occluded site. Comparison of the data obtained for two independent proteins—gene32 protein (2) and HIV-1 integrase (11)—both of which are capable of binding non-selectively to ssDNA, indicate that the reactivity of each base may be ordered (relative to $dT = 1$) as T (1) >> C (0.04/0.08) > A (0.03/0.04) >> G (~ 0). Thus, as a general rule-of-thumb, pyrimidine residues are more reactive than their purine counterparts, with *d(T)* by far the most efficient at producing protein-DNA crosslinks. Of course, for an adduct to form, a close contact must be formed between an amino acid and the reactive base. While most amino acids appear capable of crosslinking, the photochemistry is poorly understood and the influence of the side-chains on the efficiency of the reaction cannot be predicted for a complex nucleoprotein structure [10]

Least-squares analysis of graphical representations of the degree of crosslinking as a function of ligand concentration (Figure 3B) then allows the fractional; saturation to be derived. This in turn provides access to the binding isotherm from which the apparent equilibrium constant K_{obs} and apparent stoichiometry may be obtained (11–13).

2.3.3 Identification of photoreactive bases by primer extension

We describe here a simple technique for establishing those bases that have undergone a specific photoreactivity (7). Following irradiation, the bases on a

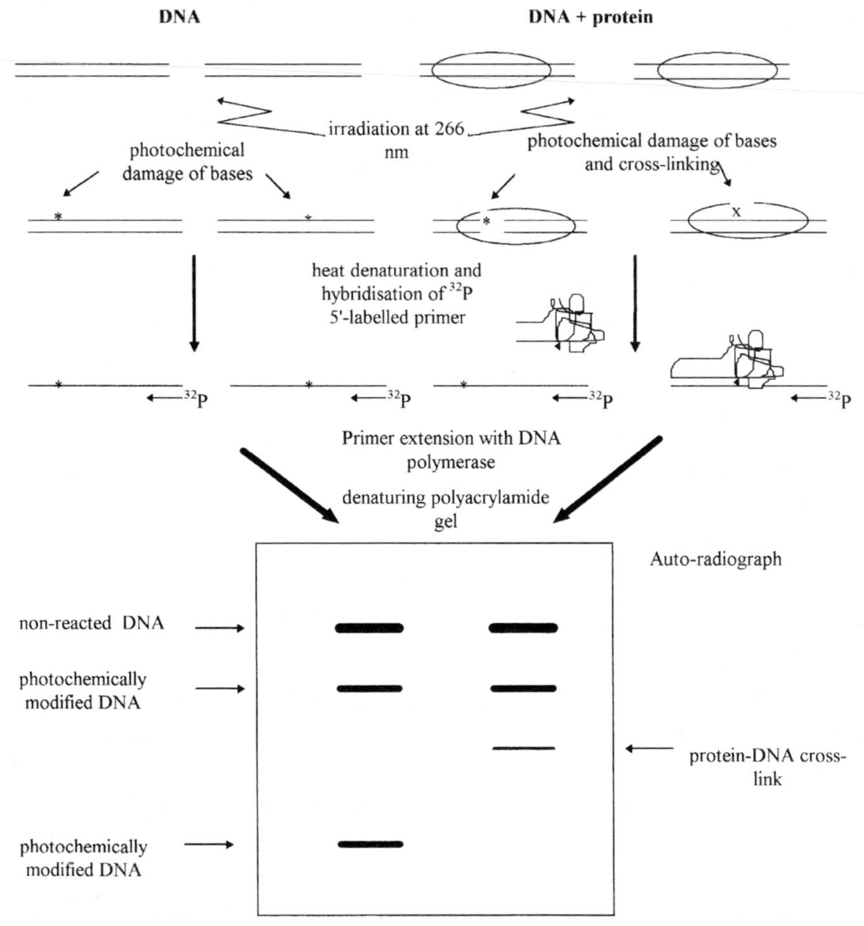

Figure 4 Primer extension protocol for the identification of crosslinked bases following UV laser irradiation of nucleoprotein complexes.

given DNA molecule will have been photomodified and the assumption is that for any given DNA molecule only one photoactive event will have taken place. *Figure 4* illustrates the methodology involved.

Protocol 3

Primer extension

Materials and reagents

- T4 PNK
- [γ-^{32}P]ATP
- A short oligonucleotide (19–25 mer) corresponding to a region complementary to the target DNA lying ≈ 20–30 bases 3′ downstream from the region of interest. The oligonucleotide should be chosen for ease of hybridization and G/C-rich 3′ extremity to facilitate Klenow fragment extension. The T_m for the oligonucleotide should be as low as possible to facilitate competitive hybridization to the template DNA. If possible, oligonucleotides should be both gel and HPLC pure
- 10× HMD buffer: 400 mM HEPES (KOH to pH 7.5), 100 mM $MgCl_2$, 10 mM DTT
- T4 PNK buffer
- 10× solution of deoxyribonucleotide triphosphates: dGTP, dTTP, dCTP and dATP (NTPs), 5 mM of each in distilled water, mixed in the same tube, stored at −20 °C
- Klenow fragment (modified DNA polymerase 1)
- 3 M sodium acetate pH 5.5
- 100% ethanol (kept at $t < -20$ °C)
- Formamide buffer

Primer extension following irradiation

1. Prepare samples and irradiate in a volume ranging from 10 to 20 µl. Freeze these samples if you are not going to carry out the extension on them immediately.
2. Label primer:
   (a) Make a dilution of the primer that is 10 µM final concentration.
   (b) Mix the following together and incubate for 30 min at 37 °C (these volumes can be modified, but maintain final concentrations): 12.5 µl H_2O, 2.0 µl 10 µM primer dilution, 2.0 µl 10× T4 PNK buffer, 2.5 µl [γ^{32}P]ATP, 1.0 µl T4 PNK.
   (c) Incubate at 65–70 °C for 20 min to inactivate the PNK.
   (d) Centrifuge briefly and freeze if you are not going to use immediately.
3. Annealing of primer to irradiated DNA sample:
   (a) Make up HMD/primer mix in bulk as follows for 10 µl (or 20 µl) samples: 2 µl (3 µl) 10× HMD buffer, 7 µl (6 µl) H_2O, 1 µl γ-^{32}P-labelled primer, 10 µl per primer extension reaction.
   (b) Add the HMD/primer mix, in 10 µl to each irradiated sample. Mix well.
   (c) Heat the tubes at 95 °C for 2–5 min (enough to fully denature the double-stranded DNA).
   (d) Place the tubes on ice immediately after heating, and allow them to sit for a further 2–5 min.

Protocol 3 continued

4 Primer extension:

(a) Make up NTP/Klenow solution in bulk as follows for 10 μl (or 20 μl) samples: 2.5 μl (5 μl) 10× dNTPs (5 mM), 0.5 μl Klenow fragment, 3.0 μl (6 μl) per primer extension reaction.

(b) Add the NTP/Klenow mix in 3 μl (6) to each sample and mix gently.

(c) Incubate this at 48 °C for 15 min.

(d) Place tubes on ice.

5 Precipitation:

(a) Add 66 μl H_2O to each tube, followed by 10 μl of 3 M sodium acetate (kept at 4 °C).

(b) Transfer this mix to a 1.5 ml Eppendorf tube.

(c) Add 325 μl 100% ethanol (kept at −20 °C) per tube.

(d) Precipitate DNA in dry ice bath, for ≈ 30 min.

(e) Spin at maximum speed in refrigerated centrifuge for 15–20 min.

(f) Remove supernatant.

(g) Add 150 μl 70% ethanol, spin again for 5 min.

(h) Remove supernatant.

(i) Let DNA pellet dry, either by sitting on the bench top for 30 min, or 5 min in a speed-vac.

6 Gel loading and migration:

(a) Pre-run gel at 50 W for 45 min.

(b) Re-suspend DNA pellets in 4–5 μl formamide buffer. Vortex well. Centrifuge briefly.

(c) Heat the samples at 95 °C for 3 min, and place immediately on ice.

(d) Carefully wash the wells of the gel just before loading.

(e) Load the entire sample and continue running the gel at 50 W.

(f) Allow the gel to run until the second dye (xylene cyanol) has migrated to the required distance, depending upon the region of interest on the DNA.

7 Dry gel normally, and either expose to film at −70 °C overnight or develop using phosphor storage densitometry.

Protocol 3 is general and can be applied to any enzyme or DNA (plasmids, fragments from restriction site or PCR).

As an example of an unusual situation we illustrate how to examine the crosslinking pattern of a catalytically active RNA polymerase.

Enzymes such as RNA polymerase, which require nucleoside triphosphates in order to actively transcribe a template strand of DNA into RNA clearly move in a two-dimensional ordered and processive manner along the DNA matrix. If this movement involves displacement of the intimate van der Waals contacts that characterize an initiating complex then UV-laser crosslinking provides a means of observing this displacement as a function of parameters such as temperature, matrix composition, NTP concentration and of course time. However, certain restrictions are introduced into the assay due primarily to the following points:

(1) The quenching effect of the nucleotides which are photo-absorbing molecules and generally in large excess (> 100 μM) compared with the template DNA (< 10 nM).

(2) Large-scale production of small RNA molecules which can hybridize to the template strand and inhibit the DNA polymerase (Klenow fragment) during primer extension.

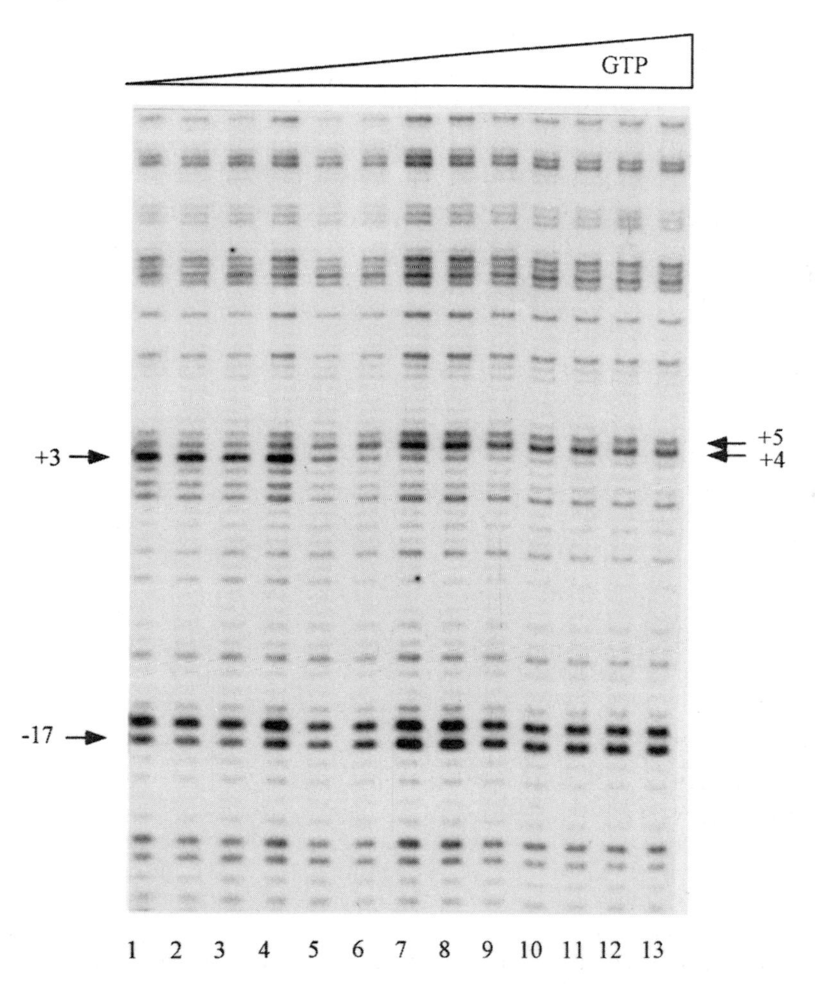

Figure 5 Primer extension pattern along the irradiated template strand of the promoter for the T7 RNA polymerase. In the absence of NTPs strong contacts are seen at positions −17 and +3 on the template strand. As the complexes are titrated with increasing concentrations of GTP, so that the RNA polymerase synthesizes a short G ladder up to around 10 nucleotides, the contact at position +3 is replaced by new contacts at +4 and +5. The contact at position −17, however, remains. This illustrates a non-symmetric movement of the RNA polymerase during the initial period of promoter clearance.

(3) In the presence of all four nucleoside triphosphates, the synthesis of abundant transcription products produces high concentrations of RNA strands which in turn can serve as templates for primer extension.

We use the following procedures to avoid these pitfalls:

(1) The total nucleoside triphosphate concentration is kept inferior to 1 mM. In other words, on average, the maximum concentration is 250 mM for each nucleotides when the four NTPs are present. This will of course depend upon the K_m of the NTPs for the enzyme under investigation. In this particular case the K_m is of the order of 50 μM.

(2) The hybridization of short newly synthesized RNA products to the template strand can be avoided by treatment with RNase H (*x* units, 37 °C, 10 min) immediately prior to the addition of the Klenow mix.

(3) Separation of the transcription products and the template prior to the annealing of the primer. For plasmid templates, exclusion size column chromatography can be used to separate high molecular weight plasmids and low molecular weight products and for shorter, linear, fragments, biotinylated DNA can be generated by PCR amplification and separated from the RNA products by use of steptavidin-treated magnetic beads (see Chapter 22 by I. K Pemberton).

We have used these modifications so as to follow an abortive cycling RNA polymerase from bacteriophage. As an example, changes in the photo-footprint of T7 RNA polymerase on a 17-bp promoter inserted in a supercoiled plasmid as the enzyme escapes from an initiating complex into elongation mode is represented in *Figure 5*.

References

1. Buckle, M. and Buc, H. (1994). In *Transcription: mechanisms and regulation* (ed. Conaway R. C. and Conaway J. W.), p. 207. Raven Press Ltd, New York.
2. Hockensmith, J. W., Kubasek, W. L., Evertsz, E. M., Mesner, L. D., and von Hippel, P. (1993). *J. Biol. Chem.* **268,** 15712.
3. Hockensmith, J. W Kubasek, W. L., Vorachek, W. R. and von Hippel, P. (1993). *J. Biol. Chem.* **268**, 15721.
4. Hockensmith, J. W., Kubasek, W. L., Vorachek, W. R., Evertsz, E. M. and von Hippel, P. (1991). *Methods Enzymol.* **208** 211.
5. Hockensmith, J. W., Kubasek, W. L., Vorachek, W. R., and von Hippel, P. (1986). *J. Biol. Chem.* **261**, 3512.
6. Buckle, M., Frotsch, A., Roux, P., Geiselmann, A., and Buc, H. (1991). *Methods Enzymol.* **208**, 236.
7. Buckle, M., Geiselmann, A., Kolb, A. and Buc, H. (1991). *Nucl. Acids Res.* **19,** 833.
8. Masnyk, T. W., Nguyen, H. T., and Minton,. K. W. (1989). *J. Biol. Chem.* **264,** 2482.
9. Spassky, A. and Angelov, D. (1997). *Biochemistry* **36**, 6571.
10. Shetlar, M. D., Christensen, J., and Hom, K. (1984). *Photochem. Photobiol.* **39**, 125.
11. Pemberton, I.K., Buckle, M., and Buc, H. (1996). *J. Biol. Chem.* **271**, 1498.
12. Brunel, F., Zakin, M. M., Buc, H. and Buckle, M. (1996). *Nucl. Acids Res.*, **24** 1608.
13. Kouznetzoff, A., Buckle, M., and Tordo, N. (1998). *J. Gen. Virol.* **79** 1005.

Chapter 15
In vivo UV-laser footprinting

Frédéric Boccard, Sylvie Déthiollaz,
Manuel Engelhorn, and Johannes Geiselmann
Laboratoire du Contrôlede l'Expression Génique, Université Joseph Fourier, CNRS, UMR5575, F–38041, Grenoble, France

1 Introduction

This chapter describes the use of UV-laser footprinting for detecting protein–DNA interactions within a living bacterial cell. The principle of UV-laser footprinting is described in Chapter 14, and we concentrate on protocols that specifically address the extension of the technique to *in vivo* experiments. DNA-binding proteins exert their function when they associate with a specific DNA target. The biological activity of such a protein is therefore a direct function of the saturation of the specific DNA-binding site with the protein. The question that we want to answer, using UV-laser footprinting *in vivo*, is: 'to what extend is a specific site on the DNA saturated by the protein that binds this sequence?' This question can be asked under different physiological conditions or in different genetic backgrounds. As an example, we describe the analysis of the interaction of the IHF of *Escherichia coli* with specific binding sites during exponential and stationary phases of growth.

2 The scope of UV-laser footprinting

2.1 Detecting DNA–protein interactions *in vivo*

A number of methods have been developed for measuring DNA–protein interactions *in vivo*. All of them involve the reaction of a footprinting reagent (in the large sense) with DNA, and the subsequent localization of the DNA modification. The footprinting reagent has to be delivered into the cell, which limits the number of chemical footprinting techniques to the use of $KMnO_4$ and DMS and excludes the use of, for example, DNase I. These reagents produce comparable signals *in vitro* and *in vivo*, allowing an easy interpretation of the experiment in structural terms (derived from *in vitro* experiments). However, the acquisition of the signal necessitates prolonged incubation (usually several minutes) of the cell with a chemical that generally reacts with many components of the cell. The creation of artefacts can therefore not be excluded. Other techniques use intracellularly produced footprinting reagents. For example, the expression of a methylase and the subsequent analysis of the DNA reveals the sites on the DNA that are protected from methylation, therefore presumably bound by protein (1). While this is a very gentle way of footprinting, the results can not be easily compared with *in vitro* experiments and the signals are limited to sequences that are recognized by the methylase. A very general drawback of these techniques is the extended incubation time necessary for signal acquisition. Apart from artefacts produced by damaging certain components of the cell, multiple reactions of the footprinting reagent within the target site will shift the equilibrium of the complex to be probed. Such experiments should therefore be extrapolated to zero incubation time; at the very least, single-hit conditions, which inevitably reduce the signal strength, have to be established.

Some of these problems are circumvented by UV-laser footprinting. The footprinting reagent (UV-light) easily enters the cell and the signal is acquired within microseconds, effectively freezing the equilibrium and preventing modifications of the structure to be probed by reactions elsewhere on the DNA. In other words, the signal is acquired before the complex has a chance to rearrange (typical DNA–protein transactions proceed on the millisecond time-scale). Furthermore, the footprinting procedure is identical for an *in vitro* and an *in vivo* experiment, which allows the comparison of easy to characterize complexes *in vitro* with their *in vivo* counterpart. It should be noted that the footprinting signal is not predictable; since the signal does *not* rely on the accessibility of the DNA by the footprinting reagent it is possible that a protein–DNA complex *does not modify* the DNA in a way which alters the photoreactivity of the DNA. A pilot experiment using purified components must be performed to assess the feasibility of the approach for each system of interest.

2.2 Principle of the reaction

The irradiation of DNA with ultraviolet light excites the nucleotide bases and leads to a wide variety of photoreactions. These photoreactions are very sensitive to the local environment of the DNA and a change in the photoreactivity therefore reflects a change in the local environment of the DNA (see chapter 14 Buckle *et al.* this volume). For a DNA–protein complex the environment of the DNA is largely determined by the presence of the protein. It is reasonable to assume that the interactions between a protein and its binding site on the DNA are identical *in vitro* and *in vivo*. We can therefore use *in vitro* UV-laser footprinting to obtain a characteristic reactivity pattern of the DNA in the complex and compare this pattern with the one observed in an *in vivo* experiment. An identical pattern is a strong indication for the presence of the protein on its binding site *in vivo*. Control experiments can verify the assumption of identical binding modes *in vitro* and *in vivo*, and hint at missing factors of the *in vitro* experiment.

3 Practical considerations

3.1 Equipment

UV-laser footprinting necessitates the use of laser light in order to limit the 'incubation time' to several nanoseconds. The most commonly used lasers are YAG lasers which emit infrared light of 1064 nm. Two consecutive passes through a frequency-doubling crystal yield high-intensity light of 266 nm, which is sufficiently close to the absorption maximum of nucleic acid bases. Such equipment is standard in any spectroscopy laboratory and can easily be used for UV-laser footprinting.

3.2 Calibration of the UV-footprinting signals *in vitro*

3.2.1 Irradiation of samples

In general, we want to know to what extent a given binding site is saturated *in vivo*. It is therefore necessary to correlate site occupancy with signal strength of the footprinting experiment. Assuming that the binding mode of the protein to DNA is identical *in vitro* and *in vivo* we can use the *in vitro* experiment to calibrate the binding signal. The general procedure for obtaining UV footprints *in vitro* are described in Chapter 14. We perform such experiments at increasing concentrations of the protein in order to deduce a binding isotherm. In this way we obtain a one-to-one correspondence between signal strength and occupancy of the binding site. *Protocol 1* illustrates the procedure for the measurement of the binding of IHF to the site located at the right extremity of the insertion sequence *IS1*.

Protocol 1

Calibration of UV-footprinting *in vitro*

Equipment and reagents

- 0.7 ml micro test tubes (e.g. Eppendorf 3813)
- IHF
- Proteinase K
- Microcentrifuge
- Binding buffer (50 mM Tris-HCl pH 7.5, 70 mM KCl, 7 mM $MgCl_2$, 3 mM $CaCl_2$, 1 mM EDTA, 10% glycerol, 200 mg/ml BSA, 1 mM β-mercaptoethanol)

Method

1. Prepare a series of Micro test tubes. The large section of these tubes reduces the height of the binding reaction and ensures total irradiation of the reaction.
2. Mix 5 nM of supercoiled plasmid DNA with increasing amounts of IHF (0 nM, 5 nM, 10 nM, 25 nM, 50 nM, 100 nM, 200 nM) in 40 μl of Binding buffer. To ensure that the primer extension pattern originates from modifications in the template following the irradiation, it is necessary to include a negative control which is not irradiated.

Protocol 1 continued

3 Briefly centrifuge the tubes to collect the reaction at the bottom of the tube and incubate for 20 min at 25 °C.

4 Open the tubes and irradiate immediately the 40-μl reaction with a single 5 ns pulse of UV-laser light.

5 Close the lid of the tubes. If a large number of samples need to be irradiated, it is possible to freeze the samples at this step in a dry-ice bath, keep the samples at −20 °C, and perform the following steps later.

6 Add 2 μl of a solution of proteinase K at 1 mg/ml and incubate for 10 min at 50 °C.

7 Extract the samples twice with 20 μl of an equal mix of Tris 0.1 M pH 8 saturated phenol and chloroform (2).

8 Add to the aqueous phase 4 μl of 3 M sodium acetate and 100 μl of ethanol.

9 Incubate for 10 min at −20 °C and centrifuge at full speed (≈ 12000 **g**) in a microcentrifuge for 10 min.

10 Decant the supernatant, wash the pellet with 100 μl of 70% ethanol.

11 Resuspend the pellet in 18.5 μl of H_2O.

3.2.2 Detection of the photoreaction

Two methods can be used to detect photoreactions on the DNA. Both methods rely on the property of a polymerase (DNA or RNA polymerase) to stall before a damaged base. The first procedure is analogous to the one described in Chapter 14 and uses T7 DNA polymerase in conjunction with an appropriate primer to detect the photodamage.

Protocol 2

Primer extension using T7 DNA polymerase

Equipment and reagents

- Annealing buffer (1 M Tris-HCl pH 7.6, 100 mM $MgCl_2$ and 160 mM DTT)
- Deaza G/A ^{T7}Sequencing Mixes (Pharmacia)
- Enzyme dilution buffer (20 mM Tris-HCl pH 7.5, 5 mM DTT, 0.1 mg/ml BSA, 5% glycerol)
- Stop solution (95% formamide, 20 mM EDTA, 0.05% Bromophenol Blue, 0.05% xylene cyanol FF)

T7 DNA polymerase

Method

1 Add 2.5 μl of 0.2 μM of radiolabelled primer to 18.5 μl of DNA.

2 Heat to 100 °C for 3 min and chill immediately on ice for 5 min.

3 Add 2.5 μl of annealing buffer (1 M Tris-HCl pH7.6, 100 mM $MgCl_2$, and 160 mM DTT) and heat to 50 °C for 3 min.

Protocol 2 continued

4 Incubate on ice for 5 min.
5 Add 2 μl of a mix containing 2 U of T7 DNA polymerase, 2.4 mM of each deoxyribonucleotide, 3 mM Tris-HCl pH7.5, 0.75 mM DTT, 15 mg/ml of BSA, 0.75% glycerol, and incubate the reaction for 10 min at 37 °C.
6 Add 150 μl of ethanol and incubate for 10 min at −20 °C
7 Centrifuge for 10 min in a microcentrifuge, dry the pellet and resuspend in 10 μl of a 1 : 1 mix of H_2O and stop solution.
8 Analyse the reactions by denaturing PAGE on an 8% sequencing gel along with sequencing reactions performed with the same radiolabelled primer.
9 Dry the gel at 80 °C for 40 min, visualize and quantify the reactions with a Phosphorimager.

To determine precisely the location of the footprinting signals, generate a reference ladder by sequencing the same plasmid DNA with the same radiolabelled primer.

1 Denature 15 nM of plasmid DNA in a volume of 8 μl by heating at 100 °C for 2 min
2 Chill on ice for 5 min.
3 Add 1 μl of radiolabelled primer (0.25 μM) and 1 μl of annealing buffer.
4 Incubate for 3 min at 50 °C.
5 Chill annealing reaction on ice for 5 min.
6 Add 2.8 μl of each Deaza G/A Sequencing Mixes in four termination tubes and prewarm at 37 °C.
7 Dilute 1 μl of T7 DNA polymerase with 4 μl of enzyme dilution buffer.
8 Add 2 μl of diluted T7 DNA polymerase to the annealing reaction.
9 Place 2.8 μl of annealing reaction in each of the termination tubes.
10 Incubate 5 min at 37 °C.
11 Add 4 μl of stop solution.

Protocol 3

Transcription with T7 RNA polymerase for detecting photoreactions

Equipment and reagents

- Stop solution (95% formamide, 20 mM EDTA, 0.05% Bromophenol Blue, 0.05% xylene cyanol FF)
- Incubation buffer (80 mM Tris-HCl, pH 7.9, 12 mM $MgCl_2$, 4 mM spermidine, 20 mM DTT, 250 mM of each ATP, CTP, GTP and UTP, 5 μCi [α-^{32}P]UTP, 10 U RNAase inhibitor and 25 U T7 RNA polymerase
- Transcription buffer (40 mM Tris-HCl, pH 7.9, 6 mM $MgCl_2$, 2 mM spermidine, 10 mM DTT, 125 mM of each ATP, CTP, GTP and UTP, 10 μCi [α-^{32}P]UTP, 30 μM 3′-dATP, 10 U RNAase inhibitor and 40 U of T7 RNA polymerase)

Protocol 3 continued

Method

1. Linearize the plasmid DNA with a suitable restriction enzyme in 40 μl; this procedure will limit the size of the full length transcripts and thus prevents any significant consumption of the pool of nucleotides.
2. Extract the samples twice with 20 μl of an equal mix of Tris 0.1 M pH8 saturated phenol and chloroform.
3. Add 2 μl of 3 M sodium acetate and 50 μl of ethanol, and incubate for 10 min at −20 °C.
4. Centrifuge for 10 min in a microcentrifuge and resuspend the pellet in 10 μl of water.
5. To 5 μl of digested DNA, add 5 μl of incubation buffer.
6. Incubate for 30 min at 37 °C.
7. Add 1 μl of 3 M sodium acetate and 25 μl of ethanol, and incubate for 10 min at −20 °C.
8. Centrifuge 10 min in a microcentrifuge and resuspend the pellet in 10 μl of a 1 : 1 mix of H_2O and stop solution.

To determine precisely the location of termination sites in the template, generate a reference RNA ladder by adding to the transcription reaction 3′-dATP as RNA chain terminator (RNA products ending with a 3′-deoxyribose migrate as if they were one-half nucleotide shorter).

1. Transcribe 500 ng of the digested plasmid DNA in 20 μl of transcription buffer.
2. Add 2 μl of 3 M sodium acetate and 50 μl of ethanol, and incubate for 10 min at −20 °C.
3. Centrifuge for 10 min in a microcentrifuge and resuspend the pellet in 10 μl of a 1 : 1 mix of H_2O and stop solution.
4. Analyse the reactions by denaturing PAGE on an 8% sequencing gel.
5. Dry the gel at 80 °C for 40 min, visualize and quantify the reactions with a Phosphorimager.

An alternative procedure uses T7 RNA polymerase (3) in conjunction with a plasmid that carries a T7 promoter, such a pBluescript, near the site studied. The advantage of this procedure is that the irradiated DNA is read several times, which results in a corresponding increase of the signal strength.

We have probed, by primer extension, the interaction of IHF with a number of *ihf* sites (sites located near *yjbE* gene, in the *gyrB* BIME-1, at the IS*1* right extremity, and the H′ site of λ *att*P), and the strongest signal common to all sites is an increase in band intensity of the guanine residue in the consensus pentamer sequence 5′-TTGAA-3′ (4, 5). This signal is not due to a covalent crosslink between IHF and the DNA, but most likely corresponds to the formation of a pyrimidine dimer on the opposite strand (the template for the primer extension). *Figure 1* shows primer extensions obtained with the *ihf* site located at the right extremity of IS1. In this case, two strong bands appear when the samples are irradiated in the presence of IHF; the first one (indicated by an arrow) corresponds to the central G of the consensus pentamer sequence, and the second one (indicated by a star) is located eight nucleotides downstream of the pentamer signal. In the absence of IHF, a significant reactivity of the two G-residues was observed; this is not very surprising because the formation of a pyrimidine dimer is a predominant photoreaction upon irradiation with UV light. The binding of IHF distorts the dinucleotides and further enhances the reactivity.

3.2.3 Quantification of the photoreactions

In the case of IHF the major footprinting signal corresponds to the guanine residue within the consensus binding site. As can be seen in the gel in *Figure 1* this major signal increases in intensity with increasing concentrations of IHF. We can therefore use this signal as an indicator of site occupancy. To be able to

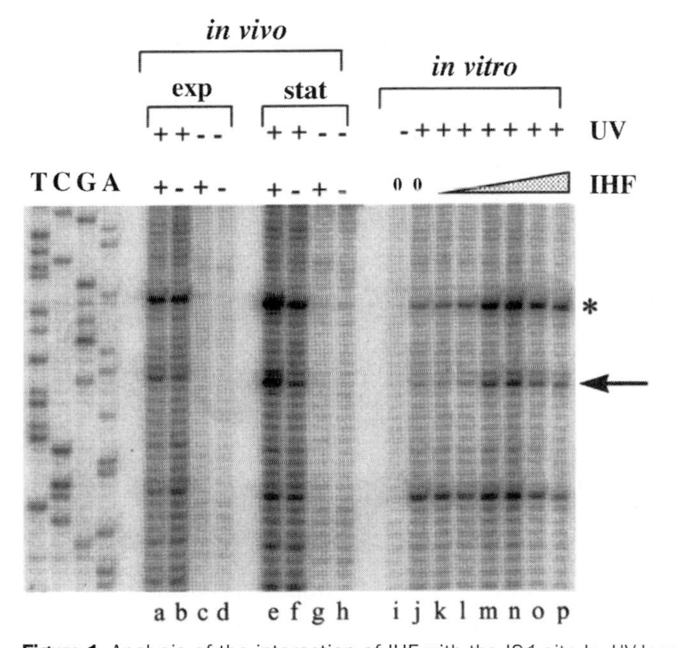

Figure 1 Analysis of the interaction of IHF with the IS*1* site by UV-laser footprinting. Plasmid DNA extracted from IHF^{+} (lanes a, c, e and g) or from IHF^{-} (lanes b, d, f and h) cells was subjected to primer extension. The cells had been irradiated during exponential growth (lanes a and b) or in stationary phase (lanes e and f). Control experiments (lanes c, d, g and h) show unirradiated samples. Plasmid DNA incubated with increasing amounts of IHF were irradiated *in vitro* and analysed by primer extension (lanes j–p). A control sample (lane i) was not irradiated. IHF concentrations were: 0 nM (lane j), 5 nM (lane k), 10 nM (lane l), 25 nM (lane m), 50 nM (lane n), 100 nM (lane o), and 200 nM (lane p). The arrow points to the main signal indicative of binding of IHF. TCGA are lanes of a sequencing reaction performed with the same primer. Two IHF footprints are seen within this site: one at the G residue of the consensus pentamer (indicated by the arrow), and another one located 8 nt downstream (indicated by an asterisk).

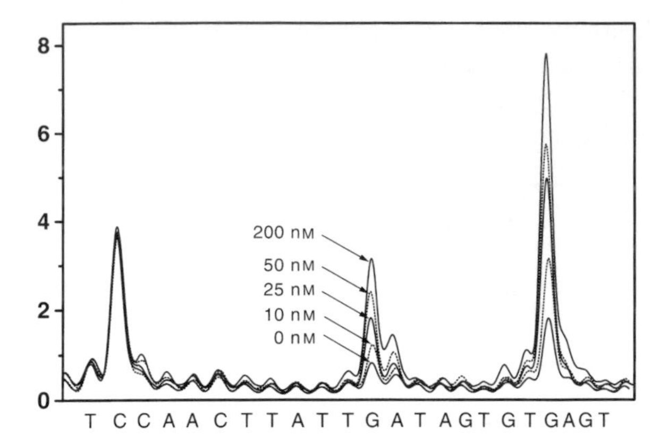

Figure 2 Quantification of the IHF UV-laser footprints. *In vitro* footprints on the IS*1* site: superposition of the scans obtained with increasing amounts of IHF, as indicated in *Figure 1*. The profiles from lanes j–p were scanned using a PhosphorImager and normalized. The nucleotide sequence of a region of 25 residues is indicated below the scans (the scans from left to right represent residues from bottom to top). Scans obtained with 5 nM and 100 nM IHF were omitted to simplify the figure.

make a quantitative comparison with the *in vivo* situation, we have to quantify the signal using a phosphor storage device.

The different reactions were scanned using a PhosphorImager (Molecular Dynamics) and the line graphs corresponding to the different concentrations of IHF were transferred to Microsoft Excel. They were normalized relative to the one obtained in the absence of IHF in such a way that the global profiles superimpose. *Figure 2* shows the superimposed scans obtained with different IHF concentrations in the IS*1* site. In the absence of IHF, the primer extension profile reveals signals induced by the modification of irradiated DNA, and the profile is specific for a given DNA fragment. The alignment of profiles allows the identification of residues that reacted differently in the presence of IHF.

For a given protein, in our case IHF, the signal strength for different binding sites must not be identical. The sequence context can modify the photoreactivity to a certain extent. We compared, for example, the G residue of the consensus pentamer for several different binding sites, all of which contain this pentamer sequence. For the IS*1* site, the two major signals described above were observed. With other *ihf* sites, it is clear from superimposed scans that other signals outside the consensus pentamer also indicate the presence of IHF. Furthermore, the absolute strength of the pentamer signal depends on the nature of the IHF-binding site. The maximal increase in signal intensity (the plateau of the fitted binding curve, see below) varies from site to site; it is 4.5 times the value obtained in the absence of IHF for the *yjbE* site, seven times at *gyrB* BIME-1 and four times baseline at IS*1*. *Figure 3* plots the peak height of the major signal versus the concentration of IHF; the data are fitted to a simple binding isotherm. This graph now relates the intensity of the footprinting signal to the saturation of the binding site with protein. These results prove that the UV footprinting method

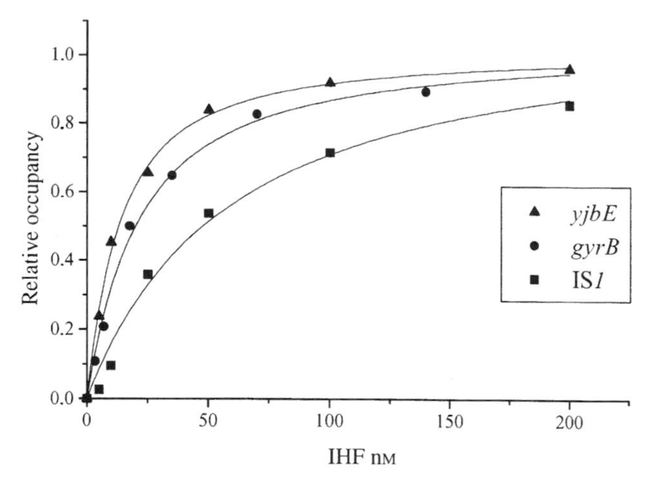

Figure 3 Binding isotherms obtained by UV laser footprinting. The normalized occupancy of the *yjbE* (▼), BIME-1 *gyrB* (●), and IS*1* (■) sites was plotted as a function of the IHF concentration. The occupancy of the sites was estimated from the intensity of the G residue of the consensus pentamer, as shown in Figure 15.2. The binding curves were fitted to a simple binding isotherm. The mathematical form of the equation is:

$$\text{Signal} = \frac{\text{Scale}}{2} \bullet \left(\text{IHF} + \text{DNA} + K_d - \sqrt{(\text{IHF} + \text{DNA} + K_d)^2 - 4.1\text{HF.DNA}} \right)$$

where IHF and DNA denote the total concentrations of IHF and the DNA-binding site, respectively, K_d is the dissociation constant, and Scale is an adjustable parameter that relates the binding signal (in arbitrary units) to site saturation.

is appropriate for analysing and predicting the fractional occupancy of IHF-binding sites (the binding isotherm obtained with the secondary signal yields a similar result as the analysis of the pentamer signal).

Protocol 4

Quantification of *in vitro* footprints

Method

1. Scan the different lanes corresponding to the different IHF concentrations using the Molecular Dynamics ImageQuant software (version 3.3) or an equivalent software.
2. Transfer the data to Microsoft Excel 97 and choose one of the lanes, e.g., the lane containing no protein, as a reference. Normalize the graphs derived from the other lanes of the gel by multiplying with the appropriate constant that makes bands outside the footprint match the corresponding bands in the reference lane. This procedure corrects for the slightly different quantities of material loaded onto different lanes of the gel. The correction factors should not exceed 20%.
3. Superimpose the different graphs and measure the peak height(s) of the IHF signal(s).
4. Plot the peak heights of the signal as a function of IHF concentration.
5. Fit the data to a simple binding isotherm

From the fit, it is possible to deduce the apparent dissociation constants, K_{dapp} (they vary from 11 nM for the *yjbE* site, 18 nM for the λ *att*P H′ site, 19 nM for the *gyrB* BIME1-site, to 56 nM for the IS*1* site). These relative affinities confirmed the classification of sites based on their K_d determined by classical bandshifts (5).

3.3 Samples for *in vivo* footprinting

The task now consists of obtaining equivalent footprinting signals from DNA within the living cell. The footprinting procedure does not change since light easily enters the cell. However, several additional constraints have to be taken into account for the sample preparation and the detection of the photoreactions. Primer extension or transcription of the irradiated template can only be performed *in vitro*. It is therefore necessary to extract the irradiated DNA from the bacteria. Our current technology allows the measurement of protein binding to specific binding sites carried on a multicopy plasmid.

The sample for *in vivo* UV-laser footprinting has to be prepared in a way such that a single pulse of the laser (typically about 30 mJ/pulse, corresponding to 4×10^{16} photons, i.e. 67 nmol of photons) delivers more photons than there are absorbing molecules in the sample. For an *in vivo* experiment, the absorbing molecules are mostly made up of cellular DNA and RNA, as well as free nucleoside phosphates. The major contribution stems from intracellular RNA nucleotides, a typical *Escherichia coli* cell containing about 630 μmol/g of dried weight (6). The total amount of nucleotide residues probably does not exceed 1 mmol/g of dried weight. Since 70% of the mass of a cell is made up of water, and assuming a cell volume of 10^{-15} l (6), there are less than 10^6 absorbing nucleotides per *E. coli* cell. Because we deliver 4×10^{16} photons/pulse and we want to ensure a great excess of photons over absorbers we can irradiate at most about 10^8 *E. coli* cells per pulse. This corresponds to about 100 μl of a suspension at 1 OD_{600}. A large number of such samples are irradiated and the cells are frozen immediately after irradiation. The plasmid DNA is extracted and subjected to primer extension or transcription by T7 RNA polymerase.

Protocol 5

Preparation of the sample

Reagents

- 0.7-ml micro test tubes (e.g. Eppendorf 3813)
- Solution I (100 mM Tris-HCl pH 7.5, 10 mM EDTA, 400 μg/ml RNase I)
- Solution II (0.2 N NaOH, 1% SDS)
- Solution III (3 M Potassium, 5 M acetate solution) (2)
- M9 medium

Method

1 Grow 50 ml cultures of *E. coli* IHF^+ and IHF^- isogenic strains carrying a plasmid with the *ihf* site of interest.

Protocol 5 continued

2. For the exponential phase assay, filter cells through a 0.22 μm filter when they reach an OD_{600} of 0.6. For the stationary phase assay, filter the cells after 16 h (OD_{600} between 4 and 5).
3. Wash the cells with M9 medium.
4. Resuspend the cells in a volume of M9 medium sufficient to obtain a cell density at 1 OD_{600}.
5. Keep the cells at 37 °C.
6. Distribute the cells in aliquots of 50 μl into 0.7-ml micro test tubes and irradiate 40–60 samples with a single 5 ns laser pulse.
7. Freeze the cells in a dry ice bath immediately after irradiation.
8. Collect 2–3 ml of cells by centrifugation in a microcentrifuge at 5000 **g** for 5 min.
9. Resuspend the cells in 100 μl of Solution I.
10. Add 100 μl of Solution II.
11. Add 100 μl of Solution III and mix by inversion several times.
12. Centrifuge the tubes at full speed for 5 min.
13. Transfer the supernatant to a clean tube, add 210 μl of iso-propanol, and incubate 5 min at room temperature.
14. Centrifuge the tubes at full speed for 10 min, discard the supernatant, and wash the pellet with 70% ethanol.
15. Resuspend the pellet in 37 μl of H_2O.
16. Use 18.5 μl of this DNA to perform a primer extension and quantify the signals as described above.

3.4 Measurement of *in vivo* binding under different growth conditions

We now have at hand all of the data necessary for measuring the *in vivo* interaction of a protein with a specific binding site. IHF, examined in *E. coli* grown under two different growth conditions, may serve as an example. Physiological studies had shown that the intracellular concentration of IHF increases by a factor of five in stationary phase, compared with exponentially growing cells. This increase in IHF concentration should lead to a corresponding increase in the occupancy of an IHF-binding site. UV-laser footprinting verifies this prediction.

In the IHF deletion mutant grown to mid-log phase or to saturation, the overall pattern is very similar (but not identical, see below) to the one obtained *in vitro* in the absence of IHF (compare lanes b and f to lane j in *Figure 1*). The profiles corresponding to these lanes therefore superimpose well (*Figure 4A*). In the IHF^+ strain, the pattern obtained with cells grown to saturation is very

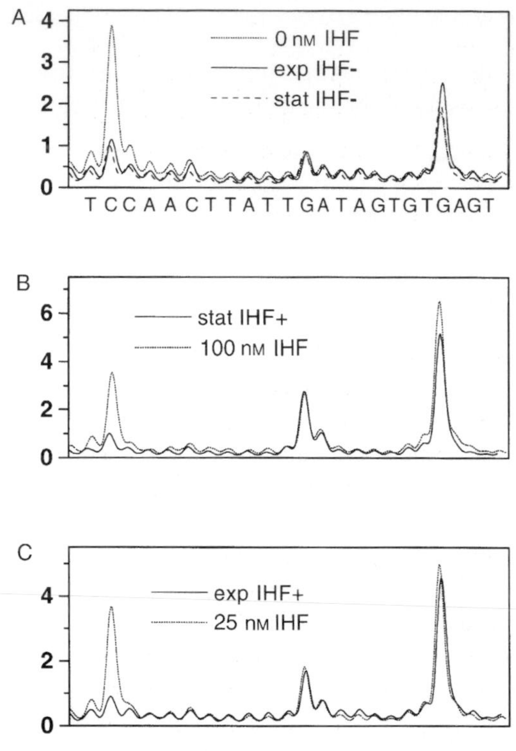

Figure 4 Quantification of the *in vivo* UV footprints obtained on the IS*1* site. (A) Scans from IHF^- cells irradiated in exponential (exp IHF^-) and in stationary (stat IHF^-) phases were superimposed onto scans obtained *in vitro* in the absence of IHF. (B) Scans from IHF^+ cells irradiated in stationary phase (stat IHF^+) were superimposed onto scans obtained *in vitro* with 100 nM IHF. (C) Scans from IHF^+ cells irradiated in exponential phase (exp IHF^+) were superimposed onto scans obtained *in vitro* with 25 nM IHF.

similar to the one of the 100 nM IHF lane *in vitro* (*Figure 4B*). The superposition of the scans obtained *in vivo* and *in vitro* allows an estimation of the occupancy of this site *in vivo*, i.e., this site is roughly 2/3 occupied *in vivo*. We performed the same experiment with cells grown to mid-log phase; the signals indicative of the presence of IHF are less apparent in exponentially growing IHF^+ cells. Superposition of the scans showed that the peaks of the *in vivo* signals are slightly lower than the ones obtained *in vitro* with 25 nM IHF (*Figure 4C*), i.e., the site is about 30% occupied in exponential phase. Equivalent experiments with this and other sites lead to the following conclusions: (i) the fractional occupancy varies with the affinity, the stronger the affinity the higher the occupancy; (ii) fractional occupancy of an IHF-binding site increases with increasing concentration of intracellular IHF; (iii) the calculated value for the concentration of free IHF (0.7 nM in exponential phase and 5 nM in stationary phase) predicts the occupancy of a different IHF-binding sites, measured by UV-laser footprinting.

Although the *in vivo* profiles superimpose well onto those obtained *in vitro*, we noted that, independently of the presence of IHF, the intensity of several signals was strikingly different in the two environments. We could not detect a clear tendency towards more or less photoreactivity *in vivo* (the differences can go either way). For example, in the case of the IS*1* site, the photoreactivity of the C residue located 10 nucleotides upstream of the pentamer signal was more pronounced *in vitro* (*Figure 4*). These results reflect the fact that the DNA is in a different environment *in vivo*. The differences probably stem from different DNA conformations, or are provoked by cellular proteins bound to DNA.

4 Conclusions and perspectives

We have shown that UV-laser footprinting reliably measures protein–DNA interactions in bacterial cells. The footprinting procedure is identical *in vitro* and *in vivo* and it is therefore easy to go back and forth between the two environments. This direct comparison can be used to verify, for a given experimental system, that the conditions and the results of *in vitro* experiments are pertinent to the biological function within the cell. In addition to qualitative comparisons between *in vitro* and *in vivo*, the technique allows the quantitative measurement of a protein–DNA interaction within the cell. The most relevant parameter, saturation of a binding site with protein, is determined directly by *in vivo* UV-laser footprinting. We therefore expect that UV-laser footprinting will be of general use for characterizing DNA–protein interactions and DNA conformations *in vivo*. The work described here has been performed on sites located on plasmids in bacterial cells. Experiments are in progress to detect DNA–protein interactions on the chromosome of *E. coli* cells, as well as in eukaryotic cells.

References

1. Tavazoie, S. and Church, G. M. (1998). *Nat. Biotechnol.* **16**, 566.
2. Sambrook, J., Fritsch, E. F., and Maniatis, T. (1989). *Molecular cloning. A laboratory manual.* Cold Spring Harbor Laboratory Press, Cold Spring Harbor.
3. Htun, H., and Johnston, B. H. (1992). *Methods Enzymol.* **212**, 272.
4. Engelhorn, M., Boccard, F., Murtin, C., Prentki, P., and Geiselmann, J. (1995). *Nucl. Acids Res.* **23**, 2959.
5. Murtin, C., Engelhorn, M., Geiselmann, J., and Boccard, F. (1998) *J. Mol. Biol.* **284**, 949.
6. Neidhardt, F. C. and Umbarger, H. E. (1996) in *Escherichia coli and Salmonella* (1996) 2nd edition (Neidhardt, F. C. *et al.* eds). ASM Press, Washington D.C., pp. 13–16.
   Bliska, J. B. and Cozzarelli, N. R. (1987). *J. Mol. Biol.* **194**, 205.

Chapter 16

Digitization and quantitative analysis of footprinting gels

Judith Smith

MRC Laboratory of Molecular Biology, Hills Road, Cambridge CB2 2QH, UK

1 Introduction

Quantitative footprinting is used in a variety of experimental procedures, including general nuclease protection, band shift analysis, drug–DNA binding, ribosome cleavage and protein component analysis. The 3-step process of quantitative gel analysis is achieved by gel electrophoresis followed by digitization of the gel, and subsequent computer analysis. Electrophoresis produces a pattern of more or less well-separated bands which correspond to the individual components of the sample. Digitization of the gel enables the pattern to be subjected to quantitative analysis and hence the determination of the contribution from each individual band. This fundamental technique is used by many experimentalists for many kinds of quantitative gel analysis, but where approximate area calculations may be sufficient for some analyses, footprinting experiments need to be able to detect minute differences between band intensities, for which a high resolution analysis is required.

Digitization can be carried out either by storage phosphor technology or by densitometry if it is a stained or autoradiographed gel. These methods require either a phosphor-imaging scanning device or a computer-driven densitometer; both systems have advantages and disadvantages.

Many software packages, commercial and otherwise, are currently available for quantitative gel analysis and the simple method of calculating peak areas by summing densities between boundaries selected at the peak minima is offered by some of them, including Geltrak (1). This approach is adequate for analyses where several bands are to be incorporated into the integration, or where approximate areas are acceptable, or where peak separation is complete. However, the tail regions of peak profiles are often obscured by those of adjacent peaks and, in this case, determination of the minima is reduced to guesswork, and the contribution to each peak from its neighbours is not taken into account. To reduce the error that this incurs, some programs offer Gaussian fitting before integration to estimate the peak tail shape and hence the peak minima (1–5). The Gaussian function has probably been the distribution of choice simply because it has been accepted as a standard in many disciplines for many years. However, it would seem that this function is not always as good a fit as has generally been believed (6). It has been found to fit the top half of gel band profiles quite closely, but where peak tails are visible, they are generally much wider than those of the Gaussian, and frequently include a degree of skewness (7–9). There has been much interest in finding a suitable mathematical function to describe gel band profiles accurately (7–10); such a function is a skew Cauchy (Lorentzian) combined with a variable shape parameter, and this provides an almost perfect fit (9).

Footprinting gels have many bands in a lane that tend to overlap and sometimes form shoulders. Most currently available programs do not separate these bands in a way which allows compensation between adjacent peaks. If experimental results are to be meaningful, a high degree of accuracy is required for area determination, and the contribution to the area for each peak by its neighbours must be accounted for. Geltrak overcomes this problem by decomposing the observed profile with a set of skew Cauchy fitted profiles combined with calculated width/height and shape/height ratio variations (9).

Many workers use profile-fitting and more primitive area-determination techniques (11–14), which may be acceptable for an estimate of band ratios where there are very few bands and differences in intensity are large. While some of these systems have the advantage that they operate in a simple and robust way, they do not address the special needs of high-resolution quantitative footprinting experiments, their highly automated approach leaving the user with little or no control over the analysis. A substantial degree of manual control is not only found to be preferred by most users, but since peak tail shapes are unpredictable it is essential for the fine adjustments needed to tailor an accurate fit. It is for this reason that Geltrak was developed as an interactive, menu-driven program with a semi-automatic approach, enabling the accurate determination of very small differences between band areas which is crucial to high-resolution quantification.

2 Implementation

2.1 Digitization

In our laboratory, densitometry of autoradiographs for high-resolution work is normally carried out on a home-built CCD gel reader which is controlled by a Macintosh computer running home-produced software. Commercially produced gel readers are available, such as the computing densitometer, Molecular Dynamics Personal Densitometer S I, which is currently used here for lower-resolution work. We also use a Molecular Dynamics Typhoon PhosphorImager to read storage phosphor plates. Both Molecular Dynamics instruments are run by the ImageQuant software, which is described in the manuals.

2.2 Analysis

Here I describe the use of a particular program, Geltrak, which was developed at the MRC Laboratory of Molecular Biology. The use of other programs will be generally similar in principle but may differ in detail. Geltrak requires the digitized image to be written in MRC Image format, and since many digitizing instruments write files in TIFF format, a conversion program called tif2mrc is available to convert the image into MRC format. Geltrak also requires the lanes to be displayed in the horizontal direction and software to rotate the image if necessary is available. The Geltrak package runs on a wide variety of UNIX computers including DEC/Alpha (DEC-UNIX), SGI (IRIX), Sun (SunOs) and PC (Linux). It requires Fortran 77 (or F2C conversion) and C compilers, graphics window system X11 release 5.0 or higher and about 6 MB of memory to run. An inexpensive Unix box with the appropriate graphics screen should be sufficient to run this software.

Although Geltrak is a general purpose program that covers many aspects of quantitative analysis, the following protocol describes a particular method for the analysis of a footprinting gel, including the calculation of difference probabilities.

Some general instructions for running this program are:

(1) Pan the image around the screen by holding down the centre mouse button while moving the mouse.

(2) Zoom an area by positioning the cursor with the mouse and clicking the right-hand mouse button. Click the centre mouse button to remove the zoom window.

(2) Mark points by positioning the cursor and clicking the left-hand mouse button.

(3) Contiguous rubber-banded lines are drawn in the following way: position the cursor at the start, then press the keyboard control key (Ctrl) and with the left-hand mouse button held down, drag the mouse to the end of the line. With the control key held down, release the mouse button, then press it again and drag to the next position and continue in this way until all the lines are drawn. At the final position release the control key and double click the mouse button.

Most gel images are compressed to fit on the screen, and the lane profile is also displayed in compressed form to match the image. However, all calculations are performed at full resolution and the original uncompressed profile is displayed for functions involving peak fitting and separation, so that peak details are easily visible. To expand the compressed profile, mark the cursor position in the centre of the area of interest. The profile is expanded to the right and left of this position.

Protocol 1

Step by step guide to using Geltrak for a footprinting analysis.

1. Enter the digitized image filename from the keyboard.
2. Enter scaling minimum and maximum. While the default is usually appropriate for scanned autoradiographs, storage phosphor images generally require a range of 0 to a fraction of the maximum displayed in the box.

Protocol 1 continued

3 Select the footprinting option. It is assumed that the first lane to be tracked will be the reference (unbound) lane.

4 Select slit size in screen pixels for left and right lane ends to allow for bandwidth variation. All of the data within the slit is averaged to produce the final density point. While it is important that as much data is included in this average as possible, the effect of neighbouring lanes must be considered and lanes where the bands are not normal to the direction of electrophoresis should have a correspondingly small slit height.

5 Track the lane by rubber-banding a series of contiguous straight lines through the lane centre, working from left to right. In order to obtain a reliable background correction, the lane should be tracked so that the extremities are well beyond the ends of the lane, even if the area of interest lies in the lane centre.

6 Enter the profile maximum. Normally the default can be used. However, in some gels the peaks of interest have been scaled down to such an extent by the presence of comparatively large peaks that they are too small to be seen properly. If this happens, enter a lower maximum. This will cut off the tops of the large peaks and correspondingly increase the height of the small peaks.

7 Track the baseline by rubber-banding. It is not essential to track between the extremities of the profile as the lines are extended automatically to the ends. However, it is recommended that the baseline is drawn as a straight line between the profile minima. If there is only one obvious minimum, position the cursor at the minimum and double click the left-hand mouse button.

8 Select 'Number the peaks'. First expand the profile around the first peak then, working from left to right, mark approximate peak positions; the program locates them precisely. Where there are 'shoulder' peaks or the program finds the wrong peak maximum, mark left- and right-hand peak boundaries. Continue marking peak positions, expanding different parts of the profile as necessary. For a comparison lane, the reference lane is also displayed and its numbers may be automatically transferred. To do this, expand the area around the first peak and mark its position. If the transfer is unsuccessful, the peak positions can be marked manually.

9 Select 'Integrate multiple peak sets by Matrix Decomposition'. Enter a file name to which calculated peak areas will be written. Select the skew Cauchy fitting function, and from the next menu choose the first output format (Peak number, distance from origin, height, area). Then a decision must be made as to how best to predict width and shape values for the whole lane. Consistency is crucial, so whichever method is used for the reference lane should also be used for the comparison lanes.

   (a) If there are at least six well-separated, regularly-shaped peaks in the profile, select 'Calculate regression coefficients by selection'. This method measures parameters from good individual peaks to calculate regression lines which predict width and shape values for overlapped and hidden peaks. Bear in mind that peaks close to the origin may look like single peaks but are often made up

Protocol 1 continued

of very closely overlapped multiple peaks and must be avoided. Expand a section of the profile around these peaks and, working from left to right, mark each peak position. The fitted profile is displayed as dashed lines superimposed over the original peak. Use the menu to refine the parameters and/or mark start and finish points to improve the fit if necessary.

(b) If all the peaks overlap, try 'Calculate width/height ratio automatically'. This method measures width and shape values from all the peaks (except those which are very small) to calculate regression lines for parameter prediction. Individual fitted profiles are displayed over the original profile. User intervention is sometimes required for very poorly shaped peaks, where the left and right sides must be marked.

(c) If the other methods fail, select 'Calculate individual ratios automatically'. This method does not calculate regression lines to predict width and shape, but fits each peak individually. User intervention may be required as in (b).

Select 'Decompose whole range'. The fitted peaks may then be examined by expanding each area and, if you have used method (c), individual peaks may be refitted When you are confident that the peaks are fitted correctly, select 'Start decomposition' The decomposition is carried out and the decomposed areas written to a disk file. Any problems are reported and if necessary one of the alternative fitting methods may be tried.

10 From the main menu select 'Track next lane' and repeat steps 5–9 for each lane to be compared with the reference lane.

11 From the main menu select 'Calculate/output difference probabilities'. Type output filename into which probabilities and log differences are written. From the menu select either 'Division by area under first peak' or 'Division by default (10000)'. Finally, type in the lane number of the first comparison lane, then the next, and so on, and 'enter' when they have all been calculated.

3 Algorithms

3.1 Peak profile-fitting function

Gel band profiles are represented by a Cauchy distribution (9), variations in peak shape and width are adjusted by the parameters σ and m in the following expression:

$$y = a\left[1 + \left(\frac{x - \mu}{\sigma}\right)^2\right]^{-m} \qquad 1$$

where a is the peak amplitude, μ is position, σ is width at half-height, m is shape.

Skewness is accommodated by the addition of a further term:

$$y = a_1\left[1 + \left(\frac{x - \mu}{\sigma}\right)^2\right]^{-m} + a_2(x - \mu)\left[1 + \left(\frac{x - \mu}{\sigma}\right)^2\right]^{-m} \qquad 2$$

3.2 Variability of peak width and shape

The effect of band broadening as a function of migration distance is used to predict the width and shape of partially obscured peaks. In the analysis of a lane of multiple overlapping peaks, a subset of well-separated peaks is selected and each peak fitted to the skew Cauchy function (*Equation 2*). The data are refined using the Levenberg–Marquardt method of non-linear least-squares profile fitting (15). If a subset of at least four peaks is not available, all the peaks in the lane are fitted, excluding only those that are very small or irregularly shaped. For each peak, width/log(amplitude) and shape/log(amplitude) ratios are calculated and used to produce a pair of regression lines against their distance (μ) from the profile origin. The significance of each regression is tested by analysis of variance where the calculated F ratio is compared at the 5% level. If the test fails, the peak whose ratio is furthest from the line is rejected and the process repeated until a satisfactory result is achieved. Both sets of regression coefficients are used to calculate constants b and c in the equations:

$$\sigma = (b_1\mu + c_1)\log(a) \qquad 3$$

to predict peak width and

$$m = (b_2\mu + c_2)\log(a) \qquad 4$$

to predict peak shape

An average weighted skew value, s, is also calculated which adds or subtracts a component to or from each point related to its distance from the mean:

$$s = a(x - \mu)\left[1 + \left(\frac{x - \mu}{\sigma}\right)^2\right]^{-m} \qquad 5$$

3.3 Area decomposition from overlapping peaks

Densities from the area of interest of adjacent peaks in the profile are stored as P_j^{obs}, where $j = 1,J$ and J is the number of densities covering the region. For each peak, amplitude (a) and position (μ) are measured directly and used to calculate peak-width (σ) and shape (m) from the regression lines (*Equations 3* and *4*). If peak fitting using parameter values derived by the regression line coefficients is unsatisfactory, each of the peaks is fitted separately and individual parameter values used. Densities between the points of inflexion of the peak are expressed as P_k^{obs}, where $k = j_1,j_2$, and the number of densities is K. The parameter values for a, μ, σ, m and s are used to calculate a set of points of a skew Cauchy curve (P_k^{calc}), over the interval -7σ to $+7\sigma$ (*Equation 2*).

This set of points is stored in correct relative alignment for the profile peak position as the matrix $S_{j,n}^{calc}$, where $n = 1,N$ and n is the peak number, N is the number of peaks in the profile and j as before. When this procedure has been carried out for all N peaks, the following decomposition is carried out to solve for coefficients A, so that given the imposed (or individual) peak width, shape and skew parameter values, the calculated skew Cauchy points fit the observed profile:

$$A = S^{calc^{-1}} P^{obs} \qquad 6$$

where S^{calc} is the set of values. This is carried out by numerical least-squares solution of the explicit component form:

$$P_j^{obs} = \sum_{n=1}^{N} S_{j,n}^{calc} A_n + \varepsilon \qquad 7$$

where ε is an error term. This method minimizes ε^2 with respect to the parameters A. For each peak, n, the final estimate of the peak area is given by:

$$AREA_n = A_n \int_n \qquad 8$$

4 Instrument and software comparison

4.1 Digitization

Figure 1 shows a comparison between three different scanning devices as carried out by Smith and Singh (9). The poor results given by scanning the autoradiograph on the Molecular Dynamics Model 300A densitometer shown in *Table 1* are probably due to the interference fringes observed on all images digitized by this instrument. Results from the home built CCD densitometer showed a con-

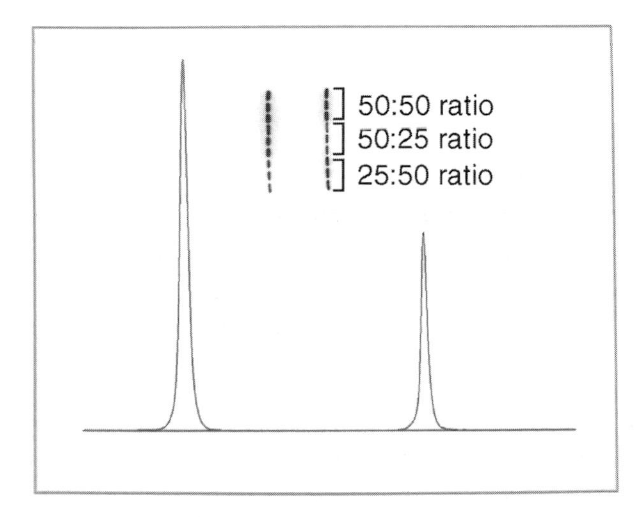

Figure 1 In an experiment to assess the accuracy of quantitative gel analysis, a perfectly clean gel with a predetermined ratio of material between the bands in a lane was produced and hence the accuracy for each digitization method was estimated. Area integration was provided by Geltrak; the baseline was drawn as a horizontal line from one extremity of the lane to the other and the areas of the two peaks calculated by integrating the densities between vertices.

Table 1 Percentage errors ca1culated between predetermined and analysed, calculated component ratios

	Short : long 50 : 50 % error	Short : long 25 : 50 % error	Short : long 50 : 25 % error	Average % error
Molecular Dynamics densitometer				
Wet gel	16	28	58	34.0
Dry gel	13	35	53	33.7
Homebuilt gel reader				
Wet gel	6	17	14	12.7
Dry gel	6	14	17	12.3
Molecular Dynamics phosphorimager				
Low exposure	5	9	6	6.7
Med exposure	4	5	5	4.7
High exposure	4	6	8	5.8

siderable improvement. This instrument does not read from the autoradiograph on a glass plate, but passes it over a light beam by means of rollers. The best results were obtained from the Molecular Dynamics Model 425S PhosphorImager, which also indicate that exposure time has an effect. It is clear that storage phosphor technology is the most accurate method of digitization; this has also been found by Johnston *et al.* (16). Exposure time is less, and both signal-to-noise ratio and dynamic range are much greater. Autoradiographs suffer from film saturation at around 1.8 optical densities, which leads to area distortion from flattened peak tops. However, storage phosphor may be the most expensive in terms of equipment and there is no 'hard copy' left for archiving purposes. Digitized image files tend to be large and expensive in terms of disk space and saving them for future use is not always easy as storage media devices are a rapidly changing technology, and their long-term reliability cannot be guaranteed. It is therefore advisable to save a duplicate copy of an important gel as an autoradiograph.

4.2 Computer analysis

4.2.1 Lane tracking and baseline selection

Of the analysis packages currently available, ImageQuant, Phoretix1D and GelExplorer (10) all require straight lanes before tracking through the centre or boxing the area to be analysed. Furthermore, they make no allowance for variations in lane width. These limitations are largely overcome by Geltrak which uses 'rubber-banding' to track the lane in a series of lines which may form a curve, averaging the data in a slit, the height of which varies along the length of the lane to include all the band information. One of the difficulties in area calculation lies in baseline selection. ImageQuant uses the lowest minimum or follows the valleys, Phoretix1D offers several automatic methods or manual selection. GelExplorer (10) subtracts an average value selected from areas of the gel with no bands. The disadvantage of a single subtraction is that no corrections are applied to allow for shading effects across the gel, which are not uncommon on autoradiographs. The selection of the baseline in Geltrak is user-dependent, but it is recommended that it be drawn as a straight line between profile minima, the slope of the line allowing for shading.

4.2.2 Peak selection and profile fitting

ImageQuant allows the user to mark boundaries between peaks before area integration. Phoretix1D automatically selects the peaks before integration from a user-controlled 'sensitivity' factor which can be modified to include small peaks. GelExplorer goes further by using multiple slices through the lane to extract multiple profiles. It extrapolates peak tail data by fitting the available part of each peak profile to a Cauchy (Lorentzian) distribution, then averages the result over all the slices. Although interesting in its approach, the method is not only computationally intensive but is entirely dependent on straight lanes of equal width for their whole length. Furthermore, it uses only three parameters, height, position, and width, to fit the band profiles. We have found the inclusion of a shape parameter in the Cauchy distribution essential for a precise fit to the form of the curve. Without this parameter, the calculated tails are considerably wider than the observed profile tails. And even when they are not sloping, bands tend to 'trail' away from the direction of electrophoresis, making them asymmetric. Geltrak uses the combination of shape and width parameters and adds skewness in the Cauchy distribution. *Figure 2* demonstrates the fitting of two three-parameter

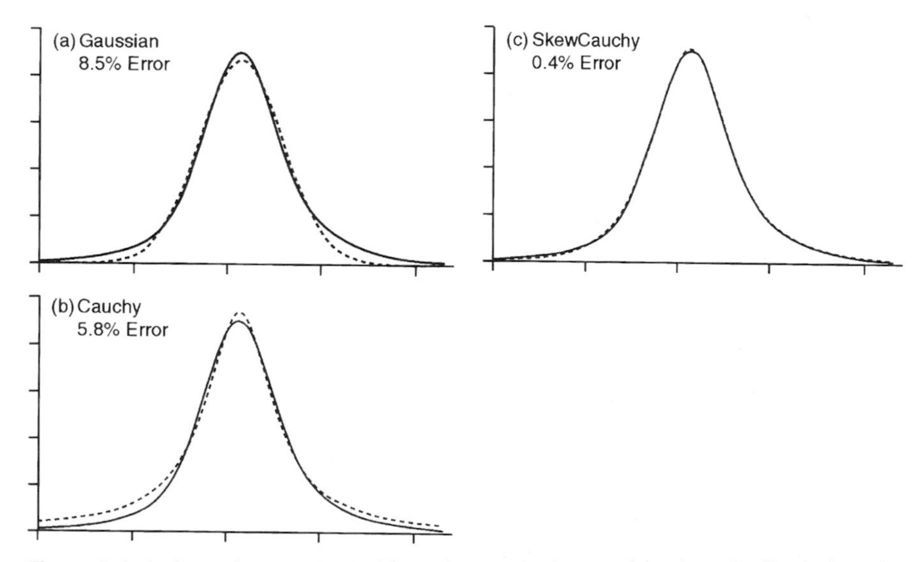

Figure 2 A single peak was extracted from the standard test gel as shown in *Figure 1*, and fitted to (a) a Gaussian, (b) a three-parameter Cauchy (refining width, height, and position) and (c) a five-parameter Cauchy (refining width, height, position, shape and skewness). The fitted peaks were integrated and their areas compared with the original peak profile, which was summed between boundaries and expressed as percentage errors.

distributions (Gaussian and Cauchy) and the five-parameter Cauchy distribution. While the Gaussian tails are too narrow, it is clear that the tailing region of the three-parameter Cauchy is much too wide. The addition of the two extra parameters clearly provide a perfect fit with an integrated area closer to the measured area by a factor greater than 10.

4.2.3 Overlapping peak separation

Separation of overlapping bands into their component areas is carried out in GelExplorer by deconvolution through curve fitting as described above, with a non-linear least-squares optimization. Geltrak first predicts width and shape values for overlapping peaks by using regression line coefficients calculated from available well-separated peaks, where width and shape are measured directly. Where there are insufficient individual peaks to calculate the regression co-efficients, all of the bands in the lane are used. Failing this, each peak is fitted separately and its individual parameter values used. All of this is under the control of the user who can examine the peak fitting and decide which option to use (*Figure 3*). The fitted peak profiles are separated into their component areas by a matrix decomposition to correct for the contribution of densities between neighbouring peaks. This method is accurate and computationally fast, as is essential for an interactive program. The accuracy achieved by a decomposition is

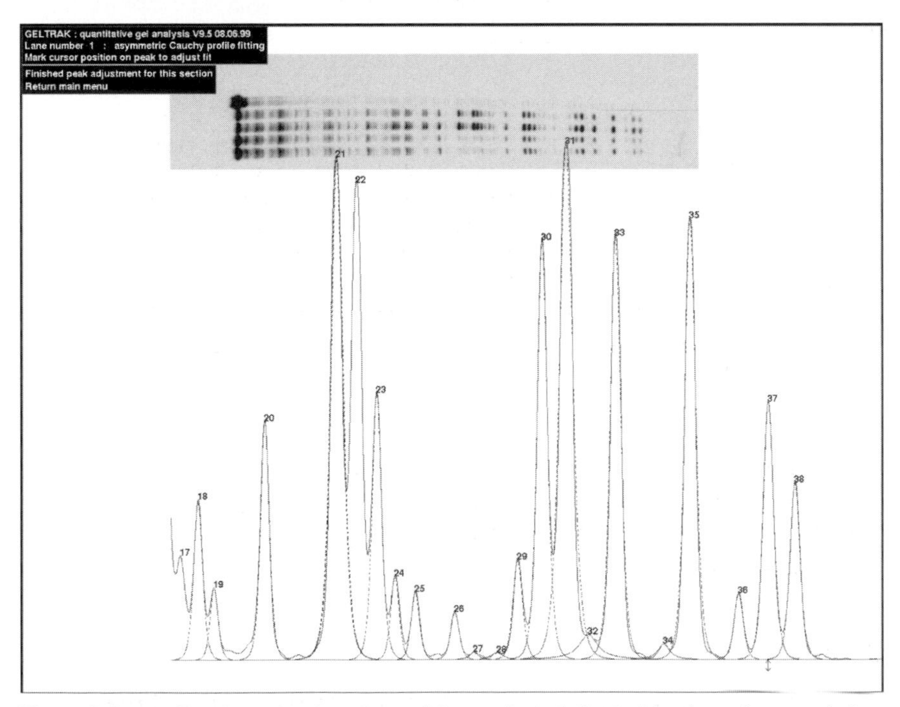

Figure 3 A set of peaks to be decomposed from a footprinting gel is shown in expanded form with the fitted skew Cauchy profiles overlaid as dashed lines. At this stage the user can try another fitting method or refit each peak individually.

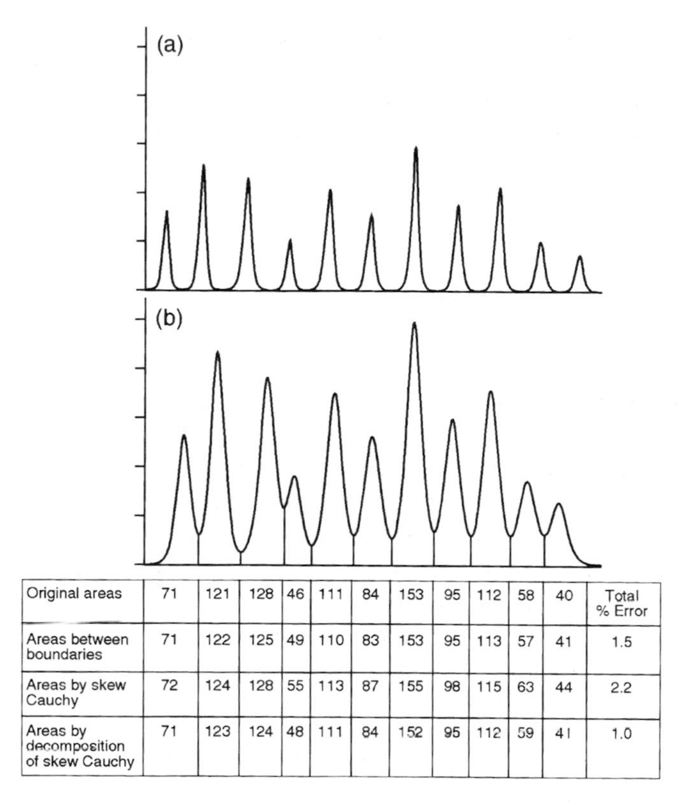

Original areas	71	121	128	46	111	84	153	95	112	58	40	Total % Error
Areas between boundaries	71	122	125	49	110	83	153	95	113	57	41	1.5
Areas by skew Cauchy	72	124	128	55	113	87	155	98	115	63	44	2.2
Areas by decomposition of skew Cauchy	71	123	124	48	111	84	152	95	112	59	41	1.0

Figure 4 To test the accuracy of extracting individual areas from fitted overlapping peaks, a set of cleven well-separated peaks were extracted from the footprinting gel illustrated in *Figure 16.3* and their areas calculated by adding the densities for each peak. Each peak was fitted to a skew Cauchy with five parameters (a) and the fitted peak integrated. Peak positions were then compressed until the peaks overlapped and their contributions added to create a profile (b) which would be typical of many gel band profiles. The peaks in the resulting profile were integrated in three ways : (i) by summing densities between minima, (II) by fitting the available part of each peak to a skew Cauchy before summation and (iii) by decomposing the fitted peaks into their individual areas. The three sets of calculated areas (c) were compared with the original areas. It can be seen that the decomposition step greatly improves the accuracy of integration.

shown clearly in *Figure 4* which illustrates a comparison between three methods of peak separation. After all the lanes have been integrated, difference probabilities may be calculated using the method described by Fairall *et al.* (17) and written with individual peak areas into a disk file.

5 Conclusion

The quality of the result of the analysis is directly related to the quality of the original gel. Errors can be introduced by any of the following: purification of the sample, pouring of the gel, making the wells, preparation of the buffer, pipetting the samples, the electrophoresis itself and the film or storage phosphor exposure. However good the resolution of the scanning device and sophisticated the mathematics in the analysis, a poor-quality gel will yield only poor-quality results. Geltrak is a general purpose analysis program and will process data from gels with crooked lanes and sloping bands, but final area calculations will be adversely affected by such data. The best results will always be achieved by gels with a high signal-to-noise ratio and straight lanes of well-separated bands. The accuracy of peak separation deteriorates with the decrease in distance between peaks, and bands which are completely overlapped will not only be lost but will artificially increase the size of the larger peak.

We have found that peaks in all gels for which a one-dimensional quantitative analysis is required are represented very closely by a skew Cauchy which includes a shape parameter. In conjunction with an estimate of both width versus position and shape versus position obtained by calculated regression coefficients, the multiple profile-fitting decomposition separates overlapping peaks accurately and efficiently. The incorporation of the footprinting option to number peaks easily and calculate difference probabilities automatically in Geltrak make it a complete high-resolution footprinting package and the clear screen display and simple menus make the program easy to use. It is rapid and interactive and gives the user a good understanding of what is happening, and has proved popular among biochemists over several years.

References

1. Smith, J. M. and Thomas, D. J. (1990). *Comput. Appl. Biosci.* **6**, 93.
2. Hansen, P., K., Christensen, J. H., Nyborg, J., Lillelund, O., and Thorgersen, H. C. (1993). *J. Mol. Biol.* **233**, 191.
3. Horgan, G. W. and Glasbey, C. A. (1995). *Electrophoresis* **16**, 298.
4. Lutter, C. L. (1978). *J. Mol. Biol.* **124**, 391.
5. Ribeiro, E. A. and Sutherland, J. C. (1993). *Anal. Biochem.* **210**, 378.
6. Press, W. H., Teukolsky, S. A., Vetterling, W. T., and Flannery, B. P. (1992). In *Numerical recipes*, p. 652. Cambridge University Press, Cambridge.
7. Galat, A. and Goldberg, I. H. (1987). *Comput. Appl. Biosci.* **3**, 333.
8. Vohradsky, J. and Panek, J. (1993). *Electrophoresis* **14**, 601.
9. Smith, J. M. and Singh, M. (1996). *Biotechniques* **20**, 1082
10. Shadle, S. E., Allen, F. A., Guo, H., Pogozelski, W. K., Bashkin J. S., and Tullius T. D. (1997). *Nucl. Acids Res.* **25**, 850
11. Morrison, T. B. and Parkinson, J. S. (1994). *Biotechniques* **17**, 922.
12. Salas, X. and Portugal, J. (1993). *Comput. Appl. Biosci.* **9**, 607.
13. Schwartz, A. and Leng, M. (1994). *J. Mol. Biol.* **236**, 969.
14. Stankus, A., Goodisman, J., and Dabrowiak, J. C. (1992). *Biochemistry* **31**, 9310.
15. Press, W. H., Teukolsky, S. A., Vetterling, W. T. and Flannery, B. P. (1992). In *Numerical recipes*, p. 678. Cambridge University Press, Cambridge.
16. Johnston, R. F., Pickett S. C., and Barker, D. L. (1990). *Electrophoresis* **11**, 355.
17. Fairall, L., Harrison, S. D., Travers, A. A., and Rhodes, D. (1992). *J. Mol. Biol.* **226**, 349.

Appendix

The Geltrak software together with tif2mrc is available on request from the authors via anonymous ftp. A licensing agreement with the Medical Research Council must first be signed. There is no handling charge to academic workers; commercial organizations must pay a licence fee.

Chapter 17
Mapping histone positions in chromatin by protein-directed DNA crosslinking and cleavage

Andrew Travers
MRC Laboratory of Molecular Biology, Hills Road, Cambridge CB2 2QH, UK

1 Introduction

The precise mapping of protein-binding sites on DNA can be facilitated by the covalent conjugation of an activatable reagent to a specific amino acid residue on the protein of interest. In this procedure, the conjugated protein is first bound to DNA and then the reagent is activated. Two general methods are currently used. The activated reagent can crosslink to DNA in van der Waals' contact and thus sensitize the backbone to alkaline hydrolysis in the immediate vicinity of the crosslink. Alternatively, on activation the conjugated reagent generates a short-lived diffusible radical, usually an hydroxyl radical, which in turn reacts with the DNA. The precision of the resulting footprint depends in both cases on the short half-life of the activated intermediate.

The initial conjugation of the reactive entity usually entails reaction with an exposed sulfhydryl group of cysteine. Such cysteines may fortuitously pre-exist in the protein at an appropriate location but in general it is necessary to use site-specific mutagenesis procedures to change a chosen amino acid to cysteine for conjugation. Conversely, cysteines at inappropriate sites may need to be mutated to another amino acid, usually serine or alanine, to facilitate the analysis. The reagents conjugated to proteins include 4-azidophenacylbromide for protein–DNA crosslinking (1) and derivatized *o*-phenanthroline (2) and EDTA for hydroxyl radical generation (3). These methods were initially developed for mapping contacts by highly sequence-specific DNA-binding proteins (1, 3) but more recently have been used to map DNA contacts by relatively non-specific chromatin-associated DNA-binding proteins including histone $H1^0$ (4), H2A (5) and H4 (6), the globular domain of histone H5 (7), the HMG domains of rat HMG1 (8) and the *E. coli* FIS protein (9). The mapping of the positioning of histone H4 on bound nucleosomal DNA is particularly useful (6). This allows the precise and unambiguous determination of the translational positions of nucleosomes reconstituted onto a DNA fragment *in vitro* (6, 10, 11).

2 Principle of the procedure

2.1 Chemistry of conjugate-directed footprinting

The chemistry of the crosslinking of 4-azidophenacyl conjugated proteins to DNA is straightforward. On activation by UV light the azido group loses nitrogen and is converted to a highly reactive nitrene. This group will, in principle, form a crosslink with any organic residue in van der Waals' contact. This reaction competes with a rapid inactivation by reaction with water (*Figure 1*)

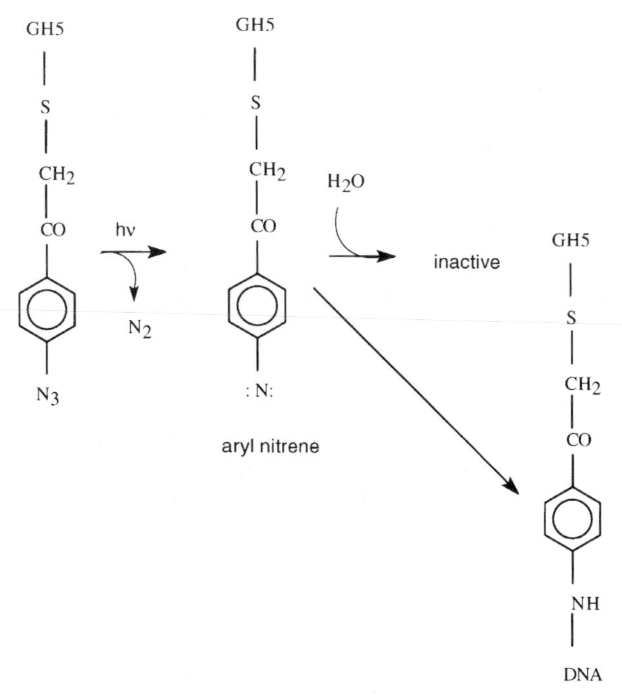

Figure 1 Mechanism of crosslinking by 4-azidophenacyl derivatives of the globular domain of histone H5 (GH5). UV irradiation of conjugated protein results in the formation of an activated nitrene with concomitant generation of nitrogen. The nitrene then reacts with any organic residue in van der Waals contact to form a covalent linkage or with water (adapted from ref. 13).

The Fenton chemistry of conjugate-directed hydroxyl-radical cleavage of DNA is as described in Chapter 12.

2.2 Limitations of the procedures

The precision of the footprint resulting from conjugate-directed reaction with DNA depends on several factors:

(1) The proximity of the chosen conjugated amino acid to DNA in the protein–DNA complex.

(2) The stability of the complex. If the protein rapidly exchanges with its DNA-binding site the probability of non-specific reaction, and hence a high back-

ground, is increased. This effect is more serious with conjugates which generate diffusible reactive entities, such as hydroxyl radical. However, when the complex is stable and the chosen conjugated amino acid is close to the DNA in the complex the reaction of the hydroxyl radical generated can be restricted to a single nucleotide. A good example of this precision is the method for the identification of the core nucleosome dyad (6) Nevertheless, other authors have attributed conjugate-directed DNA cleavage by hydroxyl radical at three successive double-helical turns to diffusion from a single dominant position of the bound protein (4). If this interpretation is correct such a phenomenon would represent a severe limitation of the procedure.

(3) The length of the conjugated reagent and its tether to cysteine. The tethered reagents are normally extend by 12–14 Å from the conjugated sulfur and thus, in principle, could vary position with respect to a bound DNA molecule by several nucleotides. This would result, as is observed in certain cases (7), in a Gaussian distribution of reaction about the principle site of reaction.

(4) Flexibility of bound protein. When a conjugated protein is precisely positioned on a DNA sequence multiple cleavage/crosslinking sites can result from the intrinsic flexibility of the region to which the conjugate is attached. A good example would be the multiple crosslinks resulting from a reagent conjugated to the *N*-terminal tails of core histones (5).

(5) The technique cannot be used when the protein contains a sulhydryl group or groups that are essential for activity. Similarly, its use is limited when the protein contains disulfide linkages since normally these would be reduced in the first step of the protocol.

3 Practical applications

3.1 Crosslinking of histones to chromatosome and core nucleosome DNA

3.1.1 Procedure for conjugating 4-azidophenacyl bromide to a cysteine residue in protein

In principle, 4-azidophenacyl bromide reacts directly and specifically with available sulfhydryl groups on mixing protein and reagent. It is first necessary to ensure that the sulfhydryl group is fully reduced by exposure to DTT. Excess DTT is then removed by gel filtration and conjugation effected by mixing the protein with the reagent overnight in the dark. After conjugation the excess 4-azidophenacyl bromide is removed by gel filtration. Following the reaction the extent of modification is checked where possible by electrophoresis in SDS gels. (*Protocol 1*)

In all cases when a conjugate has been covalently linked to a DNA-binding protein it is necessary to check that the conjugated protein is still competent to bind in the normal manner. This is usually performed using appropriate bandshift procedures (Chapter 5).

Protocol 1

Formation of 4-azidophenacyl–protein conjugates

Equipment and reagents

- 100 mM 4-azidophenacylbromide in water (kept in dark)
- Gel filtration buffer: 20 mM KCl, 20 mM Tris-HCl, pH 7.5, 0.1% Nonidet-P40
- Separating gel acrylamide stock solution: 30% acrylamide, 1% *N,N'*-methylene-bis-acrylamide in water
- Separating gel buffer: 0.375 M Tris-HCl, pH 8.8, 0.1% SDS
- Running buffer: 25 mM Tris base, 0.192 M glycine, 0.1% SDS
- SDS-PAGE gel apparatus

Method

The 4-azidophenacyl moiety is photosensitive so all procedures must be carried out in the dark.

1. Mix the protein to be modified (100–200 nM final concentration) with 2 mM DTT for 30 min on ice.
2. Remove excess DTT by Sephadex G-15 chromatography in gel filtration buffer
3. Immediately add 100 mM 4-azidophenacylbromide to a final concentration of 2–4 mM and incubate overnight at 4 °C.
4. Remove unreacted 4-azidophenacylbromide by Sephadex G-15 chromatography in the gel filtration buffer.
5. Test for the degree of modification by SDS-PAGE. For small proteins such as the globular domain of linker histone or an HMG domain of HMG1 this requires a 15 or 20% polyacrylamide gel.

3.2 Crosslinking and mapping of crosslinking sites on DNA

Crosslinking of the conjugated protein to the bound DNA requires first the reconstitution of the protein–DNA complex. The 4-azidophenacyl conjugated protein is then irradiated with long wavelength UV light for 4 s at room temperature using a transilluminator. This irradiation is sufficient for detectable crosslinking.

After crosslinking the protein to labelled DNA the crosslinked protein–DNA complexes are extracted with phenol. This removes uncrosslinked protein at the phenol–aqueous layer interface. However, care must be taken to track the crosslinked protein. Especially when the protein is small relative to the attached DNA, e.g. the complex of the globular domain of linker histone with DNA, the complex partitions into the phenol phase and is recovered from this phase by precipitation with ethanol.

The mapping of the crosslinking site on DNA depends on the sensitization of the sugar–phosphate backbone to alkaline hydrolysis at the crosslinking site and is normally accomplished by treatment with hot piperidine. The DNA is then denatured and the resulting fragment analysed by electrophoresis on DNA sequencing gels. In some complexes crosslinking may occur very close to the end of the DNA fragment. In such cases, the gel should contain a high percentage of acrylamide (normally 20%) and contain two volumes 1× TBE and one volume of 3 M sodium acetate in the lower gel-tank to ensure that any labelled small DNA fragments run as tight bands (see *Figure 2* for an example). *Protocol 2* describes the method for the crosslinking of the derivatized globular domain of linker histone H5 (GH5) but is equally applicable to crosslinking directed by derivatized core histones (5). Although in principle reagents producing hydroxyl radical can be conjugated to the same GH5 proteins the resultant cleavage patterns have a high background, possibly resulting from the rapid exchange of GH5 with the reconstituted particles.

Protocol 2

Footprinting of reconstituted chromatosomes by 4-azidophenacyl–protein conjugates

Equipment and reagents

- 312 nm transilluminator
- TBE buffer: 90 mM Tris-borate, 2mM EDTA, pH 8.3
- 20 × 40 cm gel plates and apparatus
- 1 M piperidine

Method

1. Mix 4-azidophenacyl-GH5 (10–20 pmol) with linker histone depleted chromatosomes (12) containing ^{32}P end-labelled DNA at an estimated stoichiometry of 1 : 1 in 0.5× TBE buffer.
2. The mixture is equilibrated for 10 min at 20 °C.
3. The reconstituted particles are irradiated by placing the mixture immediately above a standard 312 nm 180 W transilluminator for 4 min.
4. The volume of the crosslinking reaction mixture is adjusted to 100 μl and an equal volume of phenol–chloroform (4 : 1) added.
5. Spin for 5 min in an Eppendorf bench-top centrifuge (7000 r.p.m/5100 **g**)
6. Remove the phenol phase with a Pasteur pipette.
7. Steps 4–6 are repeated twice and the organic phases pooled.
8. Add 2 volumes of 70% ethanol in the presence of 0.4–0.8 M lithium chloride.
9. Keep at −20° for 1 h
10. Spin for 10 min in Eppendorf refrigerated bench-top centrifuge (7000 r.p.m./5100 **g**)
11. Wash pellet 3x with 70% ethanol
12. Redissolve pellet in 100 μl 1 M piperidine
13. Heat for 30 min at 90 °C.

Protocol 2 continued

14 Precipitate DNA, dissolve in formamide and heat at 100 °C for 2 min.

15 Load onto 20 × 40 cm 20% polyacrylamide denaturing gel containing 8 M urea.

16 Run gel until bromophenol blue marker is halfway down the gel.

17 Dry gel and autoradiograph.

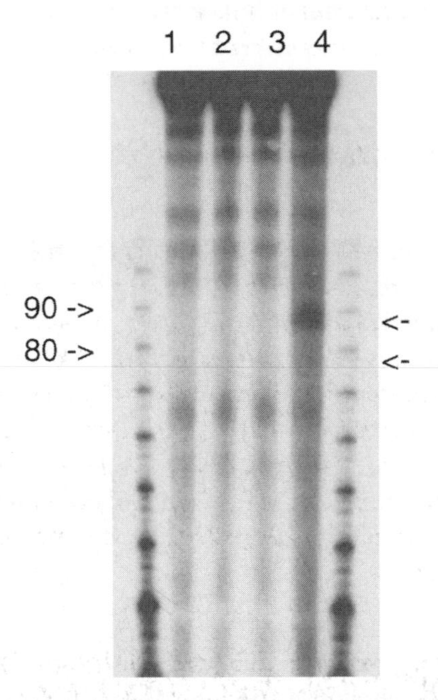

Figure 2 A 20% denaturing polyacrylamide gel electrophoresis of end-labelled linker-histone depleted chromatosomes reconstituted with native recombinant GH5 (lane1); S71C GH5 mutant before (lane 2) or after AP conjugation (lanes 3 and 4). The sample shown in lane 3 was not exposed to UV, whereas that of lane 4 was cross-linked by exposure to UV. All samples were phenol extracted, ethanol precipitated and treated with hot piperidine to cleave the DNA at the protein crosslinking point. The sizes in nucleotides are shown on the right (reproduced with permission from ref. 7)

3.3 Mapping histone positions using EDTA conjugates

The conjugation of EDTA derivatives to selected residues in histones reconstituted into nucleosomes has provided significant information about the structure of the nucleosome core particle and the translational positions adopted by the particle on a defined DNA sequence. The current method of choice to identify translational positions in *in vitro* reconstituted chromatin is to incorporate EDTA-conjugated histone H4 into the reconstituted particle (6), activate hydroxyl radical formation and then identify the translational positions by a characteristic cleavage pattern of one strong cleavage (directed by one of the two H4 histones in the particle) separated by seven nucleotides from two weaker cleavages (directed by the other H4 histone) on the same DNA strand. The position of the dyad of

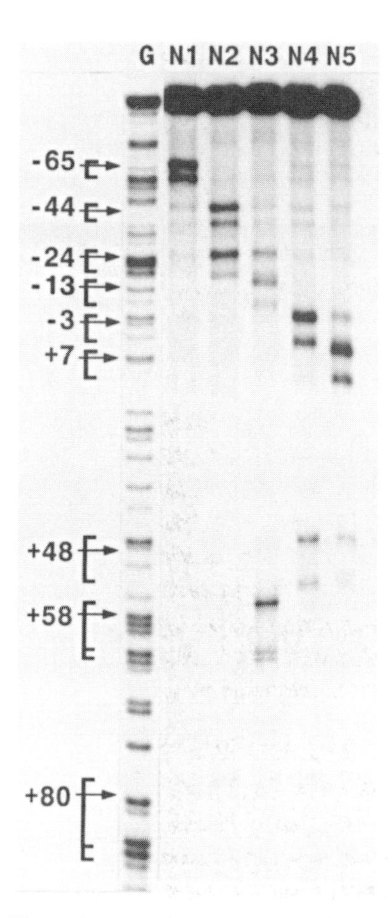

Figure 3 Nucleosome dyad mapping analysis for nucleosome core particles assembled on *Xenopus borealis* somatic 5S DNA with the coding strand radiolabelled. Nucleosome–DNA complexes were excised from a polyacrylamide gel as described in *Protocol 4* and analysed in order of decreasing electrophoretic mobility. The figure show that with decreasing mobility the dyad position is located closer to the centre of the bound DNA. G indicates the marker track showing the positions of guanine residues in the sequence.

the nucleosome core particle is then mapped to two nucleotides from the major cleavage site and five nucleotides from the minor cleavages (*Figure 3*).

The initial nucleosome reconstitution requires DNA terminally on one strand only and a mixture of core histone in which the unmodified histone, H4, is replaced by EDTA-conjugated H4 and the reconstitution effected by a salt-dilution protocol in which the mixture is dialysed against progressively lower salt concentrations.

After reconstitution the hydroxyl radical cleavage reaction is initiated by the addition of ammonium ferrous sulfate, ascorbic acid and hydrogen peroxide. For mapping of dyad positions for core nucleosomes alone the concentrations of ascorbic acid and hydrogen peroxide used are similar to those for conventional hydroxyl radical footprinting procedures (see ref. 5 and Chapter 12). However, in certain situations, for example for mapping dyad positions in the presence of the transcription factor TFIIIA the hydrogen peroxide concentration should be reduced from 0.03–05% to 0.003% to minimize damage to the factor (11). The sodium ascorbate concentration should also be reduced to 1 mM. The latter conditions are used in *Protocol 4*.

Protocol 3

Reconstitution of core nucleosomes containing EDTA-conjugated histone H4

Reagents

- Histone octamer containing EDTA-conjugated histone H4
- Radiolabelled DNA

Method

1. Mix core particles containing conjugated histone H4, DNA labelled at one terminus and mixed sequence DNA to final concentrations of 5 μM, 60 nM and 10 μM, respectively, in 20 mM Tris-HCl, pH 7.5, 1 mM EDTA at 4°C. The volume depends on the amount required.
2. Dialyse against 20 mM Tris-HCl pH 7.5, 1 mM EDTA, 2 M NaCl for 2 h at 4°C.
3. Dialyse against 20 mM Tris-HCl pH 7.5, 1 mM EDTA, 0.85 M NaCl for 2 h at 4°C.
4. Dialyse against 20 mM Tris-HCl, pH 7.5, 1 mM EDTA, 0.65 M NaCl for 2 h at 4°C.
5. Dialyse against 20 mM Tris-HCl, pH 7.5, 1 mM EDTA, 0.5 M NaCl for 2 h at 4°C.
6. Dialyse against 20 mM HEPES, pH 7.5, 0.01% Nonidet P-40 overnight at 4°C.
7. Store at 4°C prior to use.

Protocol 4

Site-directed hydroxyl radical cleavage to identify translational positions of nucleosomes

Equipment and reagents

All solutions must be freshly prepared

- Ammonium ferrous sulfate
- Ascorbic acid
- Hydrogen peroxide (this must be fresh!)
- Sequencing gel apparatus

Method

1. Mix 10 μl nucleosomes containing conjugated histone H4 (400 nM) and DNA labelled at one terminus with 5 μl 12.5 μM ammonium ferrous sulfate in a total volume of 20 μl.

Protocol 4 continued

2 Incubate at 4 °C for 15 min.

3 Add ascorbic acid to a final concentration of 1 mM and hydrogen peroxide to a final concentration of 0.003%.

4 Incubate at 4 °C for 60 min.

5 Add glycerol to 10%.

6 Load samples on a 5% (37.5 : 1 acrylamide–bis-acrylamide) acrylamide gel or a 0.7% agarose gel containing 20 mM HEPES, pH 7.5, and 0.1 mM EDTA.

7 Run the preparative acrylamide gel for 6 h at 250 V at 4 °C or the agarose gel overnight at 20 mA at 4 °C.

8 After electrophoresis the preparative gel is covered with cling film and autoradiographed at 4° (no drying).

9 The separated nucleoprotein complexes are excised.

10. Wash the gel slices with water three times for 5 min.

11 Crush the gel slices well and place in Eppendorf tubes, add Tris-HCl, pH 8.0, to 0.3 M final concentration and elute DNA for 4 h at 65 °C.

12 Precipitate DNA, dissolve in formamide and load on sequencing gel.

13 Run gel and autoradiograph.

One important caveat of this method for determining translational positions is that the modification of histone H4, Ser47Cys, and the subsequent conjugation might themselves affect the number and distribution of possible translational positions. An appropriate control for this possibility is to reconstitute core nucleosomes with wild-type H4 histone, mutated H4 histone and conjugated H4 histone in parallel and compare the number of electrophoretically separable (on 5% polyacrylamide gels—see *Protocol 4*) core nucleosome particles formed in each case. The individual bands represent core nucleosomes with different translational positions.

References

1. Pendergrast, P. S., Chen, Y., Ebright, Y. W., and Ebright, Y. H. (1992). *Proc. Natl. Acad. Sci. USA* **89**, 10287.
2. Ebright, R. H., Ebright, Y. W., Pendergrast, P. S., and Gunsekara, A. (1990). *Proc. Natl. Acad. Sci. USA* **87**, 2882.
3. Ebright, Y. W., Chen, Y., Pendergrast, P. S., and Ebright, R. H. (1992). *Biochemistry* **31**, 10664.
4. Hayes, J. J. (1996). *Biochemistry* **35**, 11931.
5. Lee, K.M. and Hayes, J.J. (1998). *Biochemistry* **37**, 8622.
6. Flaus, A., Luger, K., Tan, S., and Richmond, T. J. (1996). *Proc. Natl. Acad. Sci. USA* **93**, 1370.
7. Zhou, Y. B., Gerchmann, S. E., Ramakrishnan, V., Travers, A., and Muyldermans, S. (1996). *Nature* **395**, 402.

8. Webb, M. and Thomas, J. O. (1999). *J. Mol. Biol.* **294**, 373.
9. Pan, C. Q., Feng, J. A., Finkel, S. E., Landgraf, R., Sigman, D., and Johnson, R. C. (1994). *Proc. Natl. Acad. Sci. USA* **91**, 1721.
10. Flaus, A. and Richmond, T. J. (1998). *J. Mol. Biol.* **275**, 427.
11. Panetta, G., Buttinelli, M., Flaus, A., Richmond, T. J., and Rhodes, D. (1998). *J. Mol. Biol.* **282**, 683.
12. Segers, A., Muyldermans, S., and Wyns, L. (1991). *J. Biol. Chem.* **266**, 1502.
13. Chen, Y. and Ebright, R. H. (1993). *J. Mol. Biol.* **230**, 453.

Chapter 18

Kinetic analysis of enzyme template interactions. Nucleotide incorporation by DNA dependent RNA and DNA polymerases

Bianca Sclavi, Pascal Roux and Henri Buc
Institut Pasteur, Unité de Physicochimie des Macromolécules Biologiques (URA 1773 du CNRS), F–75724 Paris Cedex 15, France

1 Introduction

The steady-state analysis of the incorporation of nucleotide substrates by nucleic acid polymerases has been a subject of intensive research for many years. It has provided interesting rewards in its specific field, for example a better understanding of the faithful or the incorrect incorporation of nucleotides into an elongating template by a specific enzyme. It has also proved to be of great use for a more indirect purpose, namely the understanding of how, in a time-resolved manner, nucleic acid polymerases interact with their cognate templates as the copying process is initiated or proceeds at steady state. This topic is the subject of this chapter. Such assays were initially used in a rather crude, analytical manner. For example run-off transcription experiments were inserted into purification protocols to assess the degree of purity of a preparation of RNA polymerase or the extent of activation generated by a given factor. We describe here how more quantitative assays, performed with purified enzymes and templates, provide significant information on the interactions occurring between the two partners. Such assays are essential prerequisites for the establishment of a plausible kinetic scheme which will in turn be challenged by the specific molecular methods developed in other chapters of this volume. A similar approach can be adopted for the analysis of the recognition of a given nucleic acid target (here a promoter), by an enzyme (here an RNA polymerase), or for the extension of a given primer during consecutive elongation steps (here by a DNA-dependent DNA polymerase). In both cases the transition from a qualitative to a quantitative approach does not initially require very sophisticated mathematical methods of analysis. However, it does imply a similar determination of few basic parameters, stoichiometry of the partners in the final complex, measurements of equilibrium and rate constants, establishment of their dependence on enzyme and on substrate concentrations. Changes in the experimental conditions initially adopted will also follow the same goals: to ensure the maximal specificity of the recognition process, in those *in vitro* tests; to derive thermodynamic quantities associated with changes of the rate parameters; to detect and to populate some transient intermediates during the recognition process. Two examples are detailed here:

- Abortive initiation assays to probe interactions between RNA polymerases and prokaryotic promoters.
- Elongation assays. Interactions between DNA polymerases and their templates.

2 Purification of DNA fragments

High-quality DNA fragments are essential for these enzymatic and kinetic approaches. Since many of these fragments are conveniently derived either from PCR methods or by enzymatic restriction of plasmids, we suggest that purification of plasmid and fragment DNA be carried out by either size exclusion or ion exchange chromatography. We have chosen to illustrate separation by ion exchange chromatography using the SMART (Amersham Pharmacia Biotech) system (1).

Protocol 1

Ion exchange chromatographic separation of DNA fragments.

Equipment and reagents

- SMART chromatography system (Amersham Pharmacia Biotech)
- Low salt buffer: 10 mM Tris/HCl, pH 8.0, 1 mM EDTA, 300 mM NaCl
- High salt buffer: 10 mM Tris/HCl, pH 8.0, 1 mM EDTA, 600 mM NaCl
- Mono Q PC 3.2/3 ion exchange column

Separation method

ML refers to the volume at which a specific instruction will occur.

0.00 LOAD

0.00 FILL A, 1, 10, 15 000

0.00 FILL B, 1, 10, 15 000

0.00 FLOW 200

Protocol 1 continued

```
0.00   TRIPLE_WAVELENGTH 254, 280, 320
0.00   AUTOZERO
0.10   INJECT
0.10   AUTOZERO
2.00   LOAD
2.00   CONC_B 0.0
3.00   NEEDLE_POSITION DOWN
3.00   GOTO_TUBE 1
3.00   FRACTION_SIZE 200
3.90   FILL A, 1, 10, 15 000
3.90   FILL B, 1, 10, 15 000
4.00   CONC_B 50.0
10.00  FRAC_STOP
20.00  CONC_B 100
```

3 Abortive initiation assays to probe interactions between RNA polymerases and prokaryotic promoters

A kinetically competent binary complex is formed between *Escherichia coli* RNA polymerase (R) and a DNA fragment containing a promoter (P). The minimal scheme which can account for the formation of the efficient species RPo (the open complex) is:

$$R + P \underset{K_B}{\rightleftharpoons} RP^{\bullet} \underset{kr}{\overset{k_f}{\rightleftharpoons}} RP_o \qquad 1$$

The nature of the intermediate $RP^{\bullet}$ which precedes the rate limiting step (an isomerization with an associated rate constant k_f) is purposely left unspecified. This step is generally sufficiently slow so that the initial binding step corresponds to a rapid pre-equilibrium defined by an apparent association constant K_B (2).

When [R] is in large excess over [P], this scheme is analogous to the Michaelis–Menten hypothesis in enzyme kinetics and leads to a similar equation for the rate of formation of RPo.

$$v = \tau^{-1} = \frac{\partial[\text{RPo}]}{\partial t} = k_f \frac{K_B\,[R]}{1 + K_B\,[R]} \qquad 2$$

3.1 Principle of the assay (3, 4)

E. coli RNA polymerase is a very processive and accurate enzyme. At a given promoter, transcription begins by the formation of a phosphodiester bond between the first and the second nucleoside triphosphates complementary to the template (positions +1 and +2 by definition). If the next nucleotide is not provided, the aborted product is released from the stable open complex and synthesis of new products resumes. Under initial velocity conditions the number of moles of oligonucleotides released per unit time is directly proportional to the concentration of open complex RPo formed at this time (3).

In practice, the RNA start sequence must be known. A dinucleotide primer, A, which has a better affinity for the open complex than an ordinary nucleotide, is generally provided. A ribonucleoside triphosphate, B, is used as the second substrate. The concentration of A is large with respect to the corresponding apparent Michaelis constant (to maximize the fraction of open complexes which is operative), and the concentration of B is in the range of its apparent Michaelis constant (a compromise between maximal sensitivity and maintenance of initial velocity conditions for long incubation times). Two variants of the assay have been proposed:

(1) B is an α-^{32}P-labelled ribonucleoside triphosphate, complementary to the third position of the message. The labelled product is separated by ascending chromatography and counted (4) (cf. *Figure 1A*).

(2) In place of a radioactive labelling, a fluorescent probe can be used, UTP-γANS (5). In this substrate molecule, fluorescence emission is partially quenched due to the stacking of the base on to the naphthalene ring. Synthesis of the abortive product leads to the liberation of free ANS pyrophosphate, which has a quantum yield higher than the original substrate (6) (*Figure 1B*).

Three types of experiments are performed. First, the enzyme is incubated with the DNA promoter fragment so that equilibrium is reached for the formation of the open complex. Reaction is started by the addition of the two substrates, and stopped at various times by chelation of the essential Mg cation. Product release, followed as a function of the incubation time allows characterization of the enzymatic reaction (maximal turn over number, dependence on the concentration of each substrate and when the concentration of enzyme is varied, stoichiometry for the formation of a productive complex). Second, under optimal conditions derived from the first set of data, the time of incubation of RNA polymerase with the promoter is varied, to probe the rate of open complex formation at various concentrations of [R]. Third, the residence time of RNA polymerase within stable complexes is probed under poly[dI-dC] or poly[dA-dT] challenge.

We explain below how those experiments can be performed using as a reporter the incorporation of a radioactive B substrate within the abortive product. Modifications required for an optimal use of the fluorescence assay are given in section 3.6.

3.2 Steady-state assays : radioactive incorporation of an α-^{32}P-labelled NTP into the abortive product

Protocols are detailed for assays performed on the *lac*UV5 promoter. In this case, the abortive product generated corresponds to the tetranucleotide ApApUpU

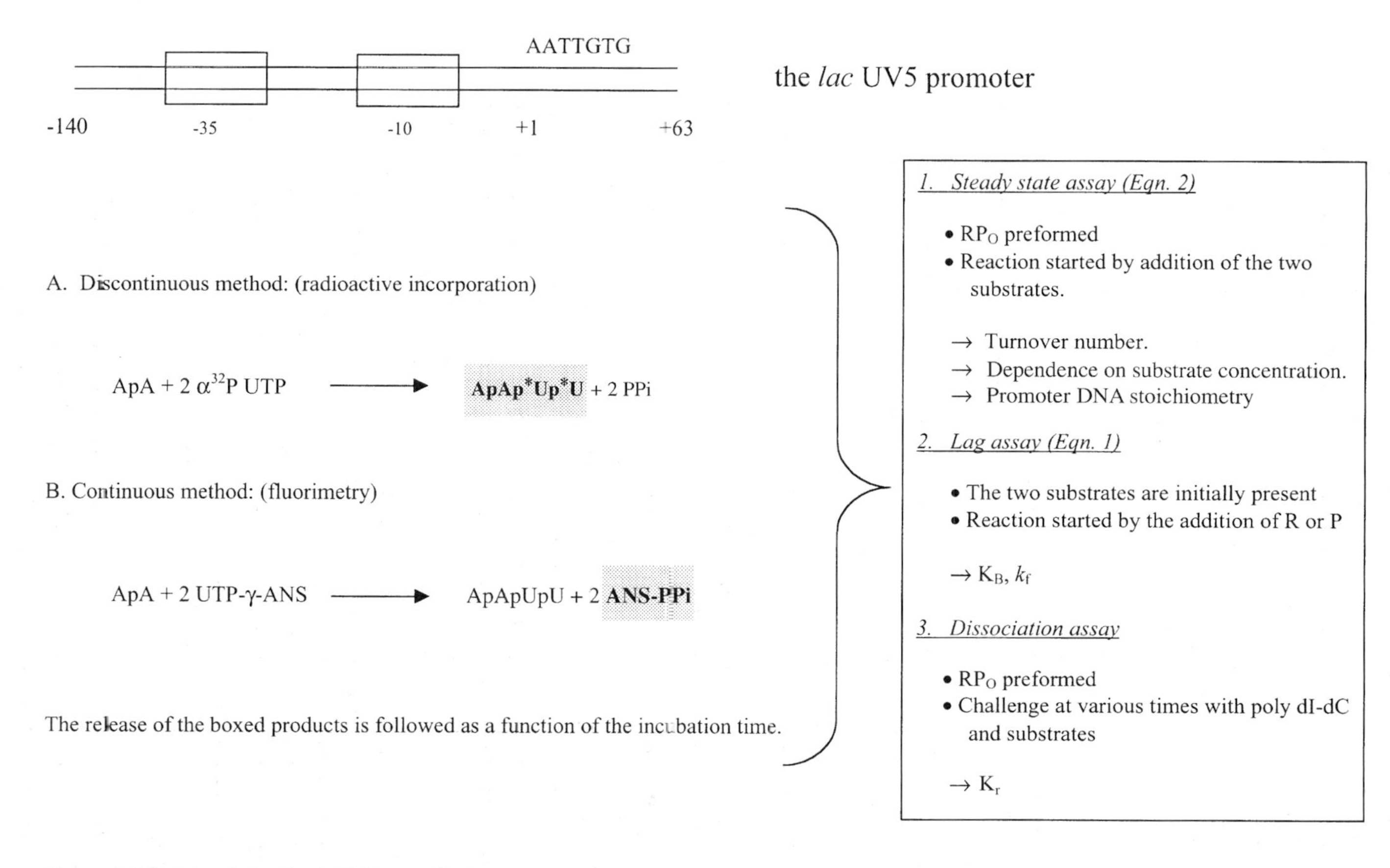

Figure 1 Principle of abortive initiation method.

(see *Figure 1*). It is separated from the radioactive UTP substrate by ascending chromatography.

Protocol 2

Abortive initiation

Equipment and reagents

- Assay buffer final composition: 0.04 M Tris-HCl, 0.10 M KCl, 0.01 M $MgCl_2$, 1 mM DTT, 0.1 mg BSA (DNase and RNase free), pH 8.0. A tenfold concentrated buffer can be stored frozen, DTT being added just before use
- *E. coli* RNA polymerase is isolated by the procedure of Burgess and Jendrisak (7) as modified by Lowe *et al.* (8). The concentrations are determined by spectrophotometry using an extinction coefficient *E*1% of 6.2/cm at 280 nm. A stock solution of ≈ 5 μM is prepared in the assay buffer before starting the experiment.
- The 203 bp fragment containing the *lac*UV5 promoter is prepared and its concentration measured as detailed in *Protocol 1*. Its final concentration in all assays will be fixed and adjusted between 1 nM and 3 nM. If measurements are performed on plasmids, two plasmids have to be prepared, one containing the *lac*UV5 fragment as an insert and another one for control experiments lacking just this insert. All experiments are now differential measurements. The plasmids should have the same superhelical density, since the degree of supercoiling of the template might affect the occupancy of the *lac*UV5 promoter and certainly affect the thermodynamic constants characterizing the association of the enzyme with its various targets on the plasmid. Specifications for this particular type of experiments are given in (9).
- A ten times stock solution for the substrates is prepared in the assay buffer. For routine assays, the final concentrations are 0.5 mM for ApA, 0.04 mM for UTP, [α-^{32}P]UTP being added to 400–1000 c.p.m./pmol. Absence of significant contamination by UMP, UDP and GTP (GTP being the next nucleotide to be incorporated into the message after UTP at this promoter), has to be checked by analytical means—for example as described in (3).
- Material for ascending chromatography. Before starting an experiment prepare 3 MM paper chromatograms. Strips 1-cm wide and 10-cm long are cut. Mark the origins and pre-spot them with a solution of 0.1 M EDTA. Ten of these strips can be attached to one cylindrical Plexiglas cover for two quart (U.S.) jars, which will contain the developing solvent (water-saturated ammonium sulfate–isopropyl alcohol, 18 : 80 : 2 by volume). This set of strips—each strip corresponding to one time point—is arranged horizontally on a blotting paper close to the water bath where the incubation will be performed. The thermostat is set at the desired temperature.

Assay

1. The various stock solutions are maintained on ice before the experiment begins (do not incubate the *E. coli* RNA polymerase and DNA fragments together at this temperature).
2. Twenty minutes before the start of the experiment, mix enzyme and promoter within the water bath, and incubate the other solutions.

Protocol 2 continued

3 At time zero, substrates are added, 20-μl aliquots are removed every 5 min and spotted at the origin of a Whatman 3 MM strip. Controls include a substrate dilution and a sample run without the DNA fragment for the longest incubation time.

4 Ascending chromatography is performed for a few hours, until the front has migrated about 15 cm. The ApApUpU product migrates with an R_F of 0.05 (at low UTP concentration the trinucleotide is also observed with a lower migration).

5 After localization of the spots corresponding to the UTP substrate and of the products on the dried chromatograms, the strips corresponding to each peak are cut, placed into scintillation vials containing water and counted, using the Cerenkov effect. Alternatively, a phosphorimager can be used.

6 The quantity of product released at each incubation time is expressed as the number of nmoles of UTP incorporated per nmole of promoter. A linear regression analysis allows determination of the initial rate of synthesis per min (v_i). This test is also used to check that, within experimental error, product accumulates linearly from time zero (ensuring that formation of the binary RPo complex was fully completed at the time origin).

3.3 Analysis of the enzymatic reaction

When required, the concentrations of both the A (ApA) and B (UTP) substrates can be independently varied. Dependence of v_i on those parameters usually fits well to hyperbolic functions corresponding to an equilibrium-ordered addition mechanism: all the substrates binding steps are at equilibrium in the steady-state; ApA binds first and in doing so, it creates for UTP a binding site competent for catalysis (3. 10). The corresponding apparent Michaelis constants for those substrates can then be derived, as well as the maximal turn over number of the reaction.

At fixed concentrations of substrates, the concentration of enzyme can also be varied, allowing establishment of a titration curve. The point of equivalence yields the fraction of active RNA polymerase within the preparation (which rarely exceeds 60%). Deviation from an irreversible reaction allows, in some cases, an estimate of an overall association constant, K_e, for the formation of the final binary complex in the presence of the two substrates.

The simplest account of those observations consists in assuming that a single catalytically competent complex is formed, which initiates transcription at a rate corresponding to the experimentally determined value of v_i. The probability of forming the corresponding open complex, initiating at +1, or the promoter occupancy, is then set equal to 1. This hypothesis is not always valid[a]. At *lac* or at *gal*, two overlapping promoters compete for the binding of R (6, 11, 12). At *malT* several binary complexes initiating at the same nucleotide can be revealed by independent methods (13). Inactive binary complexes can further lower this occupancy (14). Independent probing approaches, performed on an homogeneous

[a] As the ionic strength is decreased, the competition exerted by non-specific binding of *E. coli* RNA polymerases, in particular at the ends of the DNA fragments is considerably increased. Maintain ionic strengths higher than 0.1 M and/or use longer fragments to limit promoter occlusion by non-specific DNA binding.

set of mutated promoter sequences can be used to establish a correlation between a decreased promoter strength and a decreased occupancy of the promoter by a kinetically competent complex. In simple cases, an activator such as the cAMP–CRP complex is expected to fully populate one open complex. In this case, the corresponding turn over number is taken as reflecting full occupancy in the presence or absence of the bound activator (9, 12).

3.4 Lag assay. Quantitative characterization of the 'on' process

For *E. coli* RNA polymerases, elongation rates of the genetic message are rather slow, of the order of 50 nucleotides per second (15). The rate of synthesis of the message can only be limited at the initiation step if this process is clearly slower than the clearance time for one active RNA polymerase leaving the corresponding promoter. Such is generally the case *in vitro*: those processes are then slower than 0.1/s. They can be easily followed, in this case, after a simple modification in the steady-state protocol.

Stock solutions of enzyme DNA fragments and substrates are the same as above (in fact, those experiments are better done in parallel with those of the previous section), but this time the reaction is initiated by the addition of the enzyme solution. If this component is in large excess over the total concentration (P_T) of the promoter fragment, formation of the kinetically competent complex builds up exponentially during incubation with a characteristic time τ. For a reaction completely displaced in scheme 1 towards the formation of RPo one has:

$$\frac{[\mathrm{RPo}]}{[\mathrm{P_T}]} = 1 - e^{-t/\tau} \qquad 3$$

Accumulation of product y(t) takes place in direct proportion to the concentration of (RPo) present at a given time, or by integration of equation (3):

$$y(t) = v[\mathrm{RPo}]_t t = v_i[t - \tau(1 - e^{-t/\tau})] \qquad 4$$

After a time t equivalent to 4τ, this process occurs at the constant rate, v_i, assessed in Section 2. Extrapolation of this linear release of product back to the abscissa yields τ. A linear least-square analysis of the asymptotic behaviour is first performed and then the ratio $q = y(t)/v_i t$, a dimensionless quantity, is fitted with *Equation 4* to yield a more refined estimate of τ and to check the single exponential character of the process (16). Several programs are now available for personal computers that will do least-square fitting and allow one to enter the equations and the parameter constraints, such as Origin (Microcal Software Inc.), FigP (Biosoft), Scientist (Micromath), Magestic for Excel (Logix Consulting, Inc.), Kaleidagraph (Synergy Software).

Changing the concentration of the enzyme and repeating the experiments allows one to check the hyperbolic dependence of τ^{-1} on the concentration [R] of active enzyme (as determined in Section 2). Quantitative methods of fitting used for Michaelian kinetics can be used to determine the initial slope of the dependence of τ^{-1} on [R], ($K_B k_f$ in *Equation 1*),and the asymptotic rate at high enzyme concentrations k_f.

It should be again emphasized that those rate constants are only phenomenological and adjusted to a two-step process. They are all measured in the presence of the two substrates, which can significantly shift equilibria towards the active species. However, they provide significant constraints for the establishment of an overall multistep process for promoter recognition (16). For example, the 'on' rate constants can be directly compared with those obtained by other methods —like the potassium permanganate probing of single-stranded regions of the promoter DNA—which can also be adapted for kinetic purposes (17, G. Orsini, personal communication).

3.5 Residence time of RNAP at the promoter. Dissociation rate constants (16)

At a given temperature, complexes preformed between RNA polymerase and promoter are prepared as in Section 2, incubated for 10–20 min at the required temperature, and then challenged by a competing polynucleotide poly d(A-T) or poly(dI-dC). Since the dissociation process occurs at a rate that is independent of the enzyme concentration, a single RNAP concentration is used well above the point of equivalence. The competitor, the binary complexes, and the assay buffers are prewarmed for 10–20 min at the temperature chosen. The competitor is added at zero time. At various times ranging from 2 min to several hours, 15-μl aliquots of this solution are removed and mixed with 25 μl of the standard assay buffer. After 10 min of reaction at 37 °C, 25-μl portions of each assay are spotted on 3 MM paper strips and processed as described above. The decay of the active open complexes as the incubation time is increased is plotted on a logarithmic scale and, from the slope of this line, a dissociation rate, k_d, is obtained. In order to ensure that this rate of dissociation is strictly equal to the reverse rate constant k_r in *Scheme 1* several controls have to be performed:

(1) Dissociations are usually very slow for strong promoters. Check that the incubation times cover indeed a range of the order of $2 \cdot k_d^{-1}$.

(2) The degree by which the competitor actively displaces the open complex cannot be *a priori* predicted. Check that a change in the concentration of the polynucleotide does not affect the observed decay. Under those conditions, also check carefully whether a single rate constant accounts for the whole profile. This is one of the best tests for the unicity of the final binary complex.

(3) Measurements described in Sections 2 and 3 are performed in the presence of the two substrates, A and B. In this case, the substrates are added for a short time for each incubation point. In fact, the residence time could be affected by their final concentrations as well as by the time during which the assay is performed, since the open complex is expected to be stabilized by

the binding of A and B, and could also be affected by the abortive initiation process. It is therefore worthwhile to examine the dependence of k_r on |A| and |B| concentrations.

(3) An overall equilibrium constant K_e can be derived from those data if the final isomerization step can be assumed to be an elementary step between two species only. Then:

$$K_e = \frac{K_B k_f}{k_r} \quad 5$$

(4) Self consistency between this estimate and the response curve obtained in Section 3.2 should then be checked.

3.6 The fluorescent assay

As shown in *Figure 1B*, monitoring of abortive initiation is performed, in this case, by following the replacement of UTP-γ-ANS by ANS-PP_i as turnover proceeds. Preparation of UTP-γ-ANS is described in reference [5]. (Note that the purification protocol for the UTP-γ-ANS can be improved by first separating UTP and UTP-γ-ANS on a DEAE ion exchange column operating at atmospheric pressure; the nucleotide is then separated from its analogue by the use of reverse-phase chromatography using supports such as pepRPC from Amersham Pharmacia Biotech). As shown by Yarbrough *et al.* (5) the best differential absorbance spectrum between the two derivatives occurs at 360 nm. It is then convenient to follow the release of the pyrophosphate analogue by spectrofluorimetry (λ_{exc} = 360 nm; λ_{em} = 430 nm).

A great advantage of this method is that mixing can be efficiently operated via a stopped-flow device (6) and that product accumulation can be followed continuously. In order to do so, an injection device using two Hamilton syringes is adapted to a fluorescence cell of 600 μl thermostatically controlled within the spectrofluorimeter (originally a Jobin Yvon JY 3D spectrofluorimeter). Steady-state assays are performed by mixing the final binary complex and substrates, and determinations of on-rate constants by mixing the enzyme and the other components. The same procedures are used to analyse the data. The sole modifications which have to be introduced into the previous protocol are an absolute calibration of the number of nmoles formed (as described in ref. 5), and a careful elimination of fluorescent impurities in all the buffers. The only drawback of the method is that the signal is not specific for the initiating mRNA sequence but records any event resulting from the cleavage of the second nucleoside triphosphate. It is therefore unsuitable when several promoters compete for initiation by RNA polymerase.

3.7 Changes in mechanism for open complex formation and/or escape

There are several drastic assumptions which are introduced in the example given here. Although they are justified for a strong *E. coli* promoter, they must be re-examined in each specific case.

(1) Irreversibility in the formation of the open complex. The expression for the on-rate constant should be modified when k_r becomes significant with respect to k_f according to the equation:

$$\tau^{-1} = k_r + k_f \frac{K_B [R]}{1 + K_B [R]}$$

(2) It has been assumed that the binding step is extremely fast with respect to isomerization. Otherwise K_B should be written as an apparent Michaelis constant or:

$$K_B = \frac{k_1}{k_{-1} + k_f} \qquad 6$$

(the subscript 1 refers to the rate constants for the first step).

(3) Multistep process versus a two-step process. This point has been treated at length in references (16, 18–20).

(4) Changes in the rate limiting step. It is assumed that the formation of the open complex, binding of the substrates, transitions to the initiating and then to the elongating complexes are independent events, one of those steps being clearly rate-limiting (in our example, open complex formation). For other *E. coli* RNAP–promoter combinations, the situation might be clearly different (21, 22). A contrasting example can also be found in the case of promoter recognition and mRNA initiation by RNA polymerase from the T7 bacteriophage. In this particular case, the open complex is formed at a rate faster than 1/s, but is kinetically unstable, and requires stabilization by the incoming nucleoside triphosphates (23). The formulation given here for *E. coli* RNA polymerase-promoter complexes is inadequate. Instability of the open complex appears to be a way of achieving promoter specificity and to make those T7 promoters extremely sensitive to changes in initiating NTP concentration, a situation also encountered in the case of the ribosomal RNA promoters of *E. coli*, where a change in the concentration of NTP appears to regulate in part the growth rate dependence of rRNA transcription (22).

4 Elongation assays to study the interactions between DNA polymerases and their templates

4.1 Introduction

DNA polymerases can rapidly copy the DNA (up to 300–500 nt/s for T7 DNA polymerase) (24) with a high degree of fidelity (1 in 10^6–10^8 error for both *E. coli*'s Klenow fragment and T7) (24, 25) which cannot be explained solely by the efficiency of the correction mechanism involving the exonuclease site on the enzyme. The factors involved in the choice of the correct nucleotide have been studied extensively for a few well-known polymerases, the T7 DNA polymerases (24, 26), the Klenow fragment of *E. coli* DNA polymerase I (25, 27), and HIV reverse transcriptase (28, 29). Moreover polymerases are molecular motors, and are thus interesting models to help us understand the mechanism of translocation. A key question is whether the translocation energy comes from nucleotide hydrolysis or binding or both, and whether nucleotide binding causes a change in protein conformation resulting in forward motion or if it causes a shift in a pre-existing equilibrium between two registers in the DNA sequence (29).

In this section we outline some of the basic transient kinetic methods used to study the pathway of nucleotide incorporation; the reader should refer to references cited for additional applications of quench-flow kinetics to the study of DNA polymerases.

The steps involved in the elongation pathway for a DNA polymerase can be summarized by the following minimal scheme:

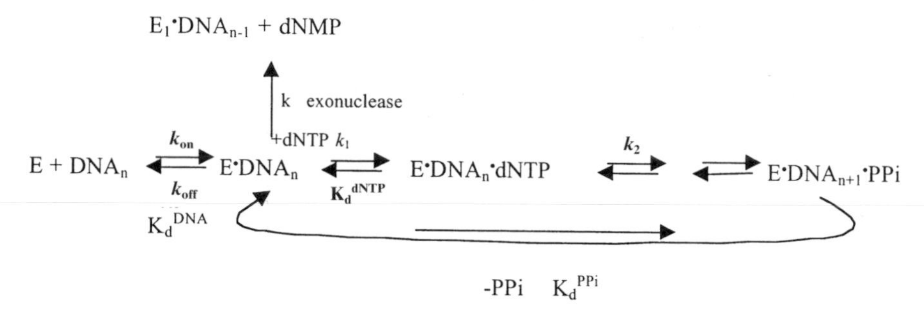

The binding of the nucleotide to the protein primer-template complex ($E^{\bullet}DNA_n$) is followed by a fast chemical step. In the case of the Klenow fragment and T7 DNA polymerases there is an additional step between nucleotide binding and chemistry, most likely a change in protein and/or DNA conformation, k_2 (31). Once the pyrophosphate moiety is released, the polymerase is ready to start a new cycle. The reverse reaction can also happen at physiological pyrophosphate concentrations (intracellular PPi concentration is about 150 μM, (32)) (25). In the case of reverse transcriptase, pyrophosphate is also a product inhibitor ($IC_{50} \approx 1$ mM) (32). If the wrong nucleotide is added, the 3′ end of the primer strand slides into the exonuclease active site where the mispaired nucleotide is cleaved off. However, the use of exo$^-$ mutant enzymes can simplify the study of the basic kinetic rates because one eliminates the possibility of the side reaction (33).

4.2 Principle of the assay

In order to study the sequence of events present in this pathway and to identify clearly the rate-limiting step one must use transient-state kinetic methods. The assay consists of using a rapid mixing quench-flow device to study the kinetics of single nucleotide incorporation to a short primer-template. By enabling one to measure directly the rate of most steps in the pathway this method can be used to determine, for example, if a given mutant lacks in binding specificity, in chemical activity or in misincorporation discrimination, or which point in the pathway is affected by the presence of an accessory protein. The methods used to study the kinetics of nucleotide incorporation by DNA polymerases have been developed by Stephen Benkovic (25) and Ken Johnson (26).

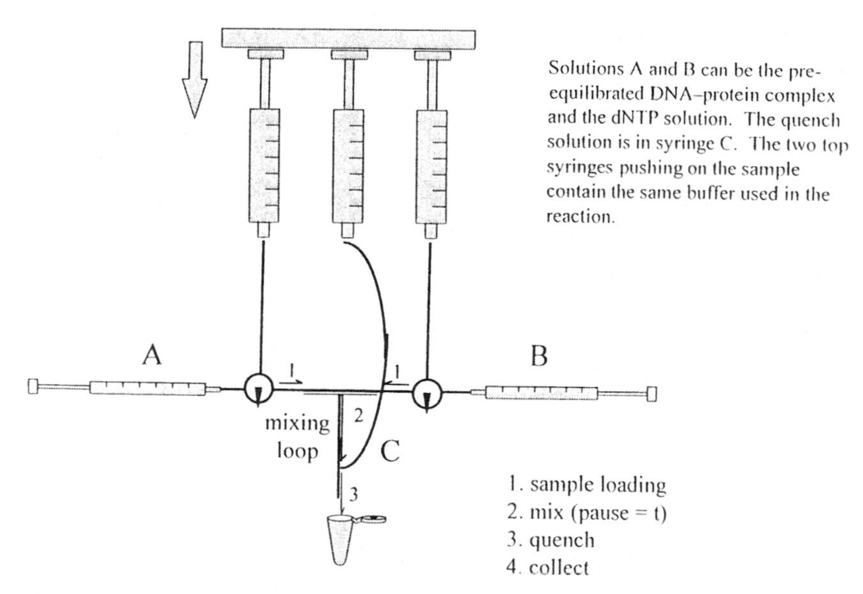

Figure 2 Schematic of mixing apparatus.

Current quench-flow devices, such as the one manufactured by Kin Tek, use only a small amount of material (15 μl per sample) because of the use of injection syringes, have a dead time of a few milliseconds and a temperature controlled chamber (*Figure 2*).

The addition of a nucleotide to the radiolabelled primer strand by the polymerase can be used to study the rate of the different steps in the reaction:

- The formation of an active enzyme–primer/template complex and its stoichiometry (k_{on}, K_d^{DNA}).
- The affinity of binding of the nucleotide substrate (K_d^{dNTP}).
- The maximum rate of incorporation (k_2)
- The nature of the chemical step
- The rate of product release (k_{off})

In many cases, the exonuclease pathway in the overall scheme can be neglected or bypassed by using an exo⁻ mutant. In the absence of pyrophosphate, the overall scheme reduces itself to a forward pathway leading to the extension of the primer:

$$E + DNA_n \underset{k_{off}}{\overset{k_{on}}{\rightleftarrows}} \underset{K_d^{DNA}}{E{\cdot}DNA_n} \underset{K_d^{dNTP}}{\overset{+dNTP\ k_1}{\rightleftarrows}} E{\cdot}DNA_n{\cdot}dNTP \overset{k_2}{\rightleftarrows} \rightleftarrows$$

Protocol 3a

Quench flow

Equipment and reagents

- TBE: 0.089 M Tris borate, 0.089 M boric acid, 0.2 mM EDTA, pH 8
- PhosphorImager (Molecular Dynamics)
- Rapid-mixing stopped flow apparatus
- Buffer specific to the reaction

Method

1. A quench-flow experiment begins with the preparation of two solutions, one with the polymerase complexed with a 5′ ^{32}P radiolabelled primer-template[a] and the other with the first nucleotide.
2. Both solutions should be in exactly the same buffer, which is specific for the system used.
3. A third syringe contains the quench solution, which can be either a highly concentrated EDTA solution or an acidic solution to denature the protein.
4. An equal volume (as small as 15 μl) of the two solutions is loaded, the mixing time is entered in the mixer's controller and the reaction is started (*Figure 3*).
5. Once the quenched sample (about 250 μl) is collected in an Eppendorf tube it is precipitated in ethanol and then resuspended in formamide loading buffer (40% formamide, 1× TBE, 0.05 % each of bromophenol blue and xylene cyanol). In the case of an acid quench the sample should go through a phenol–chloroform extraction, neutralized with a basic solution and then precipitated (26).
6. The size of the primer and template strands varies from 9/20 mer to 25/45 mer, depending on the polymerase. The product (*primer+1) can be easily separated from substrate (*primer) on a high percentage (about 16–20%, depending on the size of the DNA) denaturing acrylamide gel, and their quantities determined by using a PhosphorImager scanner and then a software such as ImageQuant (Molecular Dynamics) to determine the intensities of the bands on the gel. The intensity of the bands should be normalized for the total amount of radioactive material loaded on the gel. The amount of product formed is then:

$$\frac{I_P}{(I_P + I_S)} {}^*[S]_0 \qquad 7$$

Where I_P is the intensity of the product band, I_S is the intensity of the substrate band and $[S]_0$ is the concentration of substrate at the beginning.[b]

[a] The labelling reaction is carried out by using T4 polynucleotide kinase and [γ-^{32}P]ATP. If the DNA is an oligonucleotide which has been synthesized then it does not to be dephosphorylated at its 5′ end, otherwise calf intestine phosphatase is usually used for this step, but then it has to be inactivated and removed by phenol extraction. After labelling, the unincorporated ATP can be separated by using a chromatography column (Micro Bio-Spin 6 from BioRad, for example). The labelled primer and the cold template are then mixed in a 1 : 1 ratio and annealed by heating the mixture at 90 °C for 3 min, then placed in a 60 °C heating block and allowed to reach room temperature slowly.

[b] In an alternative method a radiolabelled nucleotide is added to a cold primer/template. The unincorporated nucleotides are removed by filter binding, and the amount of labelled primer is quantified by scintillation counting. Although this method is faster than the one described above, the noise in the data is greater since it is more difficult to normalize for the total amount of starting material.

4.2.1 Data analysis

In order to measure the rate of correct nucleotide incorporation, it is more convenient to keep primer-template and nucleotide concentration in excess. In most cases when one follows the amount of product (primer + 1) as a function of time after the addition of a nucleotide which matches the position of the primer, what is observed for DNA polymerases is a curve consisting of a burst of product formation followed by a slower increase (*Figure 3*) observed only in the case where the rate limiting step follows the chemical step. The amplitude of the burst phase is proportional to the amount of active complex [E–S]. The slower phase corresponds to the steady state rate of incorporation limited by the rate of product release.

The burst can then be fitted to the following equation:

$$Y = Ae^{(k2t)} + k_{ss}t \qquad 8$$

where A is the amplitude of the burst phase, k_2 is the rate of product formation and k_{SS} is the steady-state rate. The off-rate for the DNA, k_{off}, is obtained by dividing k_{SS} by the amplitude, A. The rate of binding of polymerase to the primer/template, k_{on}, can be determined indirectly from the values for K_d and k_{off}.

Using burst analysis, by carrying out this experiment at different nucleotide and primer/template concentrations, one can determine the K_d for binding of each of these species. In the first case one will follow a change in the rate of nucleotide incorporation (k_{obs}) that can be plotted as a function of nucleotide

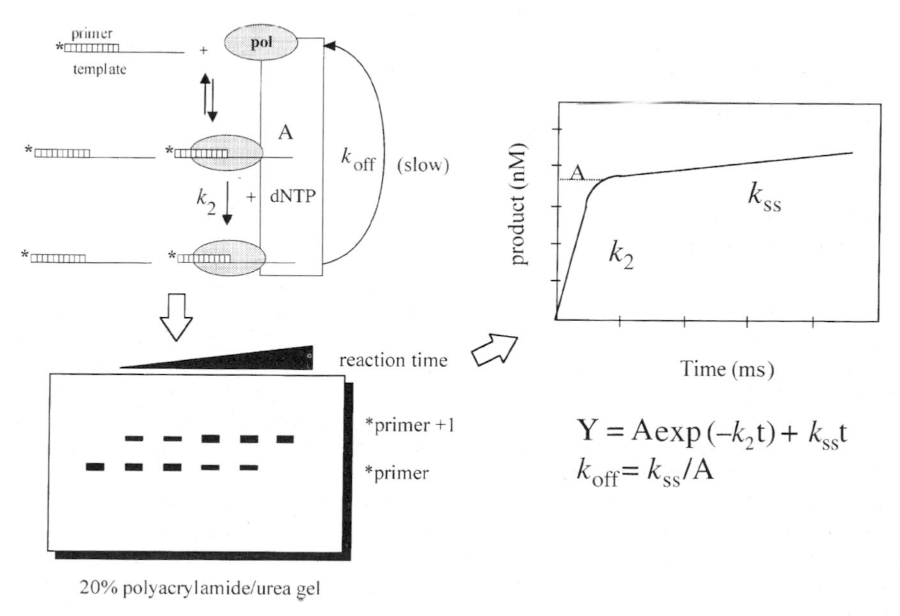

Figure 3 Kinetics of single nucleotide incorporation.

concentration and then fit to a hyperbola equation (*Equation 9*) to determine the K_d and the maximum incorporation rate, k_2.

$$k_{obs} = \frac{k_2 K_d [dNTP]}{1 + K_d [dNTP]} \quad 9$$

Two assumptions are made in this case, that the dNTP concentration is equal to the free dNTP, and that the initial binding of dNTP is a rapid equilibrium (26).

In the second case one can plot the amplitude of the burst as a function of DNA concentration. The resulting plot should be fitted to an hyperbola (*Equation 10*) if the DNA concentration [S] is always in excess of polymerase concentration [E],

$$[ES] = \frac{[E]_0 [S]_0}{[S]_0 + K_d} \quad 10$$

otherwise it should be fitted to a quadratic equation

$$[ES] = \frac{(K_d + [E]_0 + [S]_0) - \sqrt{(K_d + [E]_0 + [S]_0)^2 - 4[E]_0 [S]_0}}{2} \quad 11$$

This experiment measures the K_d for binding of DNA in an active complex only.

This methodology can then be expanded to explore other essential steps of the overall scheme, as described in (25) for *pol*I DNA polymerase, or in (24) for T7 DNA polymerase. By the use of isotope trapping experiments, pulse-chase and pulse-quench experiments, it can be shown that the chemical step of the reaction is not rate-limiting, in the first case. Conformational changes associated to the kinetic constant k_2 can then be studied by the approaches developed in other sections of this book.

By adding an excess of pyrophosphate the reaction can also be driven backwards and the corresponding rates can be measured by following the formation of labelled dATP from 3′ labelled primer (25,32).

A complete sequential scheme can then be proposed for the incorporation of a single correct nucleotide. Then, adding an incorrect nucleotide, one can study the differences in the pathway, evaluate the rate at which the primer template, mismatched at the 3′ end of the primer can be moved over to the exonuclease site, and measure the rate of excision (25,33).

4.3 Sequential incorporation of several dNTP during a run-off assay

The precise analysis of the pathway leading to the incorporation of a single deoxynucleotide within a given primer has already led to a better understanding of the mechanism of DNA polymerization. It does not necessarily mirror exactly the same elementary process when several substrates are added, allowing polymerization to proceed in a run-off manner. It is well known that the addition of a deoxynucleotide analogue, which matches the next position of a template markedly stabilizes the enzyme hybrid complex. As a consequence, several crucial rate constants of the scheme depicted above can be markedly modified. Such is indeed the case for HIV-1 reverse transcriptase interacting with a DNA–DNA template during elongation. The residence time of the enzyme is drastically modified as the substrate is bound in a pre-steady-state complex, or in a kinetically competent complex (34). It is therefore recommended that assays involving single nucleotide addition be complemented by run-off assays. For those types of kinetics a convenient gel assay has been described (35, 36) allowing identification and quantification of the *n* consecutive products and of the initial substrate as the incubation time is varied. The rates of formation and of disappearance of the various intermediate products can be simulated for a mechanism involving *n* consecutive events. In the case of HIV-1 reverse transcriptase a reasonable agreement was found between those values and the ones obtained on isolated steps of incorporation (24, 37). A similar approach has also been useful to analyse the gradual transition between an initiation mode of reverse transcription by HIV-RT and a classical elongation mode (38, 39).

Acknowledgements

B. S. is a recipient of an EMBO long term fellowship.

References

1. Nice, E., Fabri, L., Hammacher, A., Andersson, K and Helman, U. (1993). *Biomed. Chromatogr.*, **7**, 104.
2. McClure, W. R. (1980). *Proc Natl Acad Sci USA* **77**, 5634.
3. McClure, W. R., Cech, C. L., and Johnston, D. E. (1978). *J. Biol. Chem.* **253**, 8941.
4. Johnston, D. E. and McClure, W. R. (1976). In *RNA polymerase* (ed. Losick, R. and Chamberlin, M. J.), p.413. Cold Spring Harbor Laboratory Press, Cold Spring Harbor.
5. Yarbrough, L. R., Schlageck, J. G., and Baughman, M. (1979). *J. Biol. Chem.* **254**, 12069.
6. Bertrand-Burggraf, E., Lefèvre, J.-F., and Daune, M. (1984). *Nucl. Acids Res*, **12**,1697.
7. Burgess, R. R. and Jendrisak, J. J. (1975). *Biochemistry* **14**, 4634.
8. Lowe, P. A., Hager, D. A., and Burgess, R. R. (1979). *Biochemistry* **18**, 1344.
9. Malan, T. P., Kolb, A., Buc, H., and McClure, W. R. (1984). *J. Mol. Biol.* **180**, 88.
10. Menendez, M., Kolb, A., and Buc, H.(1987). A new target for CRP action at the *malT* promoter. *EMBO J.* **6**, 4227.
11. Mousse, R. E., Di Lauro, R., Adhya, S., and de Crombrugghe, B. (1977). *Cell* **12**, 847.
12. Herbert, M., Kolb, A., and Buc, H. (1986). *Proc. Natl. Acad. Sci. USA* **83**, 2807.
13. Tagami, H. and Aiba, H. (1998). *EMBO J*, **17**, 1759.
14. Melançon, P., Burgess, R. R., and Record, Jr., M. T. (1982). *Biochemistry* **21**, 4318.
15. Record, Jr., T. M., Reznikoff, W. S., Craig, M. L. McQuade, K. L., and Schlax, P. J. (1996). In Escherichia coli *and* Salomonella: *cellular and molecular biology* (ed. Neidhardt, F. C.), Vol. 1. p. 792. ASM Press, Washington, D. C.
16. Buc, H. and McClure, W. R. (1985) *Biochemistry* **24**, 2712.
17. Sasse-Dwight, S. and Gralla, J. D. (1989). *J. Biol. Chem.* **264**, 8074.
18. Roe, J.-H., Burgess, R. R., and Record, Jr., M. T. (1985). *J. Mol. Biol.* **184**, 441.
19. Tsodikov, O. V., Craig, M. L., Saecker, R. M., and Record, Jr., M. T. (1998). *J. Mol. Biol.* **283**, 757.
20. Buckle, M., Pemberton, I. K., Jacquet, M.-A., and Buc, H. (1999). *J. Mol Biol.* **285**, 955.
21. Ellinger, T., Behnke, D., Bujard, H., and Gralla, J. D. (1994). *J. Mol. Biol.* **239**, 455.
22. Gaal, T, Bartlett, M. S., Ross, W., Turnbough, Jr., C. L., and Gourse, R. L. (1997). *Science* **278**, 2092.

23. Villemain, J., Guajardo, R., Sousa, R., (1997). *J. Mol. Biol.* **273**, 958.
24. Johnson, K. A. (1993). *Annu. Rev. Biochem.* **62**, 685.
25. Benkovic, S. J. and Cameron, C. G. (1995). *Methods Enzymol.* **262**, 257.
26. Johnson, K. A., (1995). *Methods Enzymol.* **249**, 38-61.
27. Carroll, S. S. and Benkovic, S. J. (1991). *Nucl. Acids Mol. Biol.* **5**, 99.
28. Kati, W. M., Johnson, K. A., Jerva, L. F., and Anderson, K. S. (1992). *J. Biol. Chem.* **267**, 25988.
29. Reardon, J. (1993). *J. Biol. Chem.* **268**, 8743.
30. Guajardo, R. and Sousa, R. (1997). *J. Mol. Biol.* **265**, 8.
31. Dahlberg, M. E. and Benkovic, S. J. (1991). *Biochemistry* **30**, 4835.
32. Arion, D., Kaushik, N., McCormick, S., Borkow, G., and Parniak, M. A. (1998). *Biochemistry* **37**, 15908.
33. Eger, B. T., Kuchta, R. D., Carrol, S. S., Benkovic, P. A., Dahlberg, M. E., Joyce, C. M., and Benkovic, S. J. (1991). *Biochemistry* **30**, 1441.
34. Kati, W. M., Johnson, K. A., Jerva, L. F., and Anderson, K. S. (1992). *J. Biol. Chem.* **267**, 25988.
35. Boosalis, M. S., Petruska J., and Goodman M. F. (1987). *J. Biol. Chem.* **262**, 14689.
36. Richetti, M. and Buc, H. (1996). *Biochemistry* **35**, 14970.
37. Reardon, J. E. (1993). *J. Biol. Chem.* **268**, 8743.
38. Lanchy, J.-M., Keith, G., Le Grice, S. F. J., Ehresmann, B., Ehresmann, C., and Marquet, R. (1998). *J. Biol. Chem.* **273**, 24425.
39. Thrall, S. H., Krebs, R., Wöhrl, B. M., Cellai, L., Goody, R. S., and Restle, T. (1998). *Biochemistry* **37**, 13349.

Chapter 19

Kinetics of DNA interactions surface plasmon resonance spectroscopy

Björn Persson
Biacore AB, Rapsgatan 7, 75450 Uppsala, Sweden

Malcolm Buckle
Institut Pasteur, 25, rue du Docteur Roux, 75724 Paris Cedex 15, France

Peter G. Stockley
Department of Biology, University of Leeds, Leeds L52 JT, UK

1 Principles of surface plasmon resonance technology

1.1 Introduction

The procedures for biomolecular interaction analysis described in this chapter are based on biosensor technology, which offers an opportunity to study the interaction between an immobilized partner, the ligand, and an analyte molecule in solution. The interaction is monitored continuously and the binding events are studied in real time, with no need for labelling of any component. The technology is generally applicable to the study of any molecular structures which express affinity for one another, including proteins, nucleic acids, carbohydrates, lipids, low molecular weight substances, but also large particles such as vesicles, bacteria and cells. The wealth of information provided by this technology includes qualitative data such as binding specificity, but also quantitative aspects of binding like stoichiometry, affinity, concentration and kinetics and are amenable to detailed analysis (for general technical reviews see refs. 1, 2).

1.2 Principle of detection

The principle of detection relies on the optical phenomenon of surface plasmon resonance (SPR), which is sensitive to the optical properties of the medium close to a metal surface. Various configurations of instrumentation may be used to exploit the SPR effect. *Figure 1* shows the detection unit of the instrumental set up used for the Biacore (Biacore AB) system. The detection system of this SPR monitor essentially consists of a monochromatic, plane-polarized light source

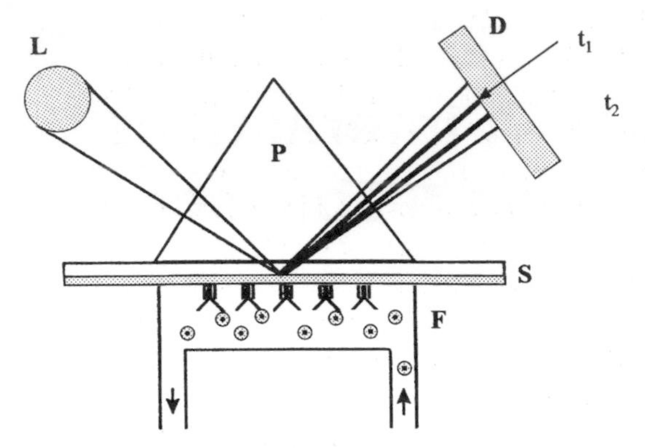

Figure 1 Surface plasmon resonance detection unit: L = light source, D = photodiode array, P = prism, S = sensor surface, F = flow cell. The two dark lines in the reflected beam projected on to the detector symbolize the light intensity drop following the resonance phenomenon at time = t_1 and t_2. The line projected at t_1 corresponds to the situation before binding of analytes to the ligand on the surface and t_2 is the position of resonance after

and a photodetector that are connected optically through a glass prism. The thin metal film positioned on the prism is in contact with the sample solution. The fan-shaped light beam incident on the back side of the metal film is totally internally reflected on to the diode-array detector.

SPR occurs when light incident on the metal film couples to oscillations of the conducting electrons, plasmons, at the metal surface. These oscillations create an electromagnetic field commonly referred to as the evanescent wave. This wave extends from the metal surface into the sample solution on the back of the surface relative to the incident light. When resonance occurs, the intensity of the reflected light decreases at a sharply defined angle of incidence, the SPR angle, which is dependent on the refractive index penetrable by the evanescent wave close to the metal surface. The SPR angle is dependent on several instrumental parameters, e.g. the wavelength of the light source and the metal of the film. When these parameters are kept constant, the SPR angle shifts are dependent only on changes in refractive index of a thin layer adjacent to the metal surface. If the SPR angle shift is monitored over time, a gradual increase of material at the surface will cause a successive increase of the SPR angle, which is detected as a shift of the position of the light-intensity minimum on the diode array.

The SPR angle shifts obtained from different proteins in solution have been correlated to surface concentration determined from radio-labelling techniques and found to be linear within a wide range of surface concentrations. The instrument output, the resonance signal, is indicated in resonance units (RU); 1000 RU correspond to a 0.1° shift in the SPR angle and for an average protein this corresponds to a surface concentration change of about 1 ng/mm^2.

1.3 Surface chemistry

The sensor chip consists of a glass substrate on to which a 50-nm thick gold film has been deposited. The gold film is covered with a long-chain hydroxyalkanethiol, which forms a monolayer at the surface. This layer serves as a barrier to prevent analyte coming into contact with the gold, but also as attachment points for the carboxymethylated dextran chains that create a hydrophilic surface to which proteins or other molecules can be covalently coupled.

In a typical kinetic analysis, one of the two interacting partners, commonly referred to as the ligand, is immobilized on the dextran. For example, substances containing primary amines can be immobilized after activation of the dextran matrix with carbodiimide/*N*-hydroxysuccinimide. Several other procedures are also possible for the immobilization of ligands, e.g. DNA is commonly synthesized with a 5′-biotin label and captured on streptavidin that is immobilized on the sensor chip by amine coupling (See Section 2.2).

An important feature of the sensor chip construction is that the dextran layer extends typically 100 nm out from the surface and thereby a small volume, penetrated by the evanescent wave, is created, where the analyte–ligand interaction can be studied.

The specificity of the analysis is determined by the identity and biological activity of the ligand.

1.4 Liquid handling system

Sample solutions containing analytes are injected into a running buffer that flows continuously over the sensor surface. A typical run in kinetic analysis consists of: (i) establishment of a baseline response in running buffer; (ii) an association phase, i.e. injection of sample in running buffer; and (iii) the dissociation phase, i.e. wash out with running buffer (See Section 2.3). For kinetic analysis of rapid interactions it is important that the flow system can perform virtually dispersion-free injections into the detector flow cell, otherwise small deviations in analyte concentration will perturb the results.

The mass transfer of analyte molecules from the bulk of sample solution in the flow cell to the ligand on the sensor surface is another crucial parameter to consider in kinetic analysis. Fast mass transfer is obtained by: (i) high linear velocity of the sample solution, thereby diminishing the thickness of the unstirred layer near the surface; (ii) a thin flow cell, facilitating a non-perturbed flow regime. It is further important to keep the ligand concentration low, otherwise depletion of analyte near the surface will perturb the effective analyte concentration and thus the kinetic constants will be underestimated.

A good baseline stability is required for kinetic analysis, e.g. when the dissociation phase of a stable complex is measured, but also in equilibrium analysis to obtain a reliable steady-state value during sample injection.

Biacore instrumentation has two (Biacore X) or four (Biacore 2000, Biacore 1000 and Biacore 3000) flow cells to address the corresponding numbers of sensor surfaces. In the multi-channel detection mode these surfaces can be addressed in serial mode, which makes it possible to run the same analyte solution over the two or four surfaces, respectively, offering an opportunity to use one surface as a reference whose signal may be subtracted from the signal of the active surface.

1.5 Data handling

1.5.1 General considerations

Interactions between molecules in Biacore are monitored over time and presented as a sensorgram, i.e. a plot of RU vs. time (s). To be able to extract kinetic parameters from the sensorgram, the curve is analysed in terms of a defined mathematical model. The Langmuir isotherm is one example of a model that describes the interaction between a monovalent analyte A and a ligand B with n independent binding sites. The fraction of the available sites on B occupied by a (θ_B) is simply:

$$(\theta_B) = \frac{[AB]}{[nB_T]}$$

The concentration of analyte is assumed to be constant during the sample injection phase and zero during the dissociation phase. There are several criteria that must be considered for the design of a proper kinetic analysis of binding.

1.5.1.1 *Multivalent analytes*

It is important to avoid multivalent analytes in kinetic analysis of binding. Antibody–antigen interactions is one such example. In the study of this type of interactions it is preferred to immobilize the bivalent antibody and use the antigen as analyte. If the antibody is used as analyte some molecules may bridge over two antigens and give rise to an avidity effect, which is not taken into account in the simple Langmuir model.

1.5.1.2 *Mass transport*

Mass transport limited binding of analyte occurs in instances where the association rate is particularly elevated and the diffusion rate of the non-immobilized molecules is not especially fast. In this situation the interaction of the free molecules with the immobilized ligand may deplete a layer of solvent immediately surrounding the immobilized ligand such that the rate-limiting step for association now becomes the rate or repletion of this layer from the bulk solvent. Fortunately, this type of effect can be identified fairly easily by injecting the analyte at different flow rates. Generally, a change of the flow from 15 μl/min to 75 μl/min should not affect rate constants more than 5–10% if no mass transfer limitation is at hand. Two practical solutions to avoid mass transfer limited binding are first to use a low immobilization density, and second to use relatively elevated flow rates ($>$ 20 μl/min). It is generally recommended to perform kinetic measurements with R_{max} values of 100–1000 RU, although lower values may be used if high quality of the signal can be attained.

1.5.1.3 *Recapture*

Recapture of analyte may occur when an analyte dissociates from the immobilized surface. The analyte may then subsequently be recruited to an adjacent molecule. The effect of this will be to decrease the numerical values ascribed to derived dissociation constants. This effect is important in systems with fast kinetics of binding and is pronounced when the immobilization level is high. Thus, in practice, the density of immobilization must be adjusted so that the dissociation rate is independent of immobilized ligand density. If this proves difficult then one can, with considerable error, extrapolate to infinite dilution. Finally, free ligand may be included during the dissociation phase in order to calculate an affinity for the competitor and thus allow an estimation of a true dissociation constant.

1.5.1.4 Fast association rates

Association rate constants for proteins are normally in the range 1×10^4–1×10^7 l/M.s and values $\approx 1 \times 10^6$ can generally be determined in Biacore. Faster rate constants can be determined by the use of non-linear regression analysis procedures such as those provided with the BiaEvaluation software 3.0. In this procedure the parameter k_t, the mass transfer constant, is fitted to the experimental curve along with the other parameters. It is related to the generally accepted mass transfer coefficient k_m by the expression $k_t \approx 10^9 \times \mathrm{Mw} \times k_m$. Values for k_t are typically of the order of 10^8 for proteins of molecular mass 50–100 kDa.

1.5.1.5 Heterogeneity

It is important to point out that heterogeneity is highly prevalent in preparations of biological molecules. This may be induced by adopted purification schemes or can be inherent in the population of native molecules. The immobilization procedure used to couple the ligand to the surface may also introduce heterogeneity.

1.5.1.6 Heterogeneity of material

If the time of injection of the analyte is varied, this should not affect the shape of the dissociation curve, provided that the interacting species are homogeneous. When heterogeneity of either ligand or analyte is at hand, different complexes with different properties will form at the sensor surface and their relative abundance will vary with time. This will affect the shape of the dissociation curve. The first remedy in this case is to consider an alternative immobilization strategy, either chemical procedures for covalent attachment or capturing, e.g. by antibodies or by other means.

1.5.2 Equilibrium analysis and kinetic binding

If a simple Langmuir model is assumed, thus the association equilibrium constant K_a is simply

$$(\theta_B) = \frac{K_a\,[A]}{1 + K_a\,[A]}$$

Since $K_a = 1/K_d$ then the equilibrium dissociation constant, K_d, can be derived from a plot of the steady-state response R_{eq} vs. A (concentration of analyte). However, to obtain a reliable K_d from R_{eq} responses, several concentrations of analyte in the range 0.1–10 times K_d must be determined. It may take many hours or even days to determine steady-state responses, while kinetic analysis is performed within minutes at concentrations of 10–1000 times $K_{d.}$ an alternative to this then is to use a kinetic analysis that determines the association and dissociation rate constants for the interaction. The analyte that is injected across an immobilized ligand on a surface should, after an infinite time, arrive at an association equilibrium giving a signal R_{eq}, and the resonance signal R at time t during this process following injection at $t = 0$ when $R = R_0$ should, in simple instances, obey the expression:

$$R(t) = R_0 + (R_{eq} - R_0)\bullet[1 - e^{(-k_{obs}\cdot t)}].$$

Similarly, for the dissociation of the bound protein, assuming that the bound molecule completely dissociates from the immobilized ligand.

$$R(t) = R_0 + (R_{eq} - R_0)\bullet e^{(-k_{off}\cdot t)}].$$

Consequently, the observed reaction rate k_{obs} for the interaction is given by:

$$k_{obs} = k_{on}[A] + k_{obs}.$$

There is thus a linear relationship between the value for k_{obs} and the total concentration of analyte [A]. The value for k_{obs} can be obtained from a direct fit of the association phase or by linear regression of a semi-log plot. It then follows that linear regression analysis of the dependence of k_{obs} on [A] allows calculation of k_{on} and k_{off}. The value thus obtained should be identical with k_d/k_a since $k_{off} = k_{on}K_d$.

1.5.2.1 Helper plots

There are several ways to establish whether the simple Langmuir model is applicable to an interaction or not. When this model is valid, all of the four plots listed below should be linear (they are all available in BiaEvaluation software provided with Biacore instruments (Biacore AB):

- Injection phase: $\ln(R_{eq}—R)$ vs. T; $\ln(dR/dt)$ vs. T
- Dissociation phase: $\ln(R/R_0)$ vs. T ; $\ln(R/R_0)$ vs. T.

For example, a plot of $\ln(dR/dt)$ *vs.* T gives a convex (upward) shape of the curve when the Langmuir model is affected by mass transfer limited binding and a concave shape for a more complex interaction.

2 SPR-assays of protein–DNA interactions

2.1 Experimental design

Traditional approaches to the assay of protein–DNA interactions rely on alterations of the properties of the partners of the complex as the interaction proceeds. For example, DNA fragments become retarded relative to unbound fragments in gel electrophoresis, or are retained on nitrocellulose filters by virtue of their interaction with proteins in filter-binding assays, or the accessibility of functional groups in the molecule becomes reduced in footprinting assays (3). These are necessarily techniques based on the separation of equilibrium mixtures or competition assays, neither of which is ideal for accurate quantification. In some cases, the molecules concerned, usually the proteins, have unique spectral properties that permit analysis of the kinetics and equilibrium distribution. However, such situations are the exception rather than the rule.

The SPR biosensors provide an excellent tool to overcome the various limitations of these traditional techniques. In many cases, it is possible to obtain stoichiometric, kinetic, thermodynamic and affinity data very rapidly with relatively small amounts of material. The first task in setting up such SPR experiments is to design the assay for the interaction of interest. For practical purposes, it is often easiest to immobilize the target DNA fragments, via biotin-streptavidin binding, and address these ligands with solutions of the test protein analyte. There are many advantages to this approach. DNA molecules are flexible and are naturally part of very large polymers which approximate to the immobilized state. There should therefore be little effect from the immobilization, although this is not so true when the DNA is used as a template for polymerases (see Section 3). Regeneration of the sensor chip is also straightforward, requiring only removal of the protein analyte which is easily achieved without denaturing the DNA ligands.

The first example that we will use describe the results obtained for the interaction of the *Escherichia coli* methionine repressor, MetJ, and a series of target DNA molecules encompassing operator sites. MetJ is a well characterized protein structurally, biochemically and genetically (4–6). It is a small homodimeric protein ($\approx$ 24 kDa), which is activated for operator binding via an electrostatic mechanism (7). The independent binding of two molecules of the co-repressor, S-adenosylmethionine (AdoMet), to each repressor dimer results in an increased positive electrostatic potential in the region of the DNA phosphodiester backbone, increasing the affinity of the repressor for its operator DNA by 100-fold. The operator regions comprise between two and five tandem repeats of an 8 bp DNA sequence, consensus = dAGACGTCT, known as a 'met box'. Each holorepressor dimer interacts with a single met box and operators are bound cooperatively with respect to protein concentration. The minimal repression complex *in vitro* comprises two repressor dimers bound to two tandem, consensus met boxes, (*Figure 2*).

2.2 DNA immobilization

Sequence-specific DNA–protein complexes characteristically have very rapid forward rate constants, reflecting the functional need to identify the DNA target against the large background of competing non-specific sites. In order to avoid problems caused by mass transport limitations, the level of immobilized DNA ligand should be relatively small ($\approx$ 100–300 RU). This can conveniently be achieved by sequential injections of the appropriate low concentrations of

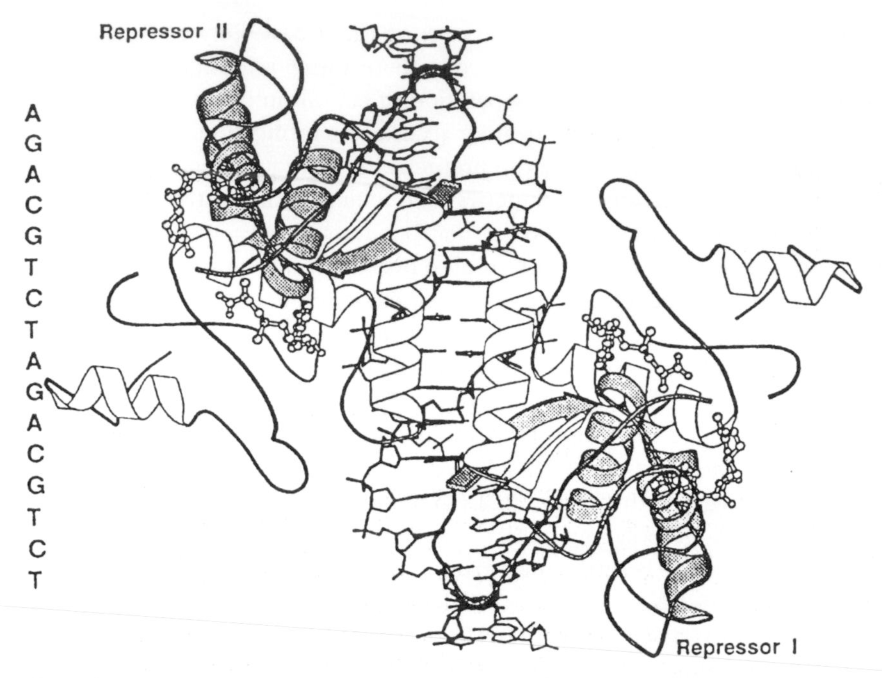

Figure 2 Representation of the X-ray crystal structure of the minimal MetJ repression complex. MetJ dimers are shown as ribbon representations, the two met box operator (sequence shown to left) as a framework model and AdoMet as ball and stick models.

biotinylated DNA ligands for differing contact times. Immobilization is usually linear during this period (*Figure 3*).

2.3 Met J sensorgrams

The MetJ–operator interaction can be followed using SPR detection, as utilized by Biacore (7). Target DNA molecules encompassing a minimal, consensus operator site (16 bp, two tandem met boxes, see *Figure 2*) were immobilized via a unique biotin tag at one end (5′) to streptavidin sensor chips (*Figure 3*). The holorepressor was then passed over the surface and the binding curves (sensorgrams) recorded (*Figure 4*). Dissociation in the absence of co-repressor was too rapid to permit kinetic analysis, therefore, dissociation was monitored in experiments in which consecutive injections of sample had been made; the first contained MetJ and AdoMet, and the second only AdoMet. Under these conditions, binding was clearly both repressor concentration- and AdoMet-dependent, as expected. Furthermore, binding saturation occurred at levels consistent with the expected stoichiometry of one MetJ dimer per met box, which is based on crystallographic and biochemical data.

It was found that although the minimal system involves the interaction of a dimer of repressor dimers on the DNA the association phase fits well to a simple 1 : 1 binding model. The dissociation phase in the presence of AdoMet, however, is clearly biphasic. We ascribed the biphasic dissociation behaviour to the in-

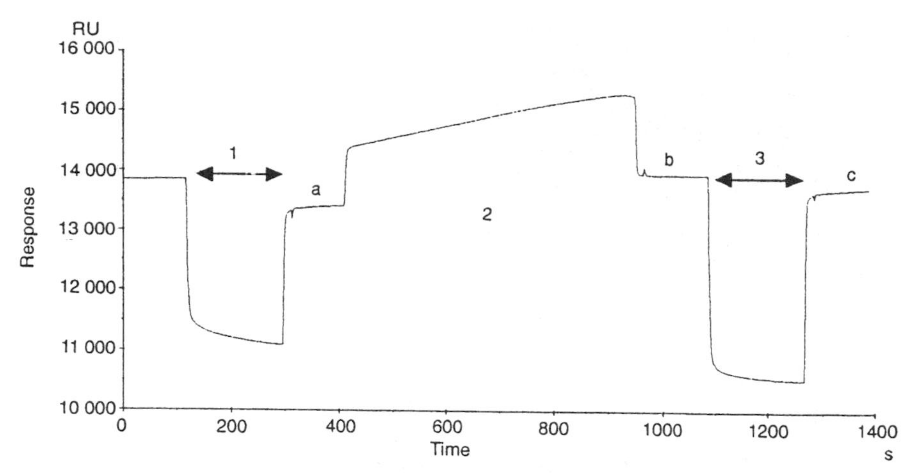

Figure 3 Sensorgram showing the different steps of the immobilization of DNA to an SA-5 sensor chip. Phase 1: drop in signal as NaOH is passed over the sensor chip surface. A new baseline signal (a) is reached at the end of this injection due to the loss of loosely bound streptavidin from the surface. Phase 2: injection and binding of biotinylated DNA to the surface. At the end of this injection an increased baseline signal is seen (b), due to the bound DNA. Phase 3: injection of SDS to remove loosely bound DNA from the surface of the sensor chip. At the end of this injection a slightly lowered baseline signal is established (c). If the baseline signal, a, is subtracted from this new reading, c, the amount of DNA immobilized onto the surface can be calculated. Taken from (8).

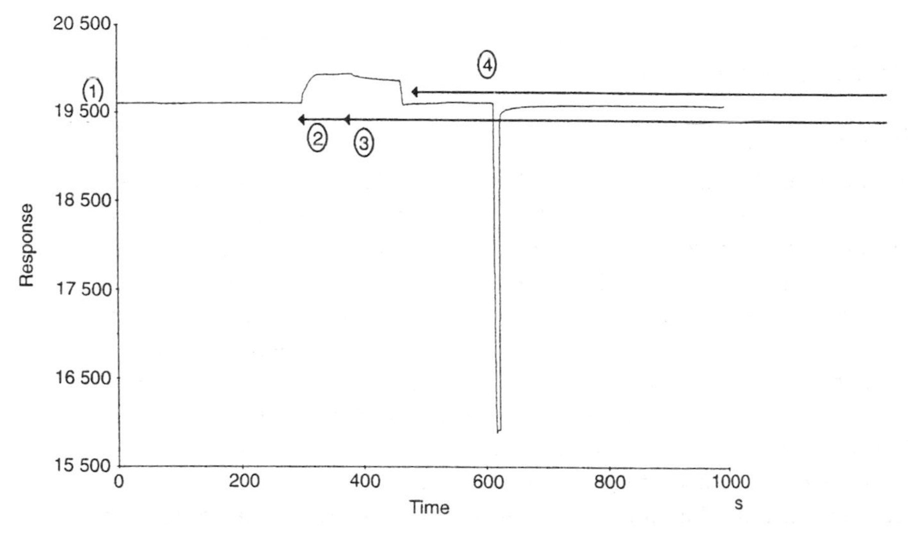

Figure 4 A typical sensorgram showing MetJ binding to and dissociating from the sensor chip surface and the regeneration of the sensor chip. Phase 1: initial baseline signal when running buffer passes over the sensor chip. Phase 2: first injection containing MetJ and effector. The change in signal is due to the binding of MetJ/effector to the chip surface. Phase 3: second injection of running buffer containing effector alone, allowing the dissociation in the presence of effector to be measured. Phase 4: regeneration of the sensor chip surface by the injection of SDS. Taken from (8).

creasing effect of operator rebinding as the concentration of free holorepressor increased. This seemed likely because of the very high forward rate constant ($\approx 10^6$/M.s). Apparent rate constants for the dissociation reaction could be calculated for two roughly linear regions of the sensorgram: the initial pseudo-first-order dissociation of the repression complex and, at later times, the slower phase consisting of both dissociation and rebinding. The calculated apparent equilibrium constant for the interaction (the ratio of the apparent kinetic constants) was in the nanomolar range, close to those determined by more traditional binding assays such as nitrocellulose filter binding and gel mobility shifts (3).

2.4 Kinetics

Data were evaluated using the linear kinetic evaluation package in the Bia-Evaluation software, as described previously (7) and as described above. Briefly, in order to calculate a value for k_{ass}, data from sensorgrams were used to plot the binding rate (dR/dT) as a function of the response (R) for each repressor concentration (*Figure 5*). The slopes (k_s) of these resulting straight lines were then calculated. By plotting these values of k_s versus concentration, k_{ass} was obtained as the slope of the resultant straight line. The value for k_{diss} was obtained from the gradient of the plot of $\ln(R_0/R_n)$ vs. $(t_n - t_0)$, where R_0 and R_n are the response values obtained along the dissociation curve at times t_0 and n.

2.5 The sensor chip surface

The very high forward rate constant observed for the binding of MetJ to its operator is typical for sequence-specific DNA–protein interactions. Care must be exercised in such cases to ensure that sensible values are obtained for the apparent rate constants (see later). Two phenomena are relevant to this discussion. The first is the rate of mass transfer of the analyte into the region of the immobilized target. As outlined above this can be increased by increasing flow rates and analyte concentrations, but the principal approach to avoid problems

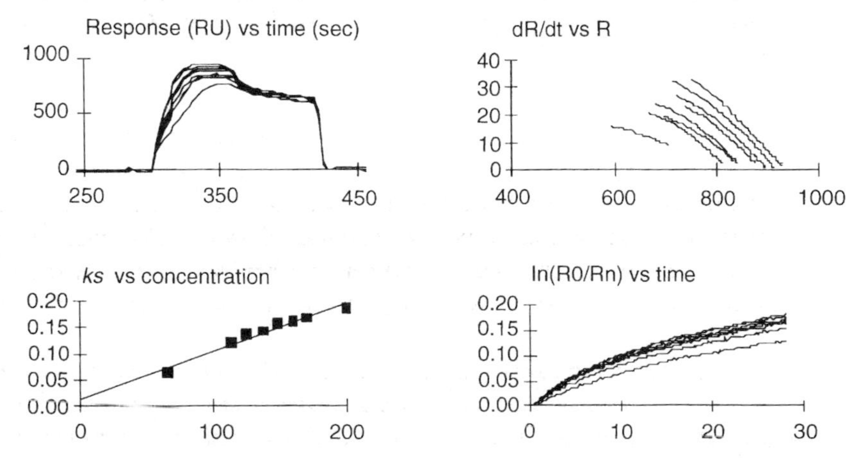

Figure 5 Calculation of kinetic constants using the linear kinetic evaluation software. Plots used for calculating the kinetic constants. Taken from [8].

caused by mass transport effects is to work at very low target concentrations, which means that the rate of analyte binding is not determined by simple mass transfer kinetics.

The second problem which has been reported is that the presence of a dextran support matrix results in hindered diffusion of the analyte within the layer, dramatically slowing the rate of target binding (9, 10). This proposal has been tested directly for the met repressor and other interactions using sensor surfaces with different lengths of dextran support or where the streptavidin is attached directly to the hydroxyalkane chain on the gold surface. There are no discernible effects from the presence of the dextran surface (11), suggesting that hindered diffusion is not a significant problem under these conditions.

These arguments emphasize the need to consider the nature of the sensor surface when designing binding assays. The ease of immobilization of the target molecule, for example, can be influenced by the net charge of the surface. Oligonucleotides are repelled by the negatively charged carboxymethyl groups on some surfaces. Such simple problems can be overcome by immobilization in high salt concentrations, etc. There is now a wide range of surfaces available from the manufacturer and customized surfaces can be designed in consultation.

3 SPR assays of polymerase action

3.1 Kinetics of RNA polymerase interactions with immobilized DNA

Many biological processes are essentially irreversible and play on an initial recognition process controlled by a binding constant followed by an isomerization event described by a rate constant. In these situations simple binding models cannot be applied. A suitable example of this is the formation of a transcriptionally active complex (RPo) between a promoter sequence on a DNA molecule (P) and its cognate RNA polymerase (R). in the case of certain prokaryotic promoters (12–14) this reaction can be described by a two-phase process going through a closed intermediate complex (RPc):

$$R + P \underset{k_{-1}}{\overset{k_1}{\rightleftharpoons}} RP_c \underset{k_{-2}}{\overset{k_2}{\rightleftharpoons}} RPo$$

Scheme 1

Since SPR will not differentiate between RPc and RPo the equilibrium is really between free R and immobilized P and RPc. Thus, in conditions where [R] >> [P] the on rate, k_{on} is given by the expression :

$$k_{on} = \frac{k_1 k_2 [R]}{k_1 [R] + k_2 + k_1} = \frac{k_2 K_B [R]}{1 + K_B [R]}$$

where

$$K_B = \frac{k_1}{k_{-1} + k_2}$$

in a similar fashion since k_{off} is k_{-2} the overall association constant K_a is not simply k_{on}/k_{off}. How then to access K_a in this situation? Since

$$K_a = \frac{k_1 k_2}{k_{-1} + k_{-2}}$$

then during the injection period there will be the transition:

$$\frac{K_B[R]}{1 + K_B[R]} \rightarrow \frac{K_a[R]}{1 + K_a[R]}$$

Thus, the initial association phase allows access to K_B but then this relatively rapidly moves into a situation which is governed by K_a. In this latter case the rate of increase in signal will be simply

$$R_{(t)} = R_{(max)} \bullet (1 - \exp^{-t/\tau})$$

where

$$\tau = \frac{1}{k_2} \bullet \left(1 + \frac{1}{K_B[R]}\right)$$

Since $k_2 >>> k_{-2}$ for the majority of promoters, then the relaxation time τ can thus be obtained from a direct fit of the association phase the values for k_2 and K_B calculated by linear least-squares analysis of the linear dependence of τ on the concentration of RNA polymerase (15). The dissociation phase following an injection period that allows full transition to open complex formation (i.e. a period in excess of $1/k_2[R]$) will give access to the dissociation constant for the open complex of k_{-2} since the decrease in response R_t is given by: $R_t = R_{eq}\exp^{(-k_d \bullet t)}$). The calculation of these values is extremely delicate given the present architecture of the Biacore apparatus. The value for the overall association rate constant $k_{obs} = k_2 K_B$ for many promoters is in excess of 10^6/M.s. Accessing the very initial phase of the association phase in which one can calculate K_B directly is often very difficult because of bulk refractive index effects and the scarcity of data points over such a short period. The dead time response of the machine (>1 s, dependent to an extent upon the flow rate) precludes access to initial events in binding studies involving rapid rearrangements. Furthermore, as is seen in the next section, delineating accurately that part of the curve where the values may be obtained is somewhat ambiguous and requires careful external controls.

This example, however, serves to illustrate that provided that care is taken and that meaningful comparisons are available with other assays on the system studied then Biacore analysis allows the acquisition of data pertaining to the binding of soluble molecules to immobilized ligands.

3.2 DNA immobilization

The next section describes the use of prepared streptavidin surfaces. One can, however, immobilize streptavidin on the standard carboxyl methylated dextran surfaces (CM5) by a simple procedure that we insert here to illustrate the programming methodology used in the Biacore system.

Protocol 1

Coupling of streptavidin

Equipment and reagents

- Streptavidin from Pierce resuspended in 0.22 μm filtered distilled water to a final concentration of 5 mg/ml. This preparation may be stored at 4 °C for up to 3 months.
- HBS buffer (10 mM HEPES, pH 7.4, 150 mM NaCl, 3.4 mM EDTA, 0.005% Biacore surfactant)
- EDC and NHS purchased from Biacore as lyophilized powders, resuspended in 0.22 μm filtered distilled water to final concentration of 100 mM each
- 1 M ethanolamine hydrochloride (pH 8.5), purchased from Biacore, stored at 4 °C
- Biacore or Biacore 2000 instrument running at 20°C with rack D in the first position and rack A in the second position
- HBS buffer, HEPES (10 mM) pH 7.4, NaCl (150 mM), EDTA (3.4 mM), surfactant P20 (0.005%)
- Sensorchip surface CM5 research grade installed in the Biacore apparatus and pre-primed with HBS buffer
- Reaction vials for the Biacore (small, plastic = 7 mm; medium, glass = 16 mm and large, glass = 2 ml) were purchased from Biacore

Method

1. Prime the apparatus with HBS buffer.
2. The thawed EDC solution in an Eppendorf tube with the top removed is placed in rack 1 position a1 (r1a1).
3. The thawed NHS solution in an Eppendorf tube with the top removed is placed in r1a2.
4. Streptavidin (5 mg/ml, 50 μl) in an Eppendorf tube with the top removed is placed in r1a3.
5. 2 ml of filtered (0.2 μm) distilled water is placed in a large glass vial in r2f7.
6. 1 M, sodium acetate buffer (1 ml, pH 4.5) is placed in a large glass vial in r2f3.
7. 1 M ethanolamine (200 μl) is placed in a large tube in r2f4
8. Two small clean plastic vials are placed in positions r2a1 and r2a2.
9. An empty large glass vial is placed in r2f5
10. The following method is programmed into the Biacore or Biacore 2000, checked for errors and run.

```
DEFINE Aprog mixing
FLOW       20
TRANSFER r1a1  r2a1   50        !rack1a1 = EDC
TRANSFER r1a2  r2a1   50        !rack1a2 = NHS
MIX            r2a1   50        !rack2a1 = EDC/NHS mix
TRANSFER r2f7  r2a2   200       !rack2f7 = distilled water
TRANSFER r2f7  r2a2   290       !rack2f7 = distilled water
```

Protocol 1 continued

```
TRANSFER r2f3  r2a2  5           !rack2f3 = 1 M acetate pH 4.5
TRANSFER r1a3  r2a2  5           !rack1a3 = streptavidin (5 µg/ml)
MIX          r2a2  50
END
DEFINE APROG bind
CAPTION activation
FLOW         20
*            INJECT     r2a1      50
-0:20        RPOINT     EDC/NHS -b
*            INJECT     r2a2      30
-0:20        RPOINT     streptavidin
*            INJECT     r2f4      35!Ethanolamine (1 M)
-0:20        RPOINT     ethanolamine
15:00        RPOINT     bound
TRANSFER r2a1  r2f5  100         !rack2f5 = waste tube
TRANSFER r2a2  r2f5  450         !rack2f5 = waste tube
END
MAIN
FLOWCELL         1
APROG        mixing
FLOWCELL         1
APROG        bind
END
```

N.B. Only surface 1 of the sensor chip will be activated using this method.

Protocol 2

Immobilization of the DNA

Equipment and reagents.

- End-biotinylated DNA suspended in HBS buffer to 10 µg/ml
- Biacore or Biacore 2000 instrument running at 20 °C with rack D in the first position and rack A in the second position
- HBS buffer
- Streptavidin activated sensor chip surface CM5 research grade installed in the Biacore apparatus and pre-primed with HBS buffer

Method

It is advisable to run this sensorgram manually so as to monitor the degree of immobilization.

1 Select a surface pre-treated with streptavidin.

2 flow HBS buffer at 20 µl/min across the surface.

Protocol 2 continued

3 Inject the DNA solution across the surface, set the baseline to the point of injection and monitor the change in RU during the injection phase. Ideally, between 20 and 50 RU of DNA should be immobilized (see later).

4 Wash the surface with a 50 μl injection of 1 M NaCl in filtered (0.2 μm) distilled water.

5 Allow the surface to equilibrate in HBS buffer to a stable baseline, the difference in RU between the beginning of the injection phase and the end of the wash period reflects the amount of DNA bound.

Table 1 DNA binding

Surface	DNA (RU)	mole DNA	[DNA] (μM)
1	226	2.2×10^{-15}	21.7
2	68	6.4×10^{-16}	6.5
3	18	1.7×10^{-16}	1.7
4	13	1.2×10^{-16}	0.8

In an experiment, four surfaces were immobilized with different concentrations of DNA. Using the empirically established relationship that 0.78 ng of DNA give a RU of 1000, and assuming that the dextran layer is 100-nm thick and presents a surface of 1 mm^2, then the effective volume is 0.1 nL and the apparent concentrations of DNA on each surface may be calculated

3.3 RNA polymerase binding to immobilized DNA

RNA polymerase is a large (465 kDa) multi-subunit enzyme that recognizes promoter regions containing a minimal recognition motif of two hexamer consensus sequences separated by an optimal 18-bp spacer region. The binding process is described in Section 3.1 and is clearly not a simple binding Langmuir isotherm. Four surfaces were immobilized with end-biotinylated DNA fragments (203 bp) containing the *lac* UV5 promoter to different degrees of immobilization and RNA polymerase (50 nM) was flowed across the surfaces at 20 μl/min at 37 °C in the appropriate buffer (10 mM HEPES pH 7.4, 150 mM NaCl, 10 mM $MgCl_2$, 0.005% P20, v/v). The resulting sensorgrams are shown in *Figure 6*.

It is somewhat surprising that each surface saturates to different extents and that the dissociation phase appeared to be different in each case. Because of the nature of the interactions involved (see Section 3.1) the evaluation procedure supplied with the Biacore could not be directly applied to the analysis of this data. If one looks at the dissociation phase it is clear that not only is it multiphasic (single-fit models do not give acceptable fit criteria) (*Figure 7*) but also that the predominant off-rate varies as a function of the DNA density.

Assuming that the true dissociation phase from an open complex is the slowest,

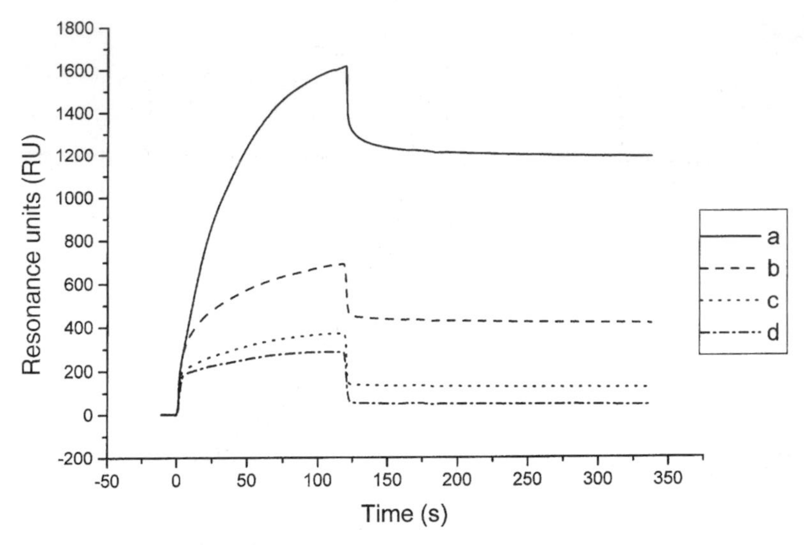

Figure 6 Sensorgrams showing RNA polymerase (50 nM) binding to different DNA densities on sensor chip surfaces. The amounts of (203 bp ds). DNA fragment bound at each surface was in resonance units (RUs): (a) 226 , (b) 68, (c) 18 and (d) 13.

then examination of the curves towards the end of the dissociation phase allows determination of a mono-exponential phase that is independent of DNA ligand density and of a value ($> 10^{-4}$/s) consistent with solution determinations for this interaction (*Figure 8*).

The association phase clearly cannot be assessed by a simple approach. As described in Section 3.1, a part of this curve contains information relating to the relaxation time (τ) for the transition to the stable open complex (whose k_{off} of 10^{-4}/s) has to be slower than the k_2 rate (*Scheme 1*, Section 3.1). a direct fit of the whole association phase using the simple expression $R = A \bullet (1 - \exp^{-t/\tau})$ (*Figure 9a*) gives a τ value of 61 s which is incommensurate with solution values of around 10 s determined under these conditions. This suggests that a large part of the association phase reflects interactions unrelated to the specific recognition of the promoter sequence or to the formation of the open complex (RPo, *Scheme 1*). It is impossible at the present moment to determine the nature of these parasite interactions and hence very difficult to account for them in experimental design or data analysis. Conversely, since we know that open complex formation has an overall on rate in excess of 10^6/M, then it is certainly rapid and thus is probably contained within the initial part of the association phase. a direct fit of this area of the sensorgram (*Figure 9b*) provides an assessment of τ as being around 9 s, i.e., in good agreement with expected values.

Obviously τ depends on the RNA polymerase concentration; however, as can be seen from *Figure 9b*, only a few data points are in fact accessible. Consequently, at higher RNA polymerase concentrations there will be a restricted region of the association phase that will provide meaningful data. Improvements in handling this problem await technological developments in SPR.

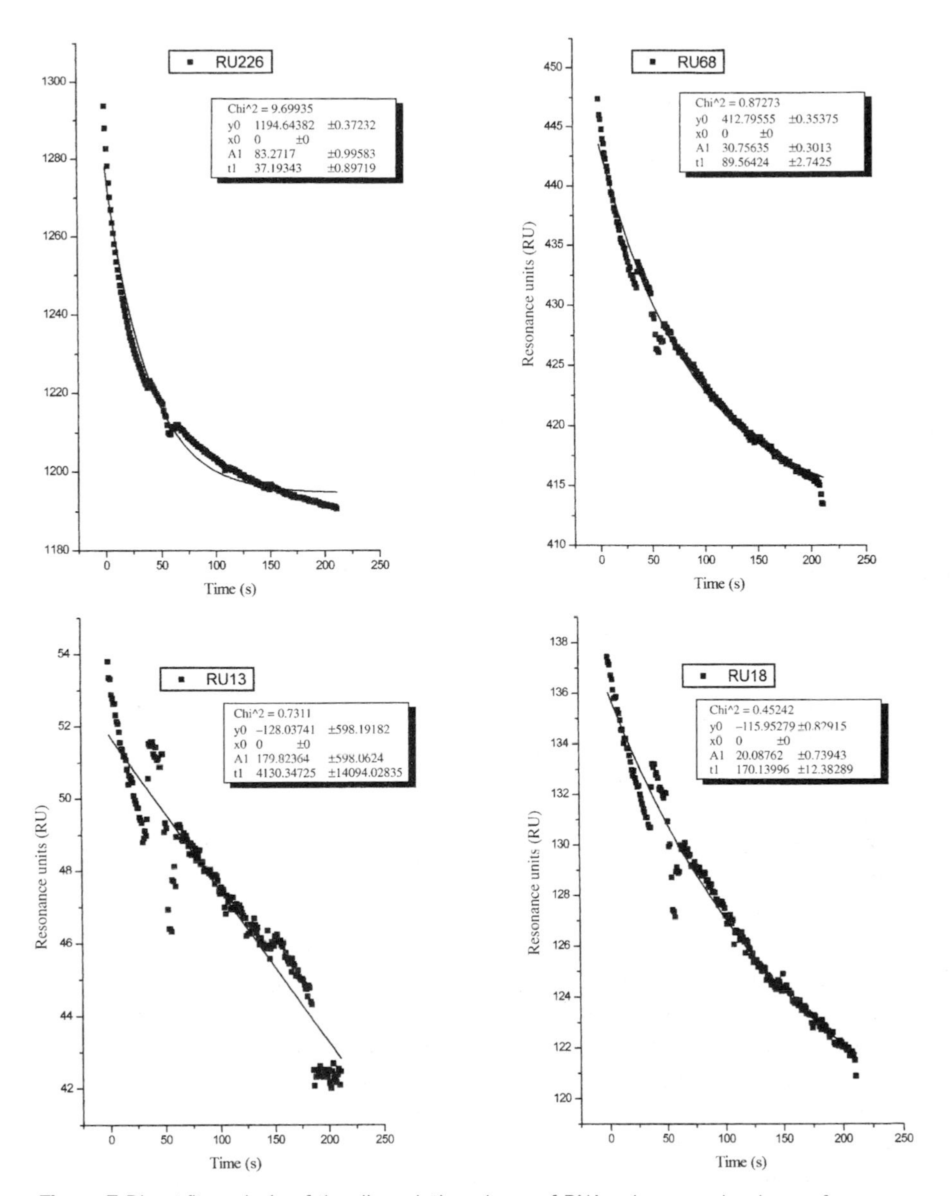

Figure 7 Direct-fit analysis of the dissociation phase of RNA polymerase leaving surfaces containing different concentrations of DNA. The analysis was carried out using a single exponential decay model, ignoring the initial bulk refractive index change subsequent to the end of the injection phase.

4 DNA hybridization

the efforts of the various genomic sequencing projects have produced enormous amounts of DNA sequence information. This is useful in basic biological and medical research but will also aid more practical applications of large diversity, such as in clinical diagnostics, the development of strategies for gene

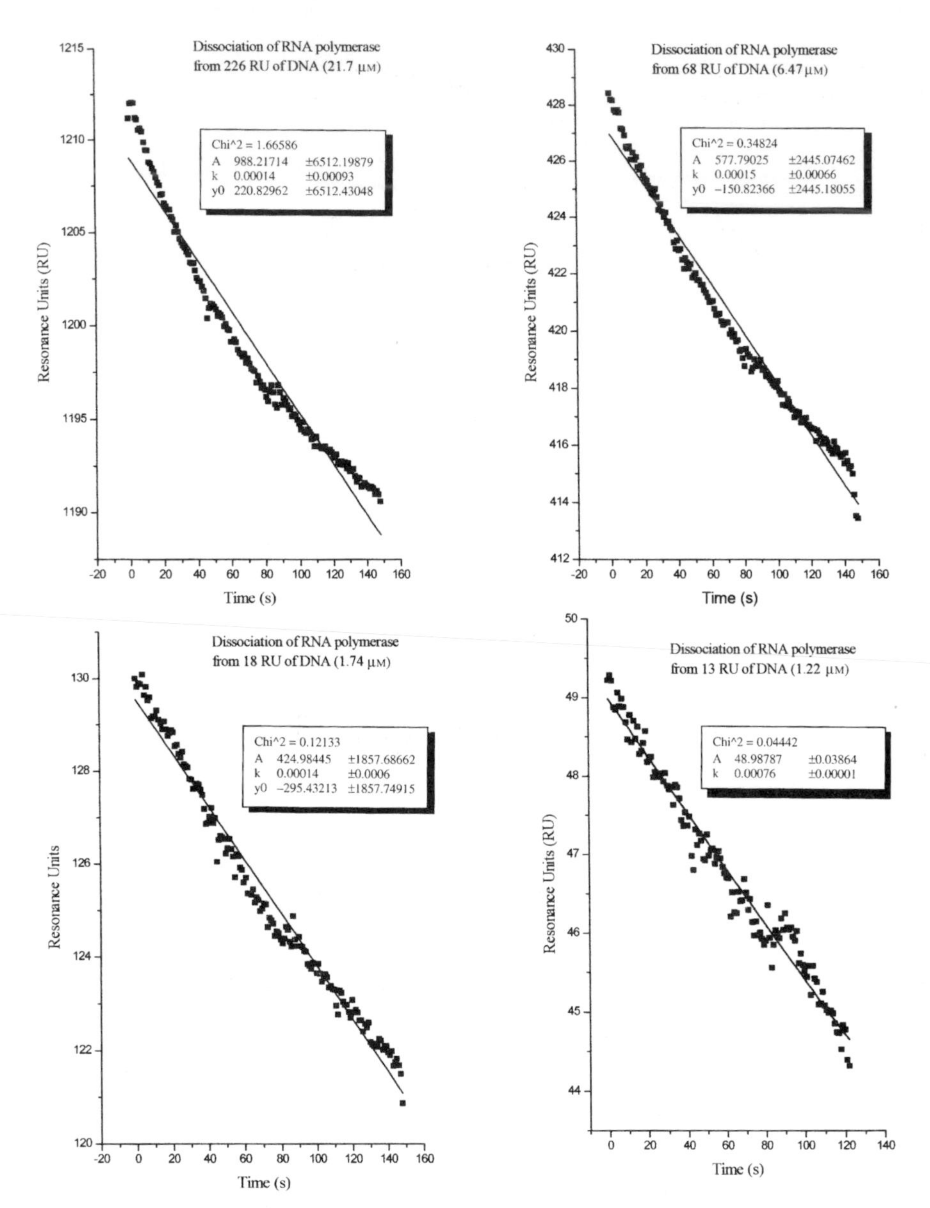

Figure 8 Direct-fit analysis of the slow phase of RNA polymerase dissociating from surfaces containing different concentrations of DNA.

therapy and the detection of food-borne pathogens. Sequencing by hybridization is one approach being developed to generate rapid sequence information. It exploits short oligonucleotide probes for the interrogation of unknown DNA samples.

This section describes how SPR can be used for the study of such hybridization reactions.

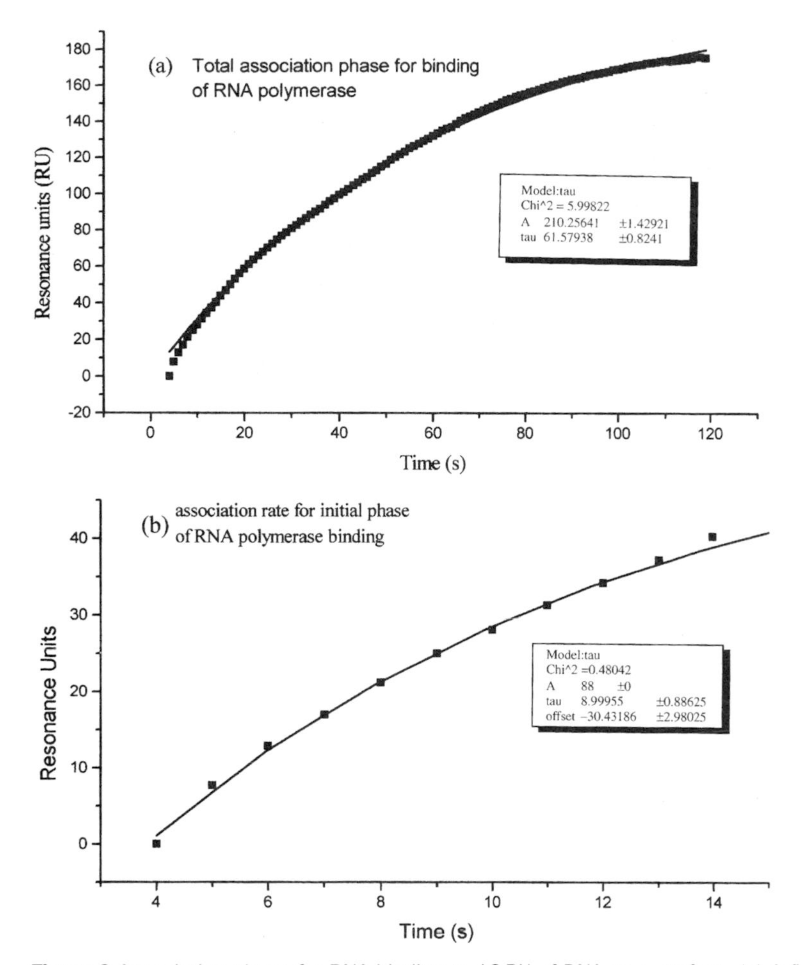

Figure 9 Association phase for RNA binding to 18 RU of DNA on a surface. (a) A fit of the whole association phase to a single exponential fit using $R = A(1-\exp^{-t/\tau})$. (b) A fit of the initial part of the association phase to the same expression.

4.1 DNA immobilization

The maximal surface concentration obtained when oligonucleotides shorter than 30 residues are captured is generally < 1500 RU. The appropriate surface ligand concentration is dependent on the application. For kinetic measurements it is always important to keep the surface concentration low, especially when the association rate constants are high. For equilibrium measurements, higher surface concentration may be used, but one must remember that this may affect the time required to reach equilibrium.

The prediction of ligand concentration is made on the estimated R_{max} value, i.e. the analyte saturation response (in RU). This corresponds to the value obtained when all binding sites on the ligand are occupied:

$$R_{max} = (R_{ligand} \times Mw_{ligand} \times n)/Mw_{analyte}$$

where R_{ligand} is the surface concentration in RU, Mw_{ligand} and $Mw_{analyte}$ are the respective molecular weights and n is the number of binding sites on the ligand. Generally, R_{max} values for kinetic measurements may be in the range 50–200 RU, but lower values can be used if sufficient quality of the response signal can be achieved.

Protocol 3

Immobilization of oligo-deoxynucleotides

Equipment and reagents

- Sensor chip SA (Biacore AB, Sweden). It is also possible to use Sensor chip CM 5 to immobilize streptavidin or avidin by amine coupling chemistry (see *Protocol 1*)
- Running buffer (HBS buffer): 10 mM HEPES, 0.15 M sodium chloride, 3.4 mM disodium EDTA, pH 7.4, with 0.005% Tween-20. Other physiological buffers may also be used
- 5′-end biotinylated oligo-deoxynucleotides
- Regeneration solution: 100 mM HCl or 50 mM NaOH. HCl works well for oligonucleotides but as the length increases, NaOH is more efficient.
- Capturing buffer: HBS buffer but with 0.5–1.0 M NaCl. Buffer substance is not critical but the salinity is important to attain efficient capture

Method

1. Oligonucleotides obtained from commercial sources are usually either delivered in high concentrations dissolved in water or lyophilized. Dilute the supplied material in capturing buffer to a final concentration of 0.1–1 μg/ml.
2. For careful control of the surface concentration of oligonucleotide it is easier if a low concentration is applied. Further, use the manual inject mode of Biacore for precise control of the immobilization level. Fill the loop with a maximum of 100 μl and thereafter inject several small injections of oligonucleotide solution and check the level of immobilization between the injections.
3. For most applications it is important to use a reference surface in order to obtain a noise-free signal from the active surface. The reference surface should be like the active surface in all respects except for the specific interaction. In this case it is achieved by immobilizing the same surface concentration of an equal-sized oligonucleotide with an irrelevant sequence in another flow cell. The analyte sample is then passed over the two flow cells using the multichannel detection mode.
4. After immobilizing the target oligonucleotide pass a few pulses of regeneration solution over the flow cells before start of the analysis. It may be useful to perform some dummy injections of analyte before starting the analysis if very high precision is needed.

4.2 Hybridization analysis

When the active surface(s), containing ligand, and the reference surface are prepared, analysis of hybridization can be started. Different buffers may be used to study the effects of chemical conditions on hybridization, such as salinity, ionic strength or presence of different ions or other additives. Effects of temperature can be studied within the range of 4–40 °C. Since the instrument is designed to give high temperature stability, it takes some time to equilibrate the unit, normally about 1 h. Thus, if several temperatures are investigated in the same run, special wait-commands in the software makes it possible to perform this kind of analysis overnight in a single run.

The efficiency of hybridization on the sensor chip is high, reflecting good availability of the immobilized partner. For example, when the hybridization of octamer nucleotides was studied, the efficiency of hybridization was estimated to be about 80% of the theoretical R_{max} value.

Protocol 4

Hybridization analysis

Equipment and reagents

- Sensor chip SA prepared according to *Protocol 1*.
- Oligo-deoxynucleotides dissolved in hybridization buffer to give a range of concentrations
- Hybridization buffer and running buffer: same buffer should be used
- Regeneration solution—100 mM HCl or 50 mM NaOH

Method

1. A kinetic analysis is performed by running a set of oligonucleotide solutions at different concentrations in the range 0.1–10 × K_d (equilibrium dissociation constant). The concentration at K_d corresponds to half-saturation of R_{max}.
2. For an equilibrium analysis, several concentrations of the analyte, spanning the complete binding curve from zero-response to R_{max}, is needed. It is recommended to run the analyte over least 10 different levels.
3. Dilute the oligonucleotides in hybridization buffer to a final concentrations according to the recommendations given above.
4. For the equilibrium analysis, check the time to reach equilibrium for the lowest concentration. The flow rate may be decreased to allow equilibrium binding—it is not necessary to run all concentrations at the same flow rate.
5. Kinetic analysis is usually performed with a flow rate > 30 μl/min.

4 After the run is finished, open the result file from the evaluation software (BIAeval 3.0). Subtract the sensorgrams obtained in the reference flow cell from those of the active flow cell(s). Adjust all corrected sensorgrams to a baseline = 0 RU. Follow the procedures of the software to perform the analysis of kinetics or equilibrium binding of hybridization.

4.3 Effects of mismatches

Detailed analysis of hybridizations have been performed with oligonucleotides ranging in size from 20 residues down to hexamers; interactions with trimers can be studied, although the responses are quite small.

Not only does the length of the fragment, as well as experimental conditions (temperature, salinity), have large effects on the affinity of the hybridizing pairs, but the base composition is also critical. Thus, when a set of octamer oligonucleotides were compared in Biacore analysis, the affinities determined spanned almost two orders of magnitude (16). The ratio of G/C to A/T bases is important, but the order in the sequence is also critical because of nearest-neighbour effects.

The advantage of using very short oligonucleotide probes is that they are very sensitive to bases in the sequence that are mismatched. When a mismatched base occurs in one of the central positions of an octamer, hybridization is almost completely abolished, and the affinity cannot be determined in the same concentration range. However, a critical situation occurs when the probe is affected by an end-mismatch. For an octamer probe the sensitivity for end-mismatches is high, but for longer probes the relative effect of an end-mismatch is much smaller. Therefore, it is advisable to use a slight nucleotide overlap for probes within the region of interrogation, i.e. use a one to three base overlap of adjacent probes.

4.4 Sequence screening

When the method for analysis of hybridization in Biacore is used for screening purposes, the idea is not to run complete binding curves containing a wide range of probe concentrations. It is enough to run only one concentration of the probe, chosen to give the best discrimination between the wild-type and the

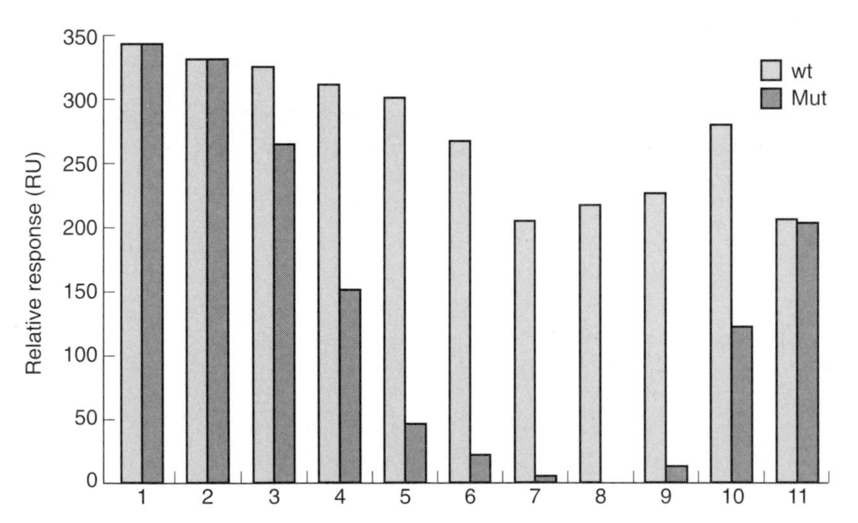

Figure 10 Octamer probes used to scan a wild-type sequence and one containing a single point mutation. Probe 2 and 9 are mismatched at the 5′- and 3′-end, respectively. Responses are corrected for bulk refractive index by subtraction of the reference flow cell signal.

mutated target. The situation for the analysis is fairly simple if the aim is to search for a single mutation at a specific site. In that case, only one probe is needed. If the mutation is to be found anywhere within a certain region, several probes must be used.

The optimization of the experimental conditions is important to get a good discriminatory power of the method. This is normally achieved by modifying the salinity (ionic strength) of the hybridization buffer. It should be pointed out that is possible to use different buffers for different probes and use a common running buffer, although it was pointed out previously that for detailed measurements it is advantageous to use the same buffer. In a screening situation, discrimination between targets may be achieved efficiently anyway. *Figure 10* shows the typical result of an experiment using octamer probes. Probes 5 and 6 have the mismatch located in either one of the two central positions.

Acknowledgements

We thank our various colleagues for their help with the SPR experiments reported here, especially Dr Isobel D. Parsons for the work on MetJ.

References

1. Fagerstam, L. G., Frostell, K. A., Karlsson, R., Persson, B. and Ronnberg, I. (1992). *J. Chromatog.* **597**, 397.
2. Jonsson, U., Fagerstam, L., Ivarsson, B., Johnsson, B., Karlsson, R., Lundh, K., *et al.* (1991). *Biotechniques* **11**, 620.
3. Stockley, P. G. (1994). *Methods Molec. Biol.* **30** 251.
4. Somers, W. S., Rafferty, J. B., Phillips, K., Strathdee, S., He, Y. Y., McNally, T., Manfield, I. *et al.* (1994). *Ann. N.Y. Acad. Sci.* **726**, 105–17.
5. Phillips, S. E., Manfield, I., Parsons, I., Davidson, B. E., Rafferty, J. B., Somers, W. S., *et al.* (1989). *Nature* **341**, 711.
6. Wild, C. M., McNally, T., Phillips, S. E., and Stockley, P. G. (1996). *Mol. Micro.* **21**, 1125.
7. Parsons, I. D., Persson, B., Mekhalfia, A., Blackburn, G. M. and Stockley, P. G. (1995). *Nucl. Acids Res.* **23**, 211.
8. Parsons, I. D. (1996). PhD thesis. University of Leeds, Leeds, UK.
9. Schuck, P. and Minton, A. P. (1996). *Anal. Biochem.* **240**, 262.
10. Schuck, P. (1996). *Biophys. J.* **70**, 1230.
11. Parsons, I. D. and Stockley P. G. (1997). *Anal. Biochem.* **254**, 82.
12. McClure, W. R. (1985). *Annu. Rev. Biochem.* **54**, 171.
13. Buc, H. and McClure, W. R. (1985). *Biochemistry* **24**, 2712.
14. Leirmo, S. and Record, M.T. Jr (1990). *Nucl. Acids Molec. Biol.* **4**. 123.
15. Adelman, K., Brody, E. N, and Buckle, M. (1998). *Proc. Natl. Acad. Sci. USA* **95**, 15247.
16. Persson, B., Stenhag, K., Nilsson, P., Larsson, A., Uhlén, M., and Nygren, P.Å. (1997). *Anal. Biochem.* **246,** 34.

Chapter 20
Quantitative DNase I kinetics footprinting

A. K. M. M. Mollah and Michael Brenowitz
Department of Biochemistry, Albert Einstein College of Medicine of Yeshiva University, 1300 Morris Park Avenue, Bronx, NY 10461, USA

1 Introduction

Nuclease protection or 'footprinting assays' have been extensively applied to the study of nucleic acid–ligand interactions and nucleic acid structure. The principal virtue of footprinting is that it can precisely localize individual sites of ligand binding or tertiary contact on nucleic acid molecules. Nucleic acid molecules hundreds or thousands of nucleotides long can be analysed. By choosing any of a number of chemical or enzymatic nucleases, changes in different aspects of the nucleic acid structure can be visualized. Quantitative footprinting protocols can also provide individual site-specific characterization of the thermodynamics and kinetics of reactions involving nucleic acids. These protocols have been recently reviewed (1).

This chapter presents a type of kinetic footprinting protocol that can be used with rapid-mixing technology, 'Quench-flow DNase I footprinting' (2). Advantages of DNase I for quantitative kinetics footprinting include millisecond time-resolution, the robust 'signal-to-noise' of protein-DNA footprints, the ready availability of DNase I and its activity over a wide range of experimental conditions (3, 4). The disadvantages of DNase I include the divalent cation requirement for activity, the lack of sequence neutrality of DNase I cleavage and the large, fairly featureless protein footprints that are generally observed for protein binding.

A technique that developed from quench-flow DNase I footprinting is 'Synchrotron X-ray footprinting' (5), a method for conducting hydroxyl radical (•OH) footprinting (6) on millisecond time-scales. Advantages of •OH footprinting include the absence of a divalent cation requirement for activity, the sequence neutrality of its cleavage, its activity over a wide range of experimental conditions and its small size. The small size of the •OH probe allows detailed structural information to be obtained within protein-binding sites (7, 8) or upon formation of tertiary contacts (9, 10) during the course of thermodynamic or kinetics experiments.

Since X-ray footprinting requires synchrotron radiation to achieve millisecond time resolution, these experiments require a significant investment in time and resources. Thus, careful consideration should be given to its applicability to a problem. The potential use of lower flux X-ray sources to reactions with slower time-scales has been discussed (11) and for reactions with time scales of at least tens of seconds, peroxonitrous acid may be a useful •OH footprinting kinetics probe (8). DNase I kinetics footprinting can provide an important foundation for the application of •OH footprinting protocols to protein–DNA interactions.

2 Quench-flow DNase I footprinting

2.1 Preliminary considerations

Our implementation of quench-flow DNase I footprinting (2) utilizes a commercially available three-syringe quench flow apparatus manufactured by the KinTek corporation (12). However, in these experiments, DNase I is loaded in the 'quench' syringe and quenching of the DNase I cleavage reaction occurs when the sample is expelled into an EDTA-containing solution present in the sample collection tube (*Figure 1*). The time of DNase I cleavage of the nucleic acid can be as short as 20 ms and is dependent upon the solution conditions used in an ex-

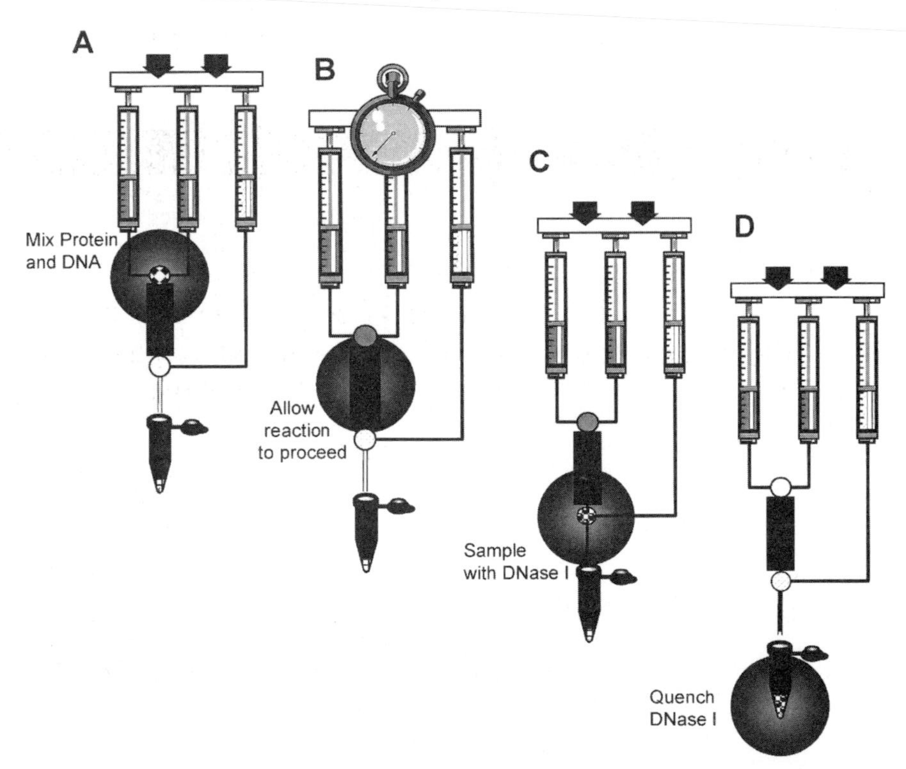

Figure 1 Schematic representation of a quench-flow DNase I kinetics experiment. (A) The protein and DNA are mixed during the first push of the syringes. (B) The reaction is allowed to proceed for the desired length of time. (C) The second push of the syringes mixes the solution containing the protein and DNA with the DNase I solution. (D) The solution is subsequently expelled into a microfuge tube containing the EDTA quench solution.

periment. A microprocessor-controlled stepping motor governs sample flow during an experiment. The parameters that control sample flow (i.e. flow rates, sample incubation time and nuclease exposure times) are readily programmed using the software provided by the manufacturer.

2.2 Reagent preparation

- Protein: quantitative studies require proteins purified to homogeneity of known specific DNA-binding activity. The protocols required to purify homogeneous preparations of proteins are idiosyncratic for each and every system. Thorough characterization of the properties of the protein (i.e. possible self-association reactions) is essential for the accurate analysis and interpretation of kinetic footprinting data.
- DNA: DNase I footprinting requires a linear DNA restriction fragment labelled at one end with ^{32}P. Ideally, a protein binding site(s) is located 30–90 bp from the ^{32}P-labelled end of the DNA, although shorter and longer spacing can be successfully analysed by using the appropriate electrophoretic conditions. The number of nucleotides from DNase I hydrolysis occluded by protein binding should not exceed 10–15% of the DNA. Primer extension is an alternative to *Protocol 1* for radiolabelling DNase I reaction products that is particularly suitable for studies of circular DNA molecules.

Protocol 1

3′ DNA Labelling

Equipment and reagents

- ^{32}P-dNTPs
- Quick Spin Column, Sephadex G-25 Fine (Boehringer Mannheim)
- TE buffer (10 mM Tris, 1 mM EDTA, pH 8.0)
- Elutip D column (Schleicher & Schuell)
- DNA polymerase I (Klenow fragment)

Method

1. Restrict 1–10 pmol of a plasmid containing the binding site(s) of interest in a 50 μl total reaction volume in order to generate a 3′ recessed end at the appropriate distance.
2. When restriction is complete, place the sample on ice and add 5 μl (16.5 pmol) of each the ^{32}P-dNTPs (specific activity ≈ 3000 Ci/mmol) complementary to the 3′ recessed end. For DNase I experiments, all the nucleotides within the 3′ recess are radiolabelled (3, 4). For X-ray footprinting only the first site is filled with a ^{32}P-dNTP. Add 10–15 units of large fragment of DNA polymerase I (Klenow fragment); incubate on ice for 30 min.
3. For DNase I substrates add 5 μl of a solution containing 5 mM of the four unlabelled dNTPs and incubate for 15 min at room temperature. Remove the unincorporated ^{32}P dNTPs using a Quick Spin Column. Purify the DNA using an Elutip D column, as described by the manufacturer and precipitate the DNA in ethanol.

Protocol 1 continued

4 Resuspend the ^{32}P-DNA in 20 μl TE buffer and restrict it with the appropriate second enzyme to generate DNA fragment labelled at one end. (Steps 1 and 4 can be combined if the second enzyme yields either a blunt or 5′ recessed end that is an unsuitable substrate for the Klenow fragment.)

5 Separate the desired restriction fragment by agarose or acrylamide electrophoresis (we use a 1% agarose gel in 1× TBE to separate fragments ranging from 200 to 900 bp). Locate the desired band by autoradiography, excise it from the gel and purify the ^{32}P-DNA using any number of protocols such as electroelution. Precipitate the DNA in ethanol, resuspend it in 100–200 μl TE and store at 4 °C. A specific radioactivity of 1×10^7 d.p.m./pmol is routinely obtained using four radioactive dNTPs.

2.3 Conducting a quench-flow DNase I 'footprinting' experiment

Prior to initiating kinetics studies of a protein–DNA interaction, qualitative DNase I analysis should be conducted to ascertain the location of the footprint and the extent of nuclease protection conferred by the protein. These qualitative studies should be followed by a thermodynamic analysis to determine the affinity of the protein for the specific sites, its specificity and the dependence of the interaction on solution variables. The decision to use a quench-flow device is dependent upon the time-scale of the reaction being studied; reactions with half-lives of minutes or greater can be simply conducted using manual pipetting (see *Protocol 3*). *Protocol 2* describes a quench-flow association experiment where the DNA is at limiting and significantly lower concentrations than the protein, i.e. the reaction is pseudo first-order in protein.

Protocol 2

Quench-flow DNase I association kinetics footprinting

Equipment and reagents

- Quench solution (50 mM EDTA prepared by dilution of 0.5 M EDTA pH 8.0 stock solution)
- Assay buffer: 25 mM Bis-tris, 5 mM $MgCl_2$, 1 mM $CaCl_2$, 2 mM DTT, 100 mM KCl, 1 μg/ml poly dG/dC, 0.01% Brij titrated to pH 7.0
- DNase I, type IV bovine pancreas (Sigma). A stock solution of ≈ 20mg/mL DNase I is prepared in 50 mM Tris-HCl, 10 mM $MgCl_2$, 1 mM $CaCl_2$, 1 mM DTT, 50% (v/v) glycerol at pH 7.2 and stored in 100-μl aliquots at −70 °C

Method

1 Thermally equilibrate the apparatus and buffers at the desired temperature.

2 The flow train of the apparatus, including the sample and reaction loops, are flushed with quench solution followed by double deionized water in order to remove residual DNase I activity.

Protocol 2 continued

3. The syringes that drive the flow of solution through the two sample loops are primed and filled with assay buffer[a] sufficient for the reaction of 25 samples (≈ 5 ml each syringe). The third 'quench' syringe is primed and filled with 5 ml of DNase I solution pre-calibrated to obtain 'single-hit' nicking of the DNA.[b] Great care must be taken in handling the high concentrations of DNase I present in this solution to prevent premature degradation of the reagent DNA.
4. Dilute the ^{32}P-DNA (≈ 25 000 c.p.m. yields rapid autoradiographic detection of each data point) and protein samples to the appropriate concentration in assay buffer. The protein solution should be twice the desired final concentration since its concentration is halved following mixing with the DNA. Draw 400–500 μl into a 1-ml disposable syringe sufficient for 16–20 injections of 20 μl each. Similarly, draw the protein solution into a second syringe. Load 20 μl of the ^{32}P-DNA containing solution into one sample loop and the same quantity of protein containing solution into the second loop. At this point, the flow control values of the quench-flow device are all turned to their 'mixing' positions.
5. A 'push–pause–push' cycling of the stepping motor is used to conduct the reaction (*Figure 1*); the calibration of the apparatus for this cycle is described elsewhere (2). The first 'push' generates turbulent flow sufficient to completely mix the DNA and protein in the reaction loop. The 'pause' allows the reaction to proceed for a defined period and is the only variable changed during an experiment. The second 'push' mixes the protein–DNA reaction mixture with the DNase I. The 'footprinting' reaction proceeds as the solution progresses down the exit tube. The flow rate and length of the exit tube determine the duration of the DNase I nicking reaction; our standard protocol results in 21 ms of DNase I nicking of the ^{32}P-DNA. The reaction is quenched when the sample is expelled into 100 μl of 50 mM EDTA, pH 8.0, present in the microfuge collection tube. The final solution volume is 356 μl.
6. Then, 250 μl of phenol–chloroform is immediately added and the solution vortexed in order to extract the DNase I from the solution.
7. After the reaction, the sample, reaction, and exit loops are cleaned by sequentially washing with quench solution, doubly deionized water and methanol (cleaning cycle). The cleaning cycle is accomplished by attaching a vacuum line to the ejection port and suctioning the three solutions in sequence through the two wash inlets. Proper precautions against the dispersal of radioactivity should be taken in the set-up and carrying out of the cleaning cycle.
8. At the conclusion of an experiment, the samples are centrifuged, the phenol-chloroform is drawn off, the ^{32}P-DNA precipitated by ethanol and prepared for electrophoresis using standard procedures (2–4,13,14).

[a]Protein–DNA interactions are highly dependent on solution conditions. Thus, the characteristics of a particular system and the goals of the study must dictate the choice of an initial assay buffer. The assay buffer often used in our laboratory as a 'reference condition' is given above. Note that magnesium and calcium ions are required for DNase I activity.

[b]The stock DNase I solution is diluted to the appropriate concentration just prior to a series of experiments. The DNase I concentration sufficient to yield single-hit kinetics must be determined empirically for each solution condition employed (3, 4).

2.3 Experimental conditions and selection of data points to be acquired

Kinetics studies are typically conducted as a function of a number of solution variables in order to probe the mechanism of the reaction. As noted in *Protocol 2*[a], both the characteristics of the system being analysed and the reaction phenomena of interest dictate the choice of experimental conditions. The experimental design must accommodate the requirement for divalent cations for DNase I activity and the diminished nuclease activity of the enzyme at high salt concentrations and low temperature.

Experiments are designed to distribute evenly the data points over the timescale of the reaction progress curve; typically 10–20 data points are collected in a random time sequence in order to minimize systematic error. Mixing the ^{32}P-DNA with buffer instead of protein solution collects the data points for 'time zero'. A control for DNase I contamination in the quench-flow device is to mix the protein and ^{32}P-DNA with buffer instead of DNase I solution in the 'quench' syringe. This control is typically conducted at the onset of an experiment.

2.4 Manual mixing DNase I kinetics footprinting

A quench-flow mixing-device is not required for reactions with half-lives on the order of minutes or hours. *Protocol 3* describes a DNase I dissociation kinetics experiment conducted by manual mixing of protein and DNA and sampling of the reaction mixture by DNase I.

Protocol 3

Manual mixing DNase I dissociation kinetics footprinting

Equipment and reagents

- Assay buffer: 25 mM Bis-tris, 5 mM $MgCl_2$, 1 mM $CaCl_2$, 2 mM DTT, 100 mM KCl, 1 μg/ml poly dG/dC, 0.01% Brij titrated to pH 7.0
- Quench solution (50 mM EDTA prepared by dilution of 0.5 M EDTA stock solution)
- DNase I, type IV bovine pancreas (Sigma). A stock solution of ≈ 20mg/mL DNase I is prepared in 50 mM Tris-HCl, 10 mM $MgCl_2$, 1 mM $CaCl_2$, 1 mM DTT, 50% (v/v) glycerol at pH 7.2 and stored in 100-μl aliquots at −70 °C

Method

1. Dilute the ^{32}P-DNA (≈ 25 000 cpm) and protein samples to the appropriate concentration in to a final concentration of 1.6 ml in assay buffer, sufficient for 16 samples of 100 μl each.
2. Allow the protein–DNA mixture to reach chemical and thermal equilibrium by incubation in a regulated water bath.
3. The fractional saturation at 'time zero' is obtained by placing 100-μl aliquots of the

Protocol 3 continued

sample into a microfuge tube containing 5 μl of a DNase I solution (whose concentration is pre-calibrated to yield single-hit kinetics[a]), mixing gently and immediately quenching the reaction by addition of 50 μl of 50 mM EDTA.

4 Then 250 μl of phenol–chloroform is subsequently added and the solution vortexed in order to extract the DNase I from the solution.

5 Dissociation progress curves for specifically bound protein are obtained by adding an excess of a competitor (such as poly dA-dT (Sigma) for the TATA binding protein) and repeating steps 3 and 4 for each time point.

6 At the conclusion of an experiment, the samples are centrifuged, the phenol-chloroform is drawn off, the ^{32}P-DNA precipitated by ethanol and prepared for electrophoresis using standard procedures (13, 14).

[b] The stock DNase I solution is diluted to the appropriate concentration just prior to a series of experiments. The DNase I concentration sufficient to yield single-hit kinetics must be determined empirically for each solution condition employed (3, 4).

2.5 Data reduction and analysis

The ^{32}P-DNA products of DNase I nicking are separated by standard denaturing gel electrophoresis. The gels are conveniently visualized using a storage phosphor screen and associated scanner to create digital images of the gel electrophoretograms (*Figures 2A* and *3A*). Densitometric analysis of the change in intensity of the protein footprints as a function of time can be conducted using a number of commercially available software packages following published protocols (3, 4, 13, 14). If multiple protein binding sites are present on the DNA, each site can and should be analysed separately (*Figures 2B* and *3B*).

Extensive discussions of the analysis of quantitative footprinting data by nonlinear least-squares analysis techniques have been published (15, 16). For progress curves composed of a single kinetic phase, the fractional saturation ($\bar{Y}$) of a protein binding site is determined from the fractional protection (p_i) of the electrophoretic bands within a protein-binding site by nonlinear least squares fitting of the data against

$$p_i = p_{i,\,lower} + (p_{i,\,upper} - p_{i,\,lower})\,\bar{Y} \qquad 1$$

and

$$\bar{Y} = 1 - e^{-k_{obs} \cdot t} \qquad 2$$

where p_i is the apparent saturation, $p_{i,\,lower}$ and $p_{i,\,upper}$ are the limits of the transition curve, k_{obs} is the pseudo first-order rate constant and t is time. For experiments conducted at less than saturating concentrations of protein, $p_{i,\,upper}$ is determined from a control sample in which a concentration of protein sufficient to saturate the binding site was allowed to come to equilibrium.

A

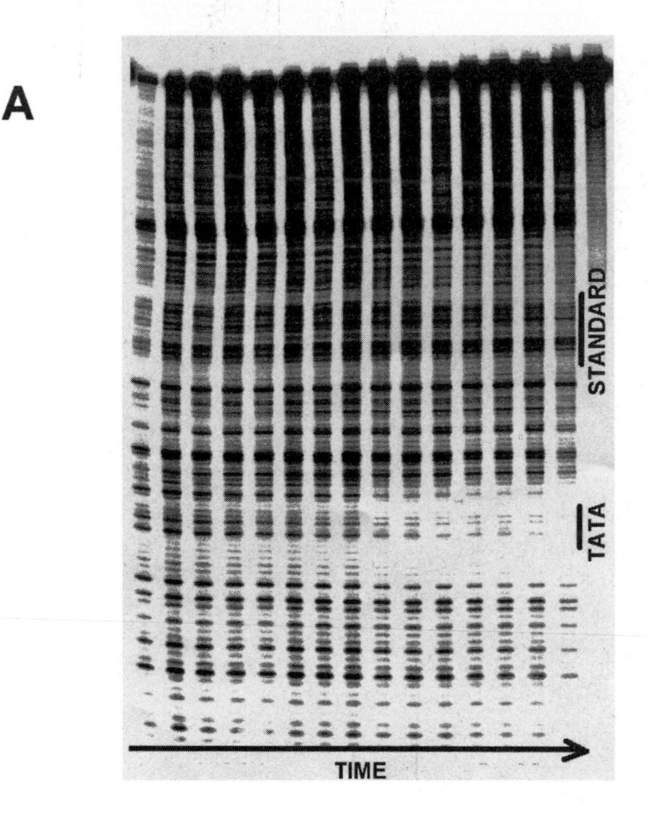

B

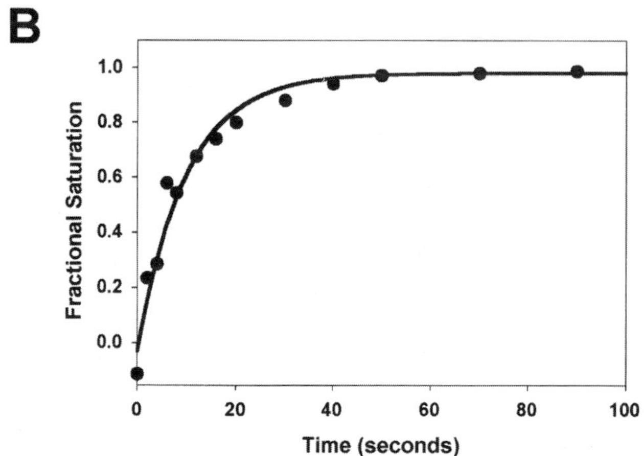

Figure 2 (A) Digital image of an electrophoretogram of a quench-flow footprinting experiment for binding of the TATA binding protein (TBP) to a single specific site (the TATA box of adenovirus major late promoter on a 282 bp restriction fragment bearing the ^{32}P label 38 bp from the binding site). The experiment was conducted in assay buffer (see *Protocol 2*) at 30 °C. The concentration of TBP was 300 nM and the ^{32}P-labelled DNA < 10 pM. (B) The individual-site progress curve determined for the experiment shown in (A). A value of $k_{obs} = 0.10 \pm 0.02$/s was determined by fitting the data to *Equations 1* and *2*.

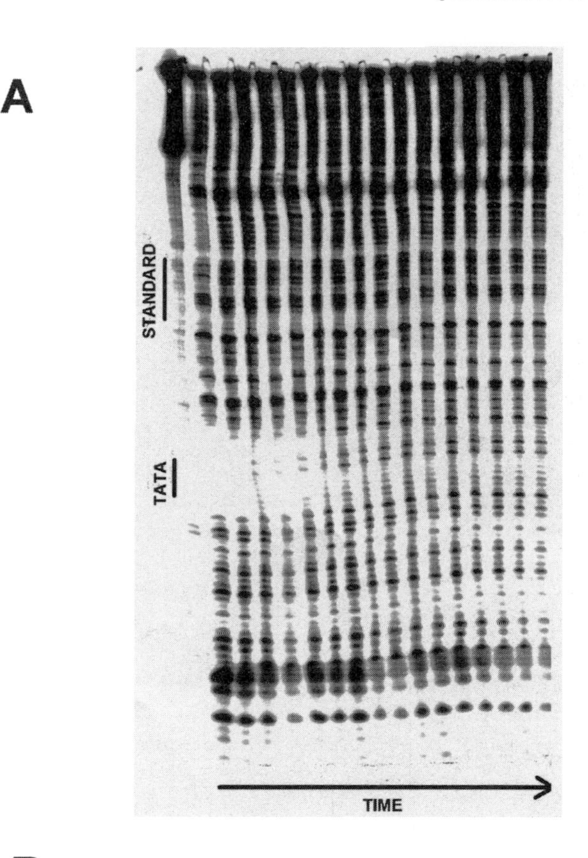

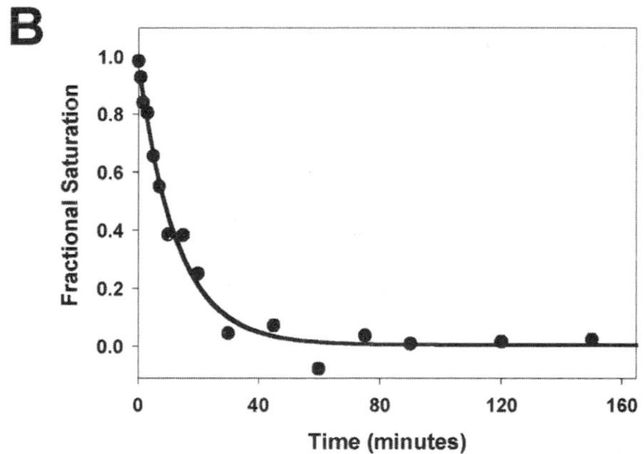

Figure 3 (A) Digital image of an electrophoretogram of a manual mixing DNase I dissociation kinetics footprinting experiment for TBP binding to a single specific site as described in the legend for *Figure 20.2*. The concentration of TBP was 100 nM and the ^{32}P-labelled DNA < 10 pM. B) The individual-site dissociation curve of the experiment shown in (A). A value of $k_{obs} = 1.3 \pm 0.2 \times 10^{-3}$/s was determined by fitting the data to *Equations 1* and *2*

2.5 Accuracy and precision of the measurements

Extensive discussions of the accuracy of quantitative DNase I footprinting have been published (2–4, 15, 16). In our laboratory, kinetics studies of the binding of the *Saccharomyces cerevisiae* TATA binding protein to promoter sequences by DNase I footprinting (17) and fluorescence (19) have been in remarkably close agreement, attesting to the accuracy of the methods. Several per cent precision in quantitative DNase I kinetics footprinting experiments have been achieved, which is comparable to that obtained in thermodynamic footprinting studies (17, 19). The precision of these experiments is dependent upon a number of factors, including the extent of protein protection of the DNA, the quality of the sample work-up and electrophoretic separation, as well as the accuracy of the reaction timing in hand-mixing experiments. It should be noted that accurate rate constants can only be obtained using purified reactants. While experiments with partly purified proteins or extracts can be conducted, only relative values can, at best, be determined. However, such relative information can provide a stimulus and a guide for the design of more rigorous experiments.

References

1. Petri, V. and Brenowitz, M. (1997). *Curr. Opin. Biotechnol.* **8**, 36.
2. Hsieh, M. and Brenowitz, M. (1996). *Methods Enzymol.* 274, 478.
3. Brenowitz, M., Senear, D. F., Shea, M. A., and Ackers, G. K. (1986). *Methods Enzymol.* 130, 132.
4. Brenowitz, M., Senear, D. F., Shea, M. A. & Ackers, G. K. (1986). *Proc. Natl. Acad. Sci. USA* **83**, 8462.
5. Sclavi, B., Sullivan, M., Woodson, S., Chance, M., and Brenowitz, M. (1998). *Methods Enzymol.* **295**, 379.
6. Dixon, W. J., Hayes, J. J., Levin, J. R., Weidner, M., Dombroski, B. A., and Tullius, T. D. (1991). *Methods Enzymol.* **208**, 380
7. Strahs, D. and Brenowitz, M. (1994). *J. Mol. Biol.* **244**, 494.
8. King, P. A., Jamison, E., Strahs, D., Anderson, V. E., and Brenowitz, M. (1993). *Nucl. Acids Res.* **21**, 2473.
9. Sclavi, B., Woodson, S., Sullivan, M., Chance, M., and Brenowitz, M. (1997). *J. Mol. Biol.* **266**, 144.
10. Sclavi, B., Chance, M., Brenowitz, M., and Woodson, S. (1998). *Science* **279**, 1940.
11. Chance, M. R., Sclavi, B., Woodson, S., and Brenowitz, M. (1997). *Structure* **5**, 865.
12. Johnson, K. A. (1986). *Methods Enzymol.* 134, 677
13. Brenowitz, M. and Senear, D. F. (1989). In *Current protocols in molecular biology* (ed. Ausubel, F. M., Brent, R., Kingston, R. E., Moore, D. D., Seidman, J. G., Smith, J. A., and Struhl, K.), Unit 12.4., pp. 12.4.1–12.4.16 John Wiley and Sons, New York.
14. Brenowitz, M., Senear, D. F., Jamison, L., and Dalma-Weiszhausz, D. D. (1993). In *Quantitative DNase I footprinting in footprinting techniques for studying nucleic acid-protein complexes (a volume of separation, detection, and characterization of biological macromolecules)* (ed. Revzin, A.), p. 1. Academic Press, New York.
15. Koblan, K. S., Bain, D. L., Beckett, D., Shea, M. A., and Ackers, G. K. (1992). *Methods Enzymol.* **210**, 405.
16. Senear, D. F. and Bolen, D. W. (1992). *Methods Enzymol.* **210**, 463.
17. Petri, V., Hsieh, M., and Brenowitz, M. (1995). *Biochemistry* **34**, 9977.
18. Parkhurst, K., Brenowitz, M., and Parkhurst, L. J. (1996). *Biochemistry* **35**, 7459.
19 Hsieh, M. and Brenowitz, M. (1997). *J. Biol. Chem.* **272**, 22092.

Chapter 21
Analysis of DNA–protein interactions by time-resolved fluorescence spectroscopy

E. H. Z. Thompson and D. P. Millar
The Scripps Research Institute, La Jolla, CA 92037 USA

1 Introduction

In addition to its uses as a probe of macromolecular structure and dynamics, fluorescence spectroscopy is emerging as a powerful tool for studies of DNA–protein interactions. Particular advantages of fluorescence-based measurements include: the ability to observe interactions under conditions of thermodynamic equilibrium; their sensitivity, allowing investigations at concentrations approaching the K_d values of many DNA–protein interactions; the ability to monitor complexation events in solution, accommodating a wide variety of conditions; and, most of all, the short time-scale of the emission process. Since the emission of a photon occurs on the same time-scale as a number of dynamic processes—including solvent relaxation, intersystem crossing, charge transfer, and quenching—it is highly sensitive to the characteristics of the local environment.

A further degree of information may be gleaned from fluorescence processes by monitoring *time-resolved* fluorescent events. The rapid time-scale for photon emission allows for direct observation of various types of molecular motion; among these are overall rotational diffusion and internal motion, properties which will typically be affected by protein–DNA complexation. Measurements of these properties can therefore provide a means of monitoring binding processes.

There are two types of experiments which can be performed using time-resolved techniques: fluorescence lifetime decay and fluorescence anisotropy decay. These parameters provide information on different properties of the probe environment, as will be explored in more depth in the next section.

2 Time-resolved fluorescence techniques

Since DNA has no measurable intrinsic fluorescence in solution at room temperature, fluorescence studies of DNA–protein interactions require the attachment of fluorescent probes to DNA. This process, and the required instrumentation for these studies, will be discussed in more detail in Sections 3.2 and 3.3.

The formation of a complex between labelled DNA and a protein may be monitored by changes in emission intensity and/or polarization anisotropy of the extrinsic probe, both in steady-state and time-resolved fluorescence experiments. While steady-state experiments provide a relatively simple means for observing complexation, the time-resolved versions can allow for a probe of dynamics of the interactions. In addition, analysis of time-resolved fluorescence signals can be used to separate different binding modes of a DNA–protein complex.

2.1 Fluorescence lifetime decays

2.1.1 Theoretical basis

A probe's fluorescence lifetime can be an informative parameter because of its extreme sensitivity to changes in the probe's environment. This environmental dependence arises because fluorescence lifetime is determined by a combination of factors which can dissipate excited-state energy:

$$\tau_f = \frac{1}{(k_r + k_{nr} + k_p + k_{et} + k_Q[Q])} \quad 1$$

In the above expression, k_r is the rate constant for radiative processes, k_{nr} is the rate constant for non-radiative processes, such as intersystem crossing and internal conversion, k_p is the rate constant for photochemical processes (including excited-state charge transfer and isomerization), k_{et} is the rate constant for energy transfer (when an acceptor is present), and k_Q is the rate constant for quenching by an exogenous quenching agent Q, present with concentration $[Q]$. Since each rate constant is dependent upon the environment of the probe, changes in that environment can be detected by a change in the fluorescence lifetime.

For a single fluorophore in a homogeneous environment, the fluorescence intensity decays exponentially with a single lifetime, τ_f. Systems such as labelled-DNA bound to a protein are more complex, and the decay is more commonly described as a summation of a number of such exponential decays, taking the form of *Equation 2*:

$$I(t) = \sum_{i=1}^{N} \alpha_i e^{-t/\tau_i} \quad 2$$

where α_i is the amplitude associated with lifetime τ_i. If the heterogeneity of *Equation 2* is due to distinct ground-state conformations of the labelled DNA or DNA–protein complex, the amplitudes reflect the relative populations of the various species. In such systems, effects on probe fluorescence with protein binding are evaluated in terms of the *average* fluorescence lifetime, defined as:

$$\tau_{ave} = \sum_{i=1}^{N} \alpha_i \tau_i \quad 3$$

Additional information can often be gleaned from the effect of protein binding on the amplitude factors.

2.1.2 Experimental approach

Specific experimental details (wavelengths for excitation and emission, concentrations of DNA and protein, required acquisition time, etc.) are highly dependent upon the type of fluorophore used as a probe of DNA–protein interaction. Some guidelines for probe selection are available in Section 3.1.

Fluorescence decay data should be collected with the emission polarizer set at the magic angle (54.75°) to eliminate polarization effects and fluorescence should be monitored at right angles to the exciting light to avoid measuring transmitted light.

For each decay function it is necessary to collect an instrument response function under the same conditions as the experimental run. To obtain an instrumental response, a scattering solution (which is simply a colloidal suspension such as barium sulphate, glycogen or milk) is substituted for the sample. The scattering of the excitation pulse into the detection path is measured at the excitation wavelength.

2.1.3 Data analysis

2.1.3.1 Deconvolution

The finite time resolution of the laser and detection system must be accounted for in the determination of fluorescence lifetimes. If the exciting light had an infinitely narrow time-width, to the point of being represented as a delta function, the observed decay profile would be the *actual* decay of the excited state. With a wider excitation peak and finite detection time, however, the situation becomes more complicated computationally; it is then necessary to make allowances for the time lag in emission between photons from probes which were excited at the beginning of the excitation peak and photons from probes which were not excited until the tail of this peak. This overlapping of excitation and emission is called convolution. The aforementioned instrument response function is used here; the deconvolution of the instrument response function and the observed decay curve yields the true decay function.

2.1.3.2 Data fitting

To obtain the lifetime parameters, we use the decay profile which is represented by *Equation 4*:

$$I(t) = K(t) \otimes g(t) \qquad 4$$

where $I(t)$ is the observed fluorescence decay, $K(t)$ is the theoretical ideal fluorescence decay, $g(t)$ is the collected instrument response function, and $\otimes$ denotes the convolution of the two functions. $K(t)$ is assumed to be represented by a sum of exponential decay functions, as in *Equation 2*.

For data fitting, the convolution of the trial function, $K(t)$, and the instrument response function, $g(t)$, is calculated numerically and compared with the measured decay function, $I(t)$. The individual lifetimes and amplitudes in *Equation 2* are then iteratively optimized using non-linear least-squares methods. The goodness of fit may be judged by the χ^2 value and residuals from the fit. The analysis should include only the minimum number of lifetime components necessary to achieve a best fit.

2.1.4 Results in terms of DNA–protein interactions

As mentioned previously, the fluorescence lifetime of a probe is influenced by a variety of dynamic processes that can be expected to change upon protein binding, although it is difficult to attribute a change in fluorescence lifetime to a change in a specific relaxation process. However, much information is available from fluorescence decay studies.

The binding of a protein to a labelled DNA will often result in a change in fluorescence lifetime for a carefully positioned probe; binding can give access to or protection from a quencher, may enhance or inhibit non-radiative processes, allow for or prevent energy transfer. In addition, a system with heterogeneity in *Equation 2* may show changes in the relative populations of several distinct fluorescing species, which may indicate a change in conformational preference. A binding event will therefore likely show up in the length of a given fluorescence lifetime, either shortening it or lengthening it; in some cases, it may even be possible to assign a preferential binding state, based on changes in fluorescing species populations.

2.2 Fluorescence anisotropy decay

Formation of a DNA–protein complex can also alter the motional properties of a DNA-attached probe. Rotational motions on the time-scale of picoseconds to nanoseconds can be monitored in time-resolved fluorescence anisotropy decay experiments. Since anisotropy decay monitors reorientation of the emission dipole during the excited-state lifetime, it can report on local fluorophore motion, segmental motions, or the overall rotational diffusion of DNA or a DNA–protein complex. The primary advantage of time-resolved anisotropy over steady-state measurement is that it can be used to separate the effects of protein binding into changes in global and local motions of a DNA-attached probe.

2.2.1 Theoretical basis

Upon excitation of a fluorophore with plane-polarized light, different decays will generally be observed for emission that is polarized parallel to the exciting light than for perpendicularly-polarized emission. If emission were instantaneous, the emitted light would retain the polarization of the exciting light; since emission occurs in a finite time (determined by the characteristic lifetime of the fluorophore), the polarization can decay as the probe and the labelled molecules rotate and tumble. The difference between the parallel and perpendicularly polarized light emitted is quantified in the time-dependent fluorescence anisotropy, $r(t)$, as defined in *Equation 5*:

$$r(t) = \frac{I_{\text{II}}(t) - I_{\perp}(t)}{I_{\text{II}}(t) + 2I_{\perp}(t)} \qquad 5$$

where $I_{II}(t)$ and $I_{I}(t)$ are polarized intensity decay functions observed parallel or perpendicular to the polarized excitation, respectively. In an homogeneous system, this anisotropy decays from some initial value (determined by the chromophore itself) to a final value of zero. The precise shape of the decay reflects the dynamics of the depolarizing motions coupled to the probe. For a probe attached to a DNA oligomer, the time-dependent anisotropy generally exhibits an initial rapid decay, reflecting local rotation of the probe at its point of attachment to the DNA, followed by a slower decay due to tumbling of the entire DNA molecule. Expressed mathematically, this dual-pathway depolarization is:

$$r(t) = \beta_1 e^{-t/\phi_1} + \beta_2 e^{-t/\phi_2} \qquad 6$$

where ϕ_1 and ϕ_2 are the correlation times for local probe motion and overall tumbling, respectively. The corresponding amplitude factors, β_1 and β_2, add up to the limiting anisotropy, an intrinsic property of the fluorophore.

Binding of a protein to labelled DNA will typically increase ϕ_2, the rotational correlation time, as the DNA–protein complex is larger than DNA alone and will therefore tumble more slowly in solution. In some cases, complexation with a protein may also affect the local rotation of the probe, restricting the motion and leading to a decrease in β_1 (and a corresponding increase in β_2, as the sum of β_1 and β_2 must stay constant). The correlation time for local rotation (ϕ_1) may also be affected. *Equation 6* can be extended to include additional depolarizing motions—for example, segmental motions of a branched molecule—by additional exponentials with corresponding amplitudes.

2.2.2 Environmental heterogeneity

At the next level of complexity is the case in which the labelled DNA binds to the protein in two or more distinct modes. Clearly, two or more local environments can exist for a probe on a DNA oligomer which binds to multiple sites on a protein. The examples discussed in this chapter deal with the case where there is a fraction of long-lived, slowly-rotating probes and a fraction of more rapidly quenched and freely-rotating probes, termed 'buried' probes and 'exposed' probes, respectively. For such a system, the observed time-resolved anisotropy is a superposition of the two probe behaviours:

$$r(t) = f_e(t) r_e(t) + f_b(t) r_b(t) \qquad 7$$

where $r_e(t)$ is the time-dependent anisotropy of the exposed probe and $r_b(t)$ is that for the buried probe. Each of these anisotropies may be expressed as a sum of local probe motion and overall tumbling, as given in *Equation 6*. It is important to note that, while the decay times may be similar, the amplitude factors in *Equation 6* will be significantly different for each population.

Each population's contribution to the observed anisotropy ($f_e(t)$ and $f_b(t)$, respectively) will change over time because of the differences in fluorescence lifetimes. That time-dependence may be represented as follows:

$$f_e(t) = \frac{x_e e^{-t/\tau_e}}{x_e e^{-t/\tau_e} + x_b e^{-t/\tau_b}} \qquad 8$$

where x_e and x_b are the equilibrium mole fractions, and τ_e and τ_b are the characteristic fluorescence lifetimes of the exposed and buries probes, respectively. A similar expression can be written for the fractional contribution of the buried probes. These expressions can also be generalized for multiple fluorescence lifetimes for each probe population.

As a consequence of the different fluorescence lifetimes and mobilities of the exposed and buried probes, the time-dependent fluorescence anisotropy described by *Equation 7* can exhibit a distinctive 'dip and rise' pattern, consisting of an initial rapid decline and a rising portion at intermediate times, followed by a slow decay at longer times. The shape of the anisotropy decay is strongly dependent upon the fractions of the buried and exposed probes: this unusual curve shape can render data fitting more difficult. However, successful analysis of these decays can reveal the relative amounts of probes in the various environments. In the case of DNA–protein interactions, different binding modes can be directly quantified using this approach, as discussed later in this chapter.

2.2.3 Experimental approach

As in the case of fluorescence lifetime decay, specifics of the experimental set-up for anisotropy decay studies are fluorophore-dependent.

In order for an experiment to yield data on fluorescence anisotropy, it is necessary to observe emission polarized in both the parallel (vertical) and perpendicular (horizontal) planes. As this is often accomplished using a single detector system, it is necessary to rotate the polarizer in front of the detector. It is desirable to perform these switches throughout the experiment (if all parallel data were collected at the beginning of an experiment, and all perpendicular data were collected at the end, the sample might have photo-bleached enough to give misleading anisotropy values). For consistency, it is best to measure the parallel decay for a short period of time and then switch the polarizer to measure the perpendicular decay for an equal amount of time. By repeating such a cycle, any number of counts may be collected.

2.2.3.1 Polarization bias

Often, the instrument which is monitoring the fluorescence can have a bias, responding differently to light polarized in different directions. It is then necessary to correct for the anisotropy inherent in the response. The appropriate correction factor (G-factor) can be determined experimentally, as described elsewhere (1). Alternately, a device called a polarization scrambler can be added to the instrumentation, *after* the emission polarizer but *before* the monochromator, which will scramble the polarization of the emitted light such that, whatever polarization is being collected, there will be no instrumental bias.

With a two-detector set-up, both decays may be measured simultaneously. Eliminating bias (polarization and instrumental) is crucial when measuring the decays with separate detectors: G-factors should be calculated for each experimental run.

2.2.3.2 *Instrument response function*

Depending upon the length of the excitation pulse relative to the rotational relaxation of the probe, it may be necessary to measure an instrumental response function using a suitable scattering solution, and with emission polarizers in parallel orientation.

2.2.4 Data analysis

If the excitation pulse is narrow relative to the length of the experiment and the rotational correlation times of interest, it is possible to compute the anisotropy of the system directly from the measured intensity decay curves. In such a case, *Equation 5* applies. The resulting anisotropy function may then be fitted directly without deconvolution, simply by adjusting the correlation times and amplitude factors in *Equation 6* (or, if necessary, *Equation 7*) for the best fit, using non-linear least-squares methods.

For systems where the excitation pulse causes convolution effects, the anisotropy may not be directly computed; instead, the data fitting is done using the measured intensity decays for each polarization. The parallel and perpendicular fluorescence decays for a probe may be expressed as the paired equations

$$I_{II}(t) = [(1/3)(1 + 2r(t))K(t)]\, g(t) \qquad 9$$

$$I_{\perp}(t) = [(1/3)(1 - r(t))K(t)]\, g(t) \qquad 10$$

The time-dependent anisotropy $r(t)$ is assumed to be represented as a sum of exponential decays as in *Equation 6*. Simultaneous fitting of *Equations 9* and *10* to the measured intensity decays may be performed by fixing the components of $K(t)$ (which may be determined from the isotropic decay) and varying the anisotropy amplitudes (β_i) and correlation times (ϕ_I) in *Equation 6*. The best experimental values are obtained by a global non-linear least-squares fit. The quality of the fit may be judged from χ^2 values and residuals for each decay curve.

In the more complicated case—that with probes in distinct environments—the curve fitting is more computationally involved. In such a case, the anisotropy must instead be considered a weighted summation of the anisotropy from probes in the different environments (as in *Equation 7*). *Equations 7*, *8*, *9*, and *10* are all needed for fitting. Such an analysis involves a large number of unknown parameters and it is not possible to find a unique solution by fitting to a single decay, $r(t)$, or pair of decays, $I_{II}(t)$ and $I_{\perp}(t)$. However, satisfactory results can be obtained by simultaneous analysis of several different samples, for which the fractions of exposed and buried probes are different in each sample but the underlying properties of the probes are identical. Many of the parameters can then be linked and globally optimized across all data sets, while the mole fractions (x_e or x_b) are determined separately for each sample.

2.2.5 Results in terms of DNA–protein interactions

In separating out the changes in local and global motions of a DNA-attached probe upon protein binding, these studies can provide information about specific sites of DNA–protein contact. By preparing a series of DNA oligomers with the probe placed at different nucleotide positions, it is possible to build up a map of the contact points (fluorescence footprinting). Changes in the amplitude of probe motion, reflected in the β_1 values, provide information on the steric restriction of the probe at each contact point.

Time-dependent anisotropy also is useful for studying complex DNA–protein systems, where multiple modes of binding are employed. In the best of cases, these time-resolved signals can be deconvolved into and assigned to the contributions of different binding modes, allowing quantification of various species present in solution. When used in conjunction with techniques for manipulating the structure of the DNA and/or the protein, such measurements can be used to dissect the interactions that stabilize different modes of binding and to obtain thermodynamic information on the energetics of the interactions. These possibilities are illustrated by the examples presented later in this chapter.

3 Experimental guidelines

3.1 Guidelines for fluorophore choice

In order to be amenable to time-resolved fluorescence studies, the desired samples must meet a few criteria. The first among these is that the probe chosen to label the DNA be compatible in excitation wavelength with the available laser set-up. Depending upon the system, this requirement may be more or less stringent; dye-lasers and titanium:sapphire lasers, obviously, have a wider range of wavelengths attainable than do non-tuneable lasers.

In addition, it is crucial that the time range of the experiment and capabilities of the instrumentation be matched with the characteristic decay times of the probe. Fluorescence anisotropy experiments, in particular, are most informative for cases where the characteristic fluorescence decay times (τ_f) and the rotational relaxation times (ϕ) are similar. If the fluorescence decay is much faster than the rotational relaxation, only the initial anisotropy can be measured and information on the motional behaviour of the probe is lost. If the situation is reversed, and rotational relaxation occurs much more rapidly than does fluorescence decay, an accurate determination of ϕ is problematic. Similarly, probes for fluorescence decay experiments should be chosen carefully; if the decay is particularly rapid, it is crucial to have a narrow excitation pulse and fast detection, or information may be lost in the convolution.

A third criteria in choosing a dye is that the binding of protein to DNA must not be inhibited by the bulk of the probe. *Table 1* details a number of probes typically used for fluorescence studies of DNA–protein interactions.

3.1.1 Minimizing probe artefacts

For experiments involving a double-stranded DNA oligomer and a protein, it is of primary importance that the labelled DNA is completely annealed and undegraded. This can typically be accomplished by HPLC purification of the

Table 1 Typical fluorescent probes for studies of DNA-protein interactions

Probe	Uses	Excitation λ	Emission λ	τ (ns)[a]
Dansyl chloride	Internal labelling of modified bases	330	510	13
Fluorescein isothiocyanate	End-labelling of oligonucleotides Internal labelling of modified bases	495	516	4
Ethenoadenosine	Nucleotide analogue	300	410	26
2-aminopurine	Nucleotide analogue	315	360	10

[a] Average fluorescence lifetime

labelled strand—to isolate-free dye—and by annealing the labelled strand with a small molar excess of unlabelled strand. Free dye or unannealed labelled strand will complicate the analysis of experimental results; its environment is likely to be different from that of the annealed and bound probe. Titration experiments will ensure that all annealed DNA is bound by protein; the necessary ratio of protein to DNA depends very much upon the system. It is also important to maintain a constant solution temperature throughout the fluorescence experiments because the fluorescence lifetime and anisotropy decay parameters are both highly temperature dependent.

3.2 DNA labelling

A number of methods exist for labelling synthetic DNA oligomers. The three main classes of labelling are:

(1) Labelling of an internal base in the oligomer.

(2) Labelling the 3′ or 5′ end of the oligomer.

(3) Incorporating a base analogue within the strand.

To maximize the information that a fluorescence experiment can yield, this labelling process must be site-specific. The examples detailed in this chapter deal with cases of specific internal labelling. Methods for end labelling or the use of base analogues are discussed in more detail elsewhere (2–4).

The simplest way to site-specifically label an internal base is to use a derivitizable base analogue to which a label may later be attached. One such method is the substitution of 5-aminopropyl uridine for a specific thymidine during solid-phase oligonucleotide synthesis using a suitably protected phosphoramidite monomer (5). After this base analogue has been incorporated, and after deprotection of the primary amine group, the oligonucleotide is reacted with a succinimidyl ester, isothiocyanate, or sulfonyl chloride derivative of the dye of interest (*Figure 2*) (5). Unreacted dye may be removed by size exclusion chromatography; the dye-labelled DNA may then be purified by reverse-phase HPLC. A particular advantage of this system is that the substitution occurs at the 5 position of the uracil ring, directing the probe and linker arm into the major

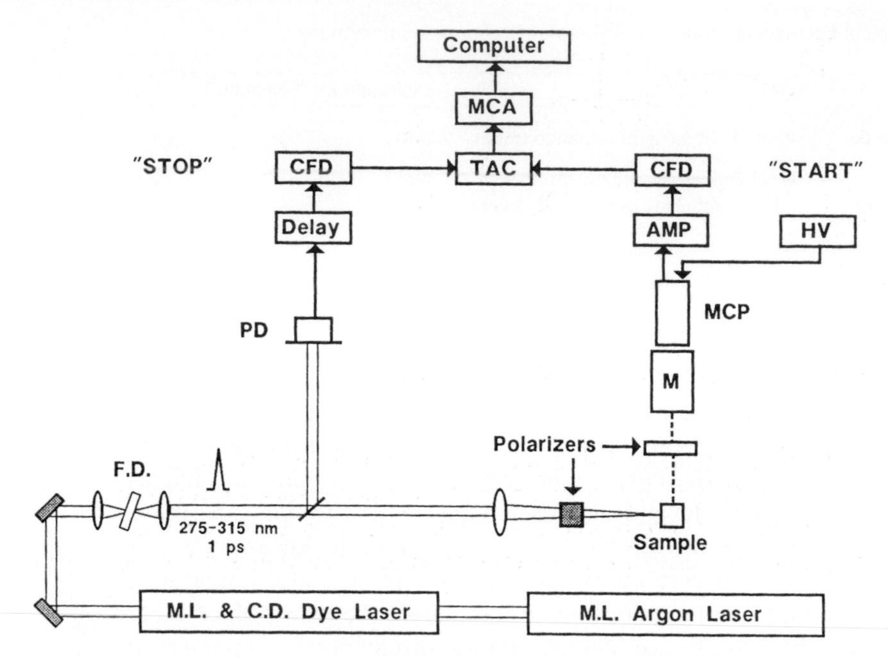

Figure 1 Picosecond laser and time-correlated single photon counting system used for fluorescence decay measurements. Excitation pulses of a few picoseconds duration are derived from a synchronously mode-locked and cavity-dumped dye laser (pumped by an actively mode-locked argon ion laser). FD = frequency-doubler for excitation in the ultraviolet region (285–340 nm); MCP = microchannel plate photomultiplier; PD = photodiode; CFD = constant fraction discriminator; TAC = time-to-amplitude converter; MCA = multichannel analyser. Time-resolved emission profiles are recorded using an 'inverted' configuration, as shown.

groove of the DNA double helix. This substitution, therefore, has only a minimal effect on base pairing or helical structure of the DNA.

3.3 Instrumentation

There are two primary methods used to record time-resolved fluorescence data: impulse-response and harmonic-response methods. The first of these involves the direct observation of the time decay of emission following a short excitation pulse; the second involves the analysis of the emission response to sinusoidally modulated excitation. In principle, the two methods should yield equivalent information, although the quality, sensitivity and resolution of the data depend on the specific instrumentation used. Here we will discuss the specifics of the impulse-response instrumental set-up used in our laboratory (*Figure 1*). Detailed descriptions of instrumentation for time-resolved fluorescence measurements can be found elsewhere (6, 7).

Excitation pulses are derived from a mode-locked dye laser, which produces picosecond pulses and is widely tuneable in the visible region of the spectrum. The laser is used in conjunction with a frequency-doubling system to provide pulses for ultraviolet excitation. Emission decay profiles are recorded by the

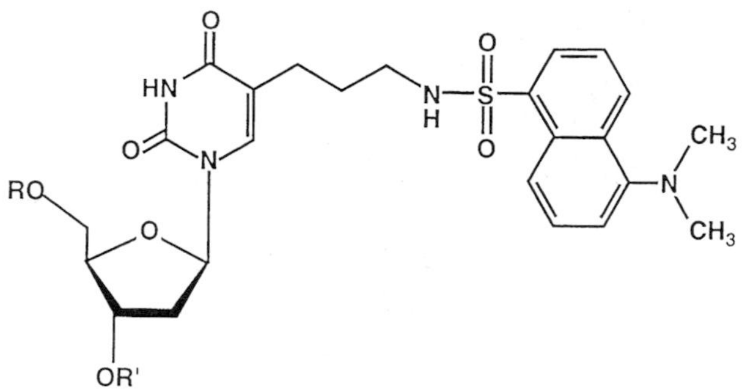

Figure 2 Fluorescent labelling of deoxyuridine with a dansyl probe.

time-correlated single-photon counting technique using a fast microchannel plate photomultiplier (6). This instrumentation allows fluorescence decays to be recorded to a very high level of precision in a relatively short collection time, while providing fast time resolution (instrument response function width of 40 ps or less).

4 Examples

There are many examples of time-resolved techniques in the literature; most of these involve applications to protein and membrane systems. In order to show some of the versatility of the technique as a probe of DNA–protein interactions, we focus on two types of studies which have been undertaken in our laboratory.

4.1 TyrR–DNA interaction: a simple example of fluorescence lifetime and anisotropy parameters

TyrR is an *Escherichia coli* regulatory protein which binds DNA at a sequence called the TyrR box and regulates the expression of enzymes involved in biosynthesis of aromatic aminoacids. Studies of the TyrR box–TyrR interaction shows how time-resolved fluorescence measurements can be used to map the binding site of a DNA-binding protein, and to infer structural information about the resultant DNA–protein complex. Bailey *et al.* examined this interaction with a fluorescein probe conjugated to various positions within and adjacent to the TyrR box (*Figure 3*) in a DNA oligomer (8). The fluorescein probe was attached to a uridine analogue similar to that shown in *Figure 2*.

The binding of TyrR to the labelled DNA resulted in significant changes in the fluorescence lifetime and rotational behaviour of the fluorescein probe. The degree of those changes was strongly dependent upon the position of the probe substitution (*Figure 4*).

In the absence of coeffectors, the TyrR protein exists as a homodimer, composed of two 57.6 kDa subunits. Upon binding its coeffector tyrosine, however, the molecule self-associates, forming a hexamer. Thus, the absence or presence of tyrosine determines the state of association of the protein; Bailey *et al.* (8)

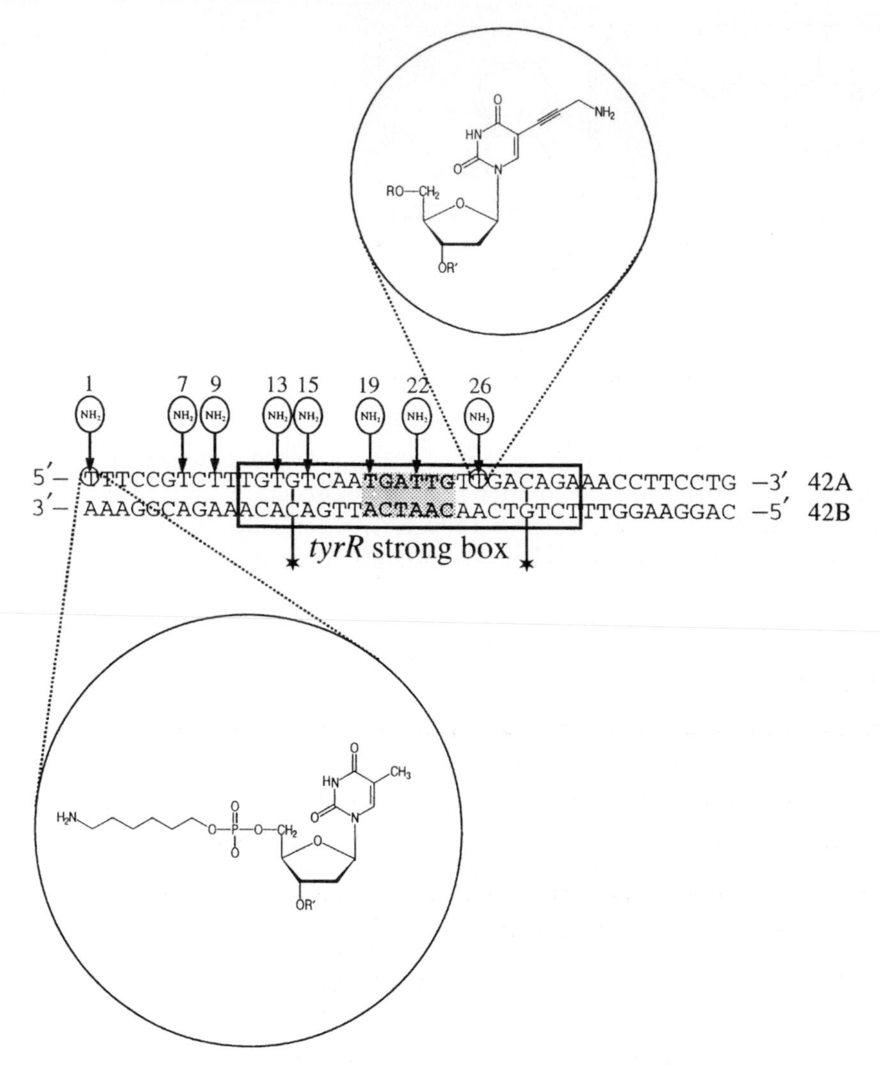

Figure 3 Design of oligonucleotides labelled at specific bases within and adjacent to the TyrR box. F denotes a fluorescein probe, and the numbers refer to the position of the labelled residue from the 5′ end. Non-conserved bases in the TyrR consensus sequence are indicated (shading), together with the positions of essential C and G residues (stars). Reprinted from *Biochemistry*, December 1995, **34**, pp. 15802–15812. with permission of the American Chemical Society.

were able to characterize the binding behaviours of the dimer and the hexamer separately. Effects of dimer binding were most significant for those oligonucleotides that were labelled at positions 13, 15, and 26 within the TyrR box: the binding was reflected in both an increase in the average fluorescence lifetime of the probe and a decrease in the amplitude of local probe motion. Both effects are consistent with close contacts between the dimer and the DNA at these positions.

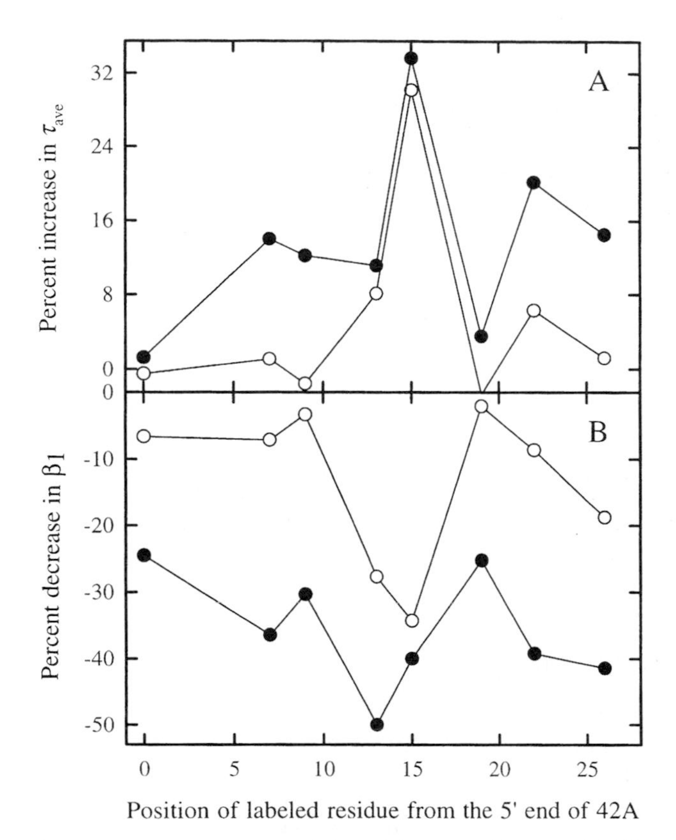

Figure 4 Fluorescence footprinting of TyrR bound to DNA. (A) Effect of the labelling position on the change in the average fluorescence lifetime of fluorescein-labelled DNA in the presence of TyrR dimer (○) or TyrR hexamer (●). (B) Change in the amplitude of the fast anisotropy decay component. The reduction in this amplitude indicates steric restriction of the fluorescein probe by TyrR. Reprinted from *Biochemistry*, December 1995, **34**, pp. 15802–15812, with permission of the American Chemical Society.

The binding of the hexamer showed those same effects, and others. In addition to positions 13, 15 and 26, hexamer-bound oligomers with the probe in positions 7, 9, 19 and 22 also showed significant changes in local motion and fluorescence lifetime. This indicates that the hexameric form of TyrR makes more extensive contacts with the TyrR box than does its dimeric counterpart. Either the hexamer simply has more protein surface for the DNA to contact, or the hexamer somehow distorts the DNA so that spatially distant regions are brought into contact with the protein surface.

4.2 Klenow fragment–DNA interaction: a complex example of time-resolved anisotropy

The Klenow fragment of *E. coli* DNA polymerase I has both polymerase and 3′–5′ exonuclease activities. The latter activity is used to edit mismatched nucleotides mistakenly incorporated during polymerization. DNA bound to the protein partitions between the two active sites ('pol' and 'exo'), which are separated by $\approx 30\,Å$.

To study DNA–polymerase interactions by fluorescence spectroscopy, Guest *et al.* (9) prepared a synthetic DNA primer-template with a dansyl probe attached to a modified uridine residue seven bases upstream of the primer 3′ terminus. Because the probe is located sufficiently far from the primer terminus to be unaffected structurally by terminal mismatches, this method is ideal for evaluating the effects of mismatches, frameshifts, and other mutagenic phenomena

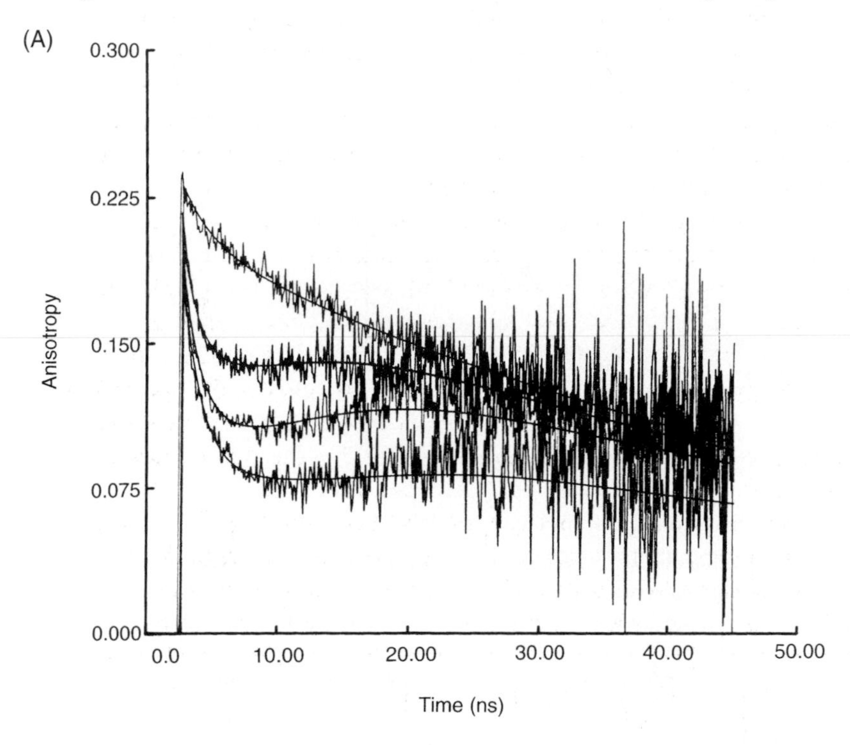

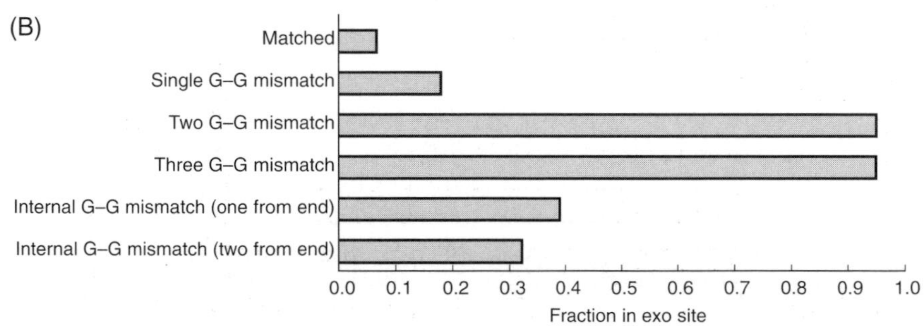

Figure 5 Partitioning of matched and mismatched DNA substrates between the Klenow exo and pol active sites. (A) Anisotropy decay profiles of dansyl-labelled primer-templates containing 0–4 mismatches bound to Klenow fragment (exo^- to prevent substrate hydrolysis). The 17-base-long primer strand contains a modified uridine residue labelled with a dansyl probe, located seven bases from the primer 3′-terminus. (B) Equilibrium partitioning obtained from analysis of anisotropy decay profiles, according to *Equations 7–10*.

on the DNA-protein interactions: changes in fluorescence behaviour are due to shifts in the probe's *environment* owing to the mismatches, etc., rather than to a change in the probe itself.

The anisotropy decay profile for the DNA-protein complex exhibited a 'dip and rise' shape indicative of two different environments for the dansyl probe; the two environments were assigned as DNA primer-templates bound at the pol or exo sites. From analysis of the anisotropy decays according to *Equations* 7 and 8, the relative fractions of primer termini bound at the two sites could be measured and the equilibrium partitioning constant K_{pe} calculated. Using a variety of matched and mismatched DNA substrates, Caver *et al.* (10) were able to quantify the effect of mismatches on partitioning of DNA between the two active sites.

As might be expected, DNA substrates with increasing numbers of mismatches partition increasingly to the exo site. *Figure* 5 shows an example of representative raw data and the calculated exo site fractions.

These observations suggest that exonuclease site partitioning is correlated with the melting capacity of the primer terminus, indicating a requirement for local melting of duplex DNA during transfer from the pol to the exo site. In addition, mismatches distant from the primer 3′-terminus result in a greater exonuclease site partitioning than do equivalent mismatches at the primer terminus itself, suggesting that unfavourable interactions between duplex DNA and the enzyme's polymerase domain can also contribute to preferential exo site partitioning (10). Similar effects have been noted with A-tract DNA and with DNA substrates containing embedded extrahelical bases (11, 12).

This time-resolved anisotropy technique has also been used to analyse the effects of protein mutations on DNA's partitioning in the Klenow fragment. Exo site residues for mutation by alanine replacement were chosen by examining X-ray crystallographic data; side-chains in close proximity to the primer 3′-terminus were selected. Each mutation was seen to have a different effect on the partitioning between the two sites (*Figure* 6).

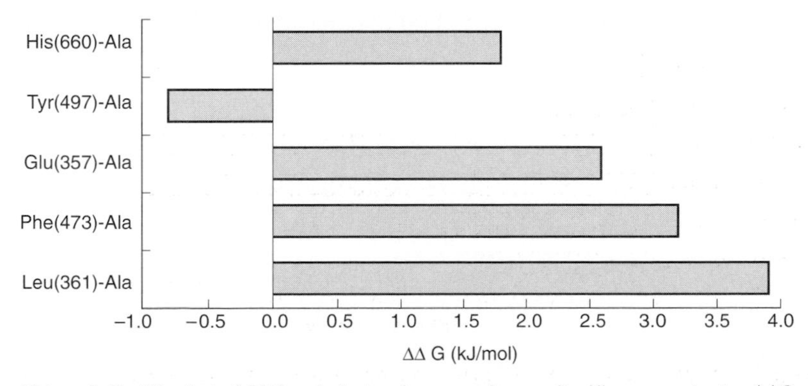

Figure 6 Partitioning of DNA substrates in exonuclease site Klenow mutants. $\Delta\Delta G$ refers to the change in the free energy of partitioning of DNA from the pol site to the 3′–5′ exo site due to the specified mutation. To encourage partitioning to the exo site, the DNA substrate contains a single G–T mismatch at the primer 3′-terminus. The mutations are examined in an exo⁻ background to prevent substrate hydrolysis.

This differential partitioning reflects the loss of binding energy due to mutation of a wild-type exonuclease site side-chain; typically, DNA substrates bind less strongly to the exonuclease site of a mutant than they did to the exonuclease site of a wild-type enzyme.

Of these mutants which caused decreased exonuclease-site binding, Leu361 and Phe473 appeared to offer the largest energetic contribution, agreeing with structural data which implicated these residues as making intimate contacts with the penultimate and terminal bases of the primer strand (13). In contrast, one mutant (Tyr497–Ala) actually showed improved exo-site binding, indicative, perhaps, of a role for the tyrosine side-chain in the wild-type enzyme in straining the DNA substrate towards the transition state of the exonuclease reaction. The results of this study demonstrate that the time-resolved fluorescence anisotropy technique can be used to quantify the energetic contributions of crystallographically defined DNA–protein contacts.

References

1. Lakowicz, J. R. (1983). In *Principles of fluorescence spectroscopy*, pp. 125–131. Plenum Press, New York and London.
2. Jameson, D. M. and Sawyer, W. H. (1995). *Methods Enzymol.* **246**, 283.
3. Waggoner, A. (1995). *Methods Enzymol.* **246**, 362.
4. Millar, D. P. (1996). *Curr. Opin. Struct. Biol.* **6**, 637.
5. Gibson, K. J. and Benkovic, S. J. (1987). *Nucl. Acids Res.* **15**, 6455.
6. Small, E. W. (1991). In *Topics in fluorescence spectroscopy*, Vol. 1 (Lakowicz, J. R., ed), pp. 97–182, Plenum Press, New York and London.
7. Lakowicz, J. R. and Gryczynski, I. (1991). In *Topics in fluorescence spectroscopy*, Vol 1 (Lakowicz, J. R., ed), pp. 293–335. Plenum Press, New York.
8. Bailey, M., Hagmar, P., Millar, D. P., Davidson, B. E., Tong, G., Haralambidis, J., and Sawyer, W. H. (1995). *Biochemistry* **34**, 15802.
9. Guest, C. R., Hochstrasser, R. A., Dupuy, C. G., Allen, D. J., Benkovic, S. J., and Millar, D. P. (1991). *Biochemistry* **30**, 8759.
10. Carver, T. E., Hochstrasser, R. A., and Millar, D. P. (1994). *Proc. Natl. Acad. Sci. USA* **37**, 10670.
11. Carver, T. E. and Millar, D. P. (1998). *Biochemistry* **37**, 1898.
12. Lam, W. C., Van der Schans, J. C., Sowers, L. C., and Millar, D. P. (1999). *Biochemistry* **38**, 2661.
13. Lam, W. C., Van der Schans, J. C., Joyce, C. M., and Millar, D. P. (1998). *Biochemistry* **37**, 1513.

Chapter 22

Analysis of protein–DNA interactions in complex nucleoprotein assemblies

Iain K. Pemberton

Unité de Physicochimie des Macromolécules Biologiques, CNRS URA 1773, Institut Pasteur, 25 Rue du Dr. Roux, F–75724 Paris Cedex, France

1 Introduction

Higher-order nucleoprotein assemblies are involved in many fundamental physiological processes. These often require the recruitment of several different proteins into large and elaborate macromolecule structures. In many cases, the protein subunit will be functional only in the context of the correctly assembled nucleoprotein. A full understanding of its contribution to the reaction mechanism inevitably requires its role to be probed within this native environment. In this chapter, a direct approach is described for the identification of proteins that participate in close-range protein-DNA contacts at specific loci within such elaborate assemblies. Although this chapter is written principally with UV-laser photocrosslinking in mind, few modifications are required to adapt the protocol to alternative protein–DNA crosslinking procedures.

2 Protocol for the transfer of a labelled nucleotide to a protein subunit by UV irradiation

As described in an earlier chapter, proteins can be coupled covalently to DNA by UV irradiation and a footprint obtained from the premature termination signals observed upon primer extension analysis to the irradiated protein–DNA complex (see Chapter 14). Since termination occurs immediately prior to the crosslinked base (1), it is thus straightforward to identify the nucleic acid involved in the contact. One can then use this knowledge to search for those proteins making these contacts. For complex nucleoproteins, one would expect the DNA fragment to be too large to allow discrimination between protein adducts on the basis of their migration on SDS-PAGE. If, however, the individual reactive nucleotide is labelled and the extraneous nucleotides are subsequently removed from the protein–DNA adduct, then the protein–nucleotide adduct should migrate close to its expected position on SDS-PAGE. This approach also facilitates the mapping of individual contacts to a defined locus on the DNA fragment. Ideally, the entire procedure is as follows:

(1) Identify contact on the DNA fragment (Protocol 3 of Buckle *et al.*, Chapter 14).
(2) Insert a radiolabel in this position.
(3) Perform crosslinking and digest the adduct with nuclease.
(4) Identify the molecular mass of the protein by SDS-PAGE.

Repeat for each nucleotide position to be scanned.

2.1 Radiolabelling of the DNA fragment at a specific nucleotide position

Figure 1 outlines the procedure to produce a DNA fragment containing a specific internally labelled nucleotide. For each nucleotide position to be studied, three PCR amplifications are performed. These will enable the reconstruction of the DNA fragment from three precise complimentary ssDNA fragments, as outlined in *Protocol 1* (N.B., sections A and B are necessary only for large DNA fragments and may be omitted for shorter DNA fragments that can be reconstructed directly from synthetic oligonucleotides, i.e. less than 100 nucleotides.

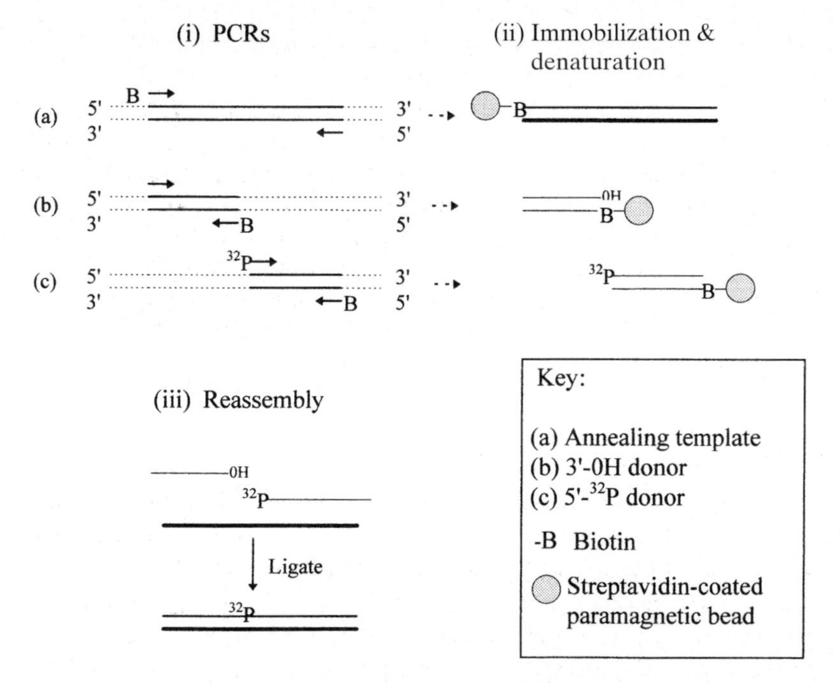

Figure 1 Construction of DNA fragments. Step (i): synthesis of annealing template, 3′-OH donor, and 5′-^{32}P donor strands in three individual PCRs (a, b and c). Step (ii): immobilization of templates on streptavidin-coated paramagnetic beads prior to alkaline denaturation to recover the non-immobilized ssDNA. Step (iii): reassembly of the internally labelled fragment by annealing and ligation of the three complimentary strands.

In each case, a biotinylated primer is used for immobilization of the PCR template on streptavidin-linked paramagnetic beads. This enables the non-biotinylated amplified strand to be removed efficiently under alkaline denaturing conditions. After neutralization, the fragments are annealed in an approximate stoichiometric ratio and the resultant 'nicked' fragment is sealed with T4 ligase. Because each step is easily controlled, the approach is simpler and more efficient than alternative primer extension/ligation approaches (a similar situation is found in site-directed mutagenesis protocols).

Protocol 1

Insertion of a ^{32}P-radiolabel at a specific nucleotide position within a DNA fragment by ligation of single stranded polynucleotides

Equipment and reagents

- Streptavidin-coated superparamagnetic polystyrene beads (Dynabeads M-280 streptavidin, Dynal)
- Magnetic particle concentrator suitable for microcentrifuge tubes (Dynal)
- Biotinylated and non-biotinylated PCR primers (purified on denaturing polyacrylamide gels)
- PCR thermocycler apparatus
- Cloned *Pfu* thermostable DNA polymerase 2.5 U/μl (Stratagene,)
- dNTPs (25 mM dTTP, dATP, dGTP, dCTP)
- TE (50 mM Tris-HCl, pH 8.0, 10 mM EDTA)
- Bead wash buffer: 1 M NaCl in TE
- Strand separation solution: 0.1 M NaOH
- Neutralization solutions: 1 M HCl and 1 M Tris-HCl, pH 7.6
- [γ-^{32}P]ATP (4500 Ci/mmol)
- Polynucleotide kinase (9.5 units/μl, Amersham Pharmacia Biotech)
- PNK buffer (10×: 0.5 M Tris–HCl, pH 7.6, 100 mM $MgCl_2$, 100 mM 2-mercaptoethanol)
- T4 DNA ligase (6.2 Weiss units/μl, Amersham Pharmacia Biotech)
- 100 mM ATP
- DNA purification spin columns and buffers (Boehringer Mannheim)
- Gel extraction kit (Jetsorb, Quantum Biotechnologies Inc.) - optional
- Calf intestinal alkaline phosphatase (20 units/μl, Boehringer Mannheim)
- General solutions: 5 M NaCl, 100 mM $MgCl_2$, 0.5 M DTT

A. Production of ligation templates by PCR

1. Perform a PCR amplification of the DNA fragment using external primers as indicated in *Figure 1(a)*. Use a 5′-biotinylated oligonucleotide as the primer for the polymerization of the *replacement* strand to allow purification of the ssDNA *annealing* template as indicated below (*Protocol 2B*). For this and the other PCR amplifications listed below, use primers at 0.5 μM final concentration. Optimize the PCR conditions if necessary to avoid spurious amplification products.[a]
2. Perform a second PCR indicated in *Figure 1(b)* between the external forward primer (non-phosphorylated) and internal reverse primer (5′-biotinylated).
3. Phosphorylate the nucleotide to be studied by 5′-end labelling of the appropriate

Protocol 1 continued

PCR primer: in 1 × PNK buffer, label 200 pmol primer with 1 μl polynucleotide kinase and 3 μl [γ-^{32}P]ATP for 30 min at 37 °C in a 20 μl reaction. The final concentration of primer is 10 μM. Heat denature the kinase for 10 min at 85 °C.

4 Perform the third PCR indicated in *Figure 1(c)* between the end-labelled and biotinylated external reverse amplification primers.

5 Purify the amplified templates on a DNA purification spin column, or alternative DNA-binding matrix, to remove enzymes and unincorporated nucleotides and primers.[b] Elute the equivalent of a 100 μl PCR reaction (≈ 25–50 pmol DNA fragment) in 50 μl sterile TE.

6 Repeat the above steps for each additional nucleotide position to be scanned.

B. Production of single stranded fragments for ligation

1 Add 12 μl 5 M NaCl to each 50 μl of the purified DNA solution, which should contain at least 20 pmol of the PCR fragment in TE.

2 Prepare 50 μl (0.5 mg) of streptavidin-linked magnetic beads[c] by resuspending in an equal volume of bead wash buffer.[d] Repeat once.

3 Withdraw the wash solution and resuspend the beads in the solution containing the PCR fragment. Incubate for 10 min at room temperature with intermittent mixing to maintain the beads in suspension.

4 Remove the liquid and resuspend in an equivalent volume of bead wash buffer.

5 Remove the liquid, resuspend and denature the fragment in 50 μl 0.1 M NaOH. Leave at room temperature for 2 min before concentrating the beads and withdrawing the liquid containing the single-stranded template.

6 Neutralize each 50 μl by adding 5 μl 1 M Tris-HCl (pH 7.6) and 5 μl 1 M HCl. Test a drop of the neutralized solution on pH indicator paper.[e]

7 Repeat the steps above for each PCR template.

C. Ligation of the DNA fragment

1 Mix the three ssDNA fragments in approximate equimolar concentration (in a final volume of 90 μl) and place in a heating block for 2 min at 95 °C. Remove the heating block and allow to cool slowly to room temperature.

2 Add 10 μl of 100 mM $MgCl_2$, 1 μl of 100 mM ATP, 1 μl 0.5 M DTT, and 1 μl T4 DNA ligase (6.2 Weiss units). Perform the ligation reaction for 2 h at 37 °C (or 4–16 h at 16 °C).

3 Denature the ligase by heating for 10 min at 65 °C. Add 1 μl calf intestinal alkaline phosphatase and dephosphorylate any unligated 5′-phosphate ends for 1 h at 37 °C.

4 Analyse a sample of the ligation reaction on a native polyacrylamide gel (using appropriate-sized markers). The labelled DNA fragments should only consist of full-length fragments, while unligated nicked DNA or unannealed single-stranded fragments are not visible upon autoradiography. Purify the DNA either (i) directly on a DNA purification spin column (as in step A.5) or (ii) by agarose electrophoresis on 1–2% gels and elute in sterile water.

Protocol 1 continued

5 Before proceeding further (protocol 3), check that the DNA fragment is functional in an appropriate DNA-binding assay.

[a] Optimum PCR conditions are often best determined empirically. Typical *Pfu* amplification reactions consist of 0.2–0.5 μM primers, 50–100 ng template and 200 μM dNTPs in a 50–100 μl reaction using the supplied 10× buffer at a final concentration of 20 mM Tris-HCl (pH 8.8), 2 mM $MgSO_4$, 10 mM KCl, 10 mM $(NH_4)_2SO_4$, 0.1% Triton X-100, and 0.1 mg/ml BSA. Generally, 30 cycles of 45 s at 95 °C, 45 s at 54 °C and 45 s at 72 °C are performed for a 200 bp fragment, although the optimum annealing and polymerization steps will vary with the T_m of the primers, the length of the DNA fragment and the polymerase employed. Consult the instructions supplied with the thermostable DNA polymerase.

[b] This step is essential for the later immobilization procedure as the presence of unincorporated biotinylated primers will efficiently block binding of the DNA fragment to the beads.

[c] Calculate the quantity of each PCR required on the basis of the pmol of DNA fragment bound by the magnetic beads. Roughly, 0.5 mg Dynabeads (10 mg/ml) will bind 100 pmol biotinylated primer, 20 pmol of a 300 bp fragment and 5 pmol of a 1 kb fragment. For larger preparations, increase the quantity of beads accordingly.

[d] The magnetic beads are concentrated on the magnetic particle concentrator for 30 s Withdraw liquid with a pipette and, after addition of the new solution, resuspend by flicking the tube or vortexing gently.

[e] Note that the DNA will now be in a solution of 100 mM Tris-HCl (pH 7–8) and 100 mM NaCl.

2.1.1 PCR amplification of ligation templates

Amplification of DNA fragments by PCR is now a standard laboratory technique. Nevertheless, a few points are worth mentioning. First, since the ligation of the DNA fragments relies on the precise complimentarity of the 3′ fragments, primers should be selected so that when the ssDNAs are annealed they create a nicked fragment (with the 3′-OH terminus of the upper strand lying immediately adjacent to the base bearing the radiolabelled phosphate group). Primer sets are required for each nucleotide position to be scanned, as indicated in *Figures 1* and *2*. They should be gel purified to ensure that all corresponding PCR products are full length and should be synthesized unphosphorylated at their 5′-ends. Second, if either *Taq* or another non-proof-reading thermostable polymerase is used, then a further 'blunt-ending' step will be required to remove the additional 3′-dA residues incorporated by such enzymes. To avoid this problem, it is best to use a thermostable polymerase with a 3′–5′ exonuclease activity, several of which are commercially available.

2.1.2 Production of ssDNA fragments

The immobilization of biotinylated PCR templates on streptavidin-linked paramagnetic beads facilitates the production of large quantities of the corresponding ssDNA in a matter of minutes. The amount of beads required will depend on the size of the template and the quantity of fragment required. Roughly, while 1 mg of beads will bind ≈ 40 pmol of a 300 bp fragment, each doubling in size of the DNA fragment will approximately halve the binding capacity of the beads (consult the information supplied by the manufacturers). Aim to reach the capacity of the beads, erring in favour of excess PCR template. The use of high salt will optimize the efficiency of immobilization. As the protocol demands strong alkaline strand-denaturing conditions, calibrate the neutralization step prior to the experiment. Mix the stated volumes of the respective NaOH, Tris-HCl and HCl solutions and test with pH paper. Adjust the respective volumes if necessary to attain pH 7–8.

2.1.3 Ligation and analysis of the fragment

The ssDNA strands are combined, heated to 95 °C and allowed to anneal by slow cooling to room temperature. Mix the fragments in an equimolar ratio (1 : 1 : 1). It is important to ensure that the annealing template is not present in a large excess over the other fragments, as this will reduce the ligation efficiency. Ligations may be performed overnight at 16–18 °C, although shorter periods at higher temperatures (e.g. 2 h at 37 °C) are generally sufficient. Since DNA nicks are resealed efficiently by the ligase, a high percentage (> 90%) of the exposed 5′-end labelled ^{32}P should be incorporated into the phosphatase-resistant form of the DNA phosphodiester backbone. This can be tested directly by treating the ligation reaction with calf intestinal alkaline phosphatase (a step that will also remove the residual 5′-^{32}P present on any unannealed ssDNA, which might otherwise complicate the crosslinking analysis). If low ligation efficiencies are encountered, these may be due to one or more of the following:

- Inefficient kinase reaction: ensure that the primer has been efficiently phosphorylated with ^{32}P by estimating the specific activity of the end-labelled primer after purification from the free nucleotide (i.e. by gel filtration).
- Poor annealing of ssDNAs: check that the fragments have annealed correctly by analysing the hybrids on a gel pre- and post-treatment with alkaline phosphatase.
- T4 Ligase and/or ATP is not functional: ensure that the pH of the Tris-buffered template is correct for the ligase and that the ATP (do not use dATP) and ligase are functional (use the bacteriophage T4 ligase not the *Escherichia coli* enzyme, since the latter uses NAD rather than ATP). The ligation works just as well if a 10× ligase buffer (provided with most T4 ligases) is used to supply the ATP.

The DNA fragment should now be repurified on a DNA-binding matrix prior to crosslinking studies. Gel purification is preferable when it is necessary to remove the unlabelled contaminating ssDNA prior to the DNA-binding reaction. In this case, reagents designed for extraction of DNA fragments from gels (such as Jetsorb, Quantum Biotechnologies Inc.) have proved highly efficient when used as per the manufacturers instructions. Before moving on to the crosslinking reactions, it is first worthwhile testing the purified DNA fragment in a simple DNA-binding reaction. Classical approaches such as gel retardation or nitrocellulose filter binding combined with competition assays suffice to ensure that the fragment has been reconstructed appropriately and maintains the expected binding specificity.

2.2 Identification of subunit interactions by UV irradiation and nuclease digestion

2.2.1 UV-laser irradiation of samples

The technique of UV-laser irradiation is described in Chapter 14. This approach minimizes structural perturbations to the nucleoprotein assembly, thereby increasing the likelihood of capturing real contacts, while eliminating the serious possibility of spurious or near-neighbour artefacts. The short irradiation times of the laser are also compatible with kinetic studies. These can be applied to monitor the assembly of the nucleoprotein and/or the corresponding reaction mechanism(s), offering a unique insight into the time-resolved trajectory of the nucleic acid within these complex assemblies. It should be noted, however, that the specific nature of the laser irradiation approach does impose a constraint on the yield of crosslinked adduct (i.e., at a typical quantum yield of 0.5% adduct per crosslinked nucleotide, one expects as little as 0.5 fmol adduct for a standard 10 μl sample containing 10 nM DNA fragment). In view of the low yield of product, a standard protocol for the precipitation of the adduct with trichloroacetic acid is presented, which can be employed when the adduct(s) must be concentrated from dilute solutions. Do not prolong the period the adduct remains under acid conditions, which may be deleterious to its stability. In some cases, it may be preferable to simply increase the concentration of the irradiated nucleoprotein to a level where the adduct may be detected directly.

2.2.2 Alternative UV crosslinking procedures

Alternative crosslinking procedures can be considered if a high-energy UV-laser is not available or if it would be advantageous to increase the yield of crosslinked adduct. For example, halogenated isosteric analogues of thymine, such as 5′-bromo-deoxyuridine or 5′-iodo-deoxyuridine (2), have commonly been used for such purposes. Their chromophores are excited by long-wavelength UV light ($\lambda > 300$ nM) that, by itself, does not activate or damage the DNA fragment. The nucleotides are readily available as phosphoroamidites and hence PCR primers can be synthesized with the appropriate substitution at their 5′ ends. The photoreactive nucleotide should be radiolabelled and incorporated into the DNA fragment in the same way as described above. Alternative approaches involve coupling photoreactive azide moieties, such as azidophenacyl bromide (3), into a thiophosphate introduced at specific positions within a synthetic oligonucleotide (4). While some precision may be lost in the UV crosslinking owing to structural perturbations in the complex during the prolonged course of the reaction (several minutes of UV irradiation are typically required), the substitution may, nonetheless, provide remarkable crosslinking efficiencies (often one order of magnitude higher than the quantum yield of the laser). This makes photoreactive-base substitution an attractive candidate for some purposes.

Protocol 2

Transfer of labelled nucleotide to the DNA-binding protein by UV crosslinking and nuclease digestion

Equipment and reagents

- Benzon nuclease (100,000 units/vial, Merck)[a]
- Snake venom phosphodiesterase I (1 mg/vial, Amersham Pharmacia Biotech)[b]
- SDS sample buffer (0.125 M Tris-HCl (pH 6.8), 0.2 M DTT, 4% SDS, 20% glycerol, 0.01% bromophenol blue)
- UV crosslinking apparatus (e.g. pulsed Nd:YAG laser)
- SDS polyacrylamide gel electrophoresis apparatus, gel and running buffers
- ^{14}C-labelled rainbow high molecular weight protein markers (Amersham)

Method

1. Perform the DNA-binding reaction under the appropriate conditions and irradiate the sample with a UV light source. Irradiate a protein-free sample to serve as a control.
2. To each reaction, add 0.5 μl Benzon nuclease (> 100 units) and incubate for 30 min to 1 h at 37 °C. Briefly spin down the reactions on a bench-top centrifuge (for a few seconds) to collect the liquid at the bottom of the tube.
3. Add 1 μl phosphodiesterase I, and continue the incubation for a further 30 min to 1 h at 37 °C.[c]
4. Stop the reaction by adding a 1/10 vol. of 10× SDS sample buffer and heat to 95°C for 3 min. Centrifuge the tubes again briefly to collect the liquid at the bottom of the tube.
5. Separate the UV crosslinked protein–nucleotide adduct by SDS-PAGE. Migrate the sample beside the control reaction and ^{14}C-labelled protein markers. After electrophoresis, dry the gels and visualize by autoradiography or phosphorimager densitometry.

[a] Ensure that the commercial grade purchased is free of contaminating proteases.

[b] Resuspend lyophilized phosphodiesterase at 1 mg/ml in a solution of 110 mM Tris-HCl (pH 8.9), 110 mM $NaCl_2$, 15 mM $MgCl_2$ and 50% glycerol.

[c] Omit step 3 for reactions if less precision in locating the crosslinked base is required (see text).

2.2.3. Nucleases

After formation and irradiation of the protein–DNA assembly, the DNA substrate is fragmented with nucleases so that thc adduct formed with the region of interest may be sized accurately by SDS-PAGE (*Protocol 2*). Two such nucleases are described here. Benzon nuclease is a non-specific DNAse I-type endonuclease, derived originally from *Serratia marscescens* and over-expressed and purified from

E. coli; it exhibits a high specific activity for dsDNA, cleaving it rapidly into small fragments of ≈ 3–8 nt. Snake venom phosphodiesterase I (PDE I) is an exonuclease that progressively releases 5′-mononucleotides from ssDNA in the 3′–5′ direction. While PDE I is highly efficient with ssDNA or short DNA duplexes (less than 50 nt), it will not cleave large DNA fragments efficiently. However, if the DNA is first digested into small fragments by Benzon nuclease, then a subsequent incubation with PDE I will quickly reduce these into mononucleotides.

3 Specific example of promoter recognition by the *E. coli* RNA polymerase

The *E. coli* RNA polymerase core enzyme is composed of several subunits arranged in the stoichiometry $\alpha_2\beta\beta'$. A σ subunit is also required for the recognition of a specific promoter. While several genetic studies have indicated that region 4.2 of σ^{70} is directly involved in the recognition of the −35 consensus region (Pribnow box) on the promoter, it remained uncertain whether σ^{70} was also implicated in the catalysis of promoter melting in the transcription initiation complex. The rate of appearance of a protein–DNA contact in the −35 region of the *lac*UV5 promoter, identified by primer extension analysis at −33 (5), was found to be coincident with a well-characterized kinetic intermediate of this process (Rp_i) (6). To identify the subunit involved, a 203 bp DNA fragment was prepared, with a radiolabel positioned at position −34 of the *lac*UV5 promoter (*Figure 2*).

Protocol 3

TCA precipitation of the protein–nucleotide adduct

Reagents

- 20% (w/v) TCA stored at 4 °C
- 80% acetone (stored at -20 °C)

Method

1. To the reaction, add an equal volume of 20% TCA. Place the solution on ice for 5 min.
2. Centrifuge at 10 000 × g for 10 min at 4 °C.
3. Carefully withdraw the supernatant with a pipette. Add 500 μl of 80% acetone.
4. Repeat steps 2 and 3 twice.
5. Dry the pellet on the bench or under vacuum in a centrifuge.
6. Resuspend either in 2× SDS gel loading buffer (for SDS-PAGE) or in a suitable buffered solution for further manipulations.

Complexes were formed between *E. coli* RNA polymerase (200 nM) and a 203 bp DNA fragment (10 nM) in a 10 μl volume. After 15 min at 37 °C, heparin was added to 0.1 mg/ml. After a further 5 min to allow dissociation of unstable (closed)

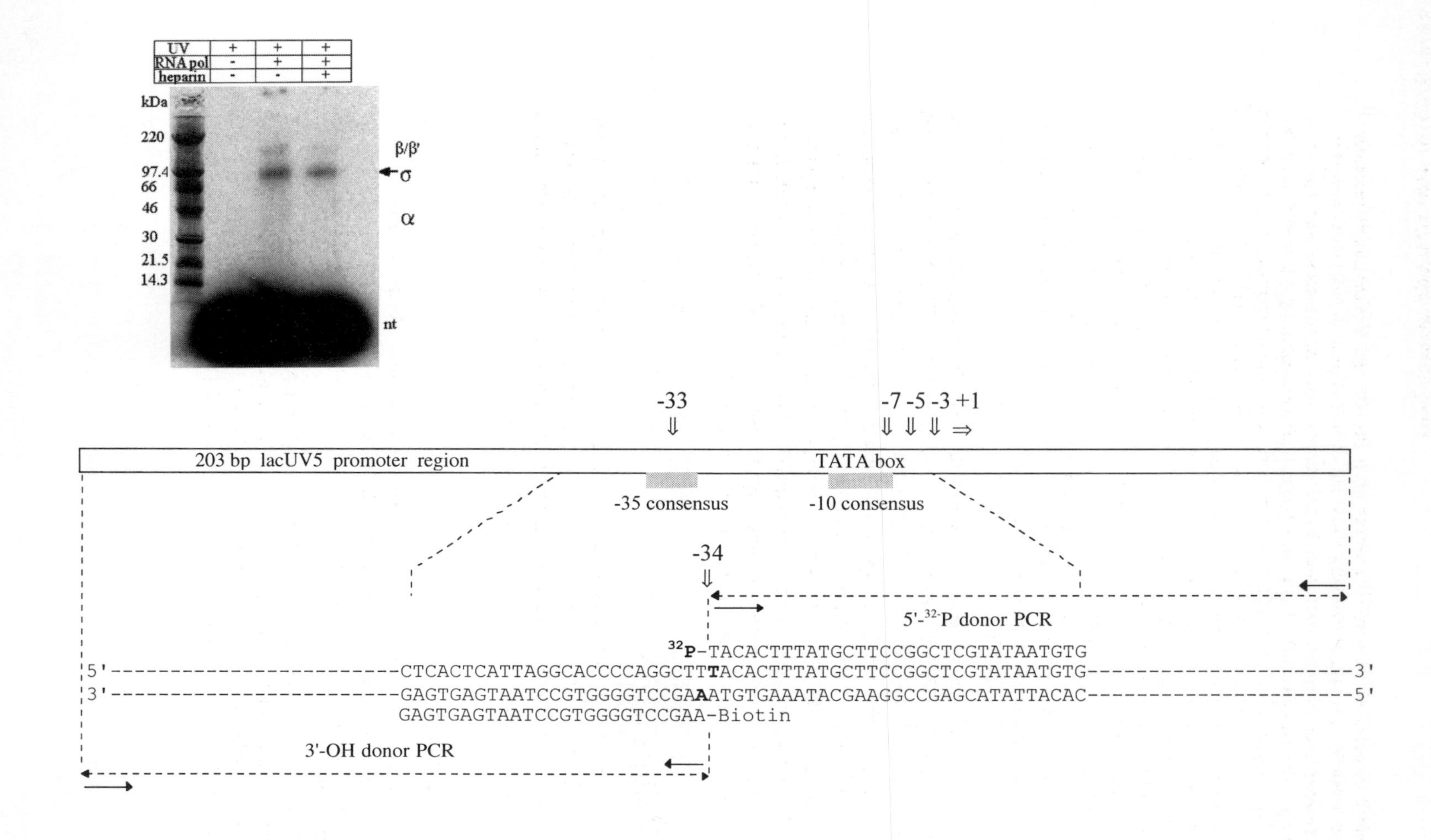

UV + + +
RNA pol - + +
heparin - - +
kDa
220
97.4
66
46
30
21.5
14.3
β/β'
σ
α
nt
-33
-7 -5 -3 +1
203 bp lacUV5 promoter region
TATA box
-35 consensus
-10 consensus
-34
5'-32P donor PCR
32P-TACACTTTATGCTTCCGGCTCGTATAATGTG
5'-CTCACTCATTAGGCACCCCAGGCTTTACACTTTATGCTTCCGGCTCGTATAATGTG-3'
3'-GAGTGAGTAATCCGTGGGGTCCGAAATGTGAAATACGAAGGCCGAGCATATTACAC-5'
GAGTGAGTAATCCGTGGGGTCCGAA-Biotin
3'-OH donor PCR

Figure 2 Detection of intimate σ^{70} contacts within the –35 region of the lacUV5 promoter. Primer extension analysis on irradiated *Escherichia coli* RNA polymerase–*lac*UV5 promoter complexes (2) identified four protein–DNA contacts at or near the two consensus regions at –35 and –10 (positions –33, –7, –5 and –3). The nucleotide occurring immediately after the –33 stop (–34) was radiolabelled and incorporated into the 203-bp fragment, as described in Figure 22.1 and the text. The sequences of the set of juxtaposed internal PCR primers required for this are given above. The reconstructed promoter (10 nM) was incubated in the presence or absence of the RNA polymerase holoenzyme (200 n) for 15 min at 37 °C prior to UV-laser irradiation and complete nucleolytic digestion with Benzon nuclease for 2 h at 37 °C. The subunits crosslinked to the –34 region were identified by electrophoresis on 8–25% SDS-PAGE and autoradiography. From left to right: lane 1, molecular weight markers; lane 2, UV-irradiated DNA control; lanes 3 and 4, UV irradiation of RNA polymerase–promoter DNA complexes before (lane 3) or after (lane 4) a 2.5-min chase period with heparin (0.1 mg/ml). The molecular weights of the marker proteins and the expected electrophoretic positions of the individual polymerase subunits are noted.

complexes, the reaction was irradiated with a single, 5 ns pulse of 266 nm UV light emitted by a Nd:YAG laser (Chapter 14). The protein–DNA adduct was then digested with Benzon nuclease (for 2 h at 37 °C) prior to analysis by SDS-PAGE. The predominant crosslinked species observed migrated close to that expected for the σ^{70} subunit. These results thus provided the first direct evidence that, in agreement with the conclusions of the former genetic studies, σ^{70} not only interacts directly with the −35 consensus sequence, but also plays a direct role in the engagement of the promoter DNA during the formation of RP_i, the forerunner to an open and transcriptionally competent complex (7).

4 Perspectives

In addition to the variety of complex transcriptional structures currently known, several other fundamental biological processes such as DNA replication, recombination and repair also involve complex nucleoprotein assemblies which should prove amenable to the approach described above.

References

1. Buckle, M., Geiselmann, J., Kolb, A., and Buc, H. (1991). *Nucl. Acids Res.* **19**, 833.
2. Blatter, E. E., Ebright, Y. W., and Ebright, R. H. (1992). *Nature* **359**, 650.
3. Willis, M. C., Hicke, B. J., Uhlenbeck, O. C., Chec, T. R., and Koch, T. H. (1993). *Science* **262**, 1255.
4. Hixson, S. H. and Hixson, S. S. (1975). *Biochemistry* **14**, 4251.
5. Yang, S. and Nash, H. A. (1994). *Proc. Natl. Acad. Sci. USA* **91**, 12183.
6. Buc, H. and McClure, W. (1985). *Biochemistry* **24**, 2712–2723
7. Buckle, M., Pemberton, I. K., Jacquet, M. A., and Buc, H. (1999). *J. Mol. Biol*, **285**, 955.

Chapter 23

Site-specific protein–DNA photocrosslinking

Tae-Kyung Kim, Thierry Lagrange and Danny Reinberg
Howard Hughes Medical Institute and Department of Biochemistry, University of Medicine and Dentistry of New Jersey-Robert Wood Johnson Medical School, Piscataway, NJ 08854, USA

Nikolai Naryshkin and Richard H. Ebright
Howard Hughes Medical Institute, Waksman Institute, and Department of Chemistry, Rutgers University, Piscataway, NJ 08854, USA

1 Introduction

Transcription and replication involve multiprotein–DNA complexes with molecular weight masses in excess of 400 kDa (1–5). Understanding transcription and replication will require understanding the arrangement of proteins relative to DNA in these complexes. High-resolution structures of components of the complexes have been determined (1–5). However, the intact, fully assembled, complexes are too large for high-resolution structure-determination by current methods. Therefore, efforts to define the arrangement of proteins relative to DNA within the intact complexes must rely heavily on biochemical methods.

Recently, we have developed a site-specific protein–DNA photocrosslinking procedure to define positions of proteins relative to DNA in multiprotein–DNA complexes (6–8). The procedure provides three types of positional information: the translational position relative to the DNA sequence, the rotational orientation relative to the DNA helix axis, and the groove orientation relative to the DNA major and minor grooves (6–8). The procedure has been applied successfully to analysis of bacterial and eukaryotic transcription complexes, including a eukaryotic transcription complex containing 27 distinct polypeptides and having a molecular mass in excess of 1700 kDa (6–9; unpublished data).

2 Procedure

2.1 Outline of procedure

The procedure has four parts (*Figure 1*):

(1) Chemical (10–12) and enzymatic (13) reactions are used to prepare a DNA

fragment containing a photoactivatible crosslinking agent and an adjacent radiolabel incorporated at a single, defined DNA phosphate.

(2) The multiprotein-DNA complex of interest is formed using the site-specifically derivatized DNA fragment, and the multiprotein–DNA complex is UV-irradiated, initiating covalent crosslinking with polypeptides in direct physical proximity to the photoactivatible crosslinking agent.

(3) Extensive nuclease digestion is performed, eliminating uncrosslinked DNA and converting crosslinked DNA to a crosslinked, radiolabelled 3–5 nt 'tag'.

(4) The 'tagged' polypeptides are identified.

This procedure is related to a procedure developed by Geiduschek and co-workers (14–17; see also 18–23), but offers important advantages. First, since the photo-activatible crosslinking agent is incorporated into DNA chemically, it can be incorporated at a single, defined site. (In the procedure of Geiduschek and co-

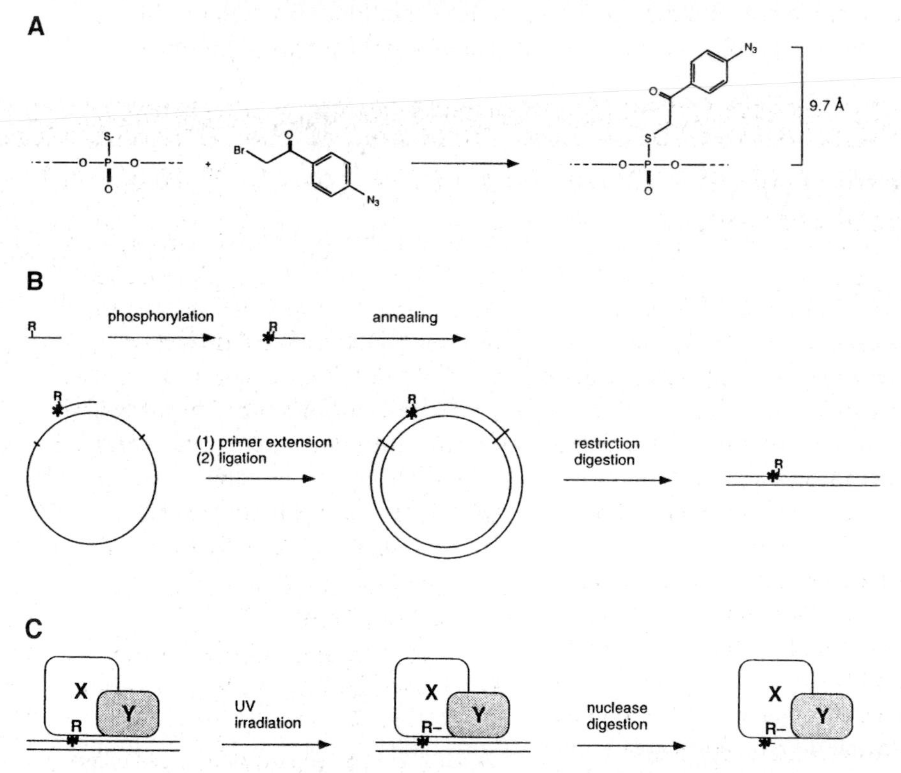

Figure 1. Site-specific protein-DNA photocrosslinking (6–8). (A, B) Chemical and enzymatic reactions are used to prepare a DNA fragment with a phenyl-azide photoactivatible cross-linking agent (R) and an adjacent radioactive phosphorus (*) incorporated at a single, defined site. Based on the chemistry of incorporation, the maximum distance between the site of incorporation and the photoreactive atom is 9.7 Å; the maximum distance between the site of incorporation and a crosslinked atom is ≈11 Å. (C) UV-irradiation of the derivatized protein-DNA complex initiates crosslinking. Nuclease digestion eliminates uncrosslinked DNA and converts crosslinked, radiolabelled DNA to a crosslinked, radiolabelled 3-5 nucleotide 'tag'.

workers, this is true only at certain DNA sequences.) Second, since the photo-activatible crosslinking agent is incorporated on the DNA phosphate backbone, it can be incorporated at any nucleotide: A, T, G or C. Third, since the photo-activatible crosslinking agent is incorporated on the DNA phosphate backbone, it probes interactions both in the DNA minor groove and in the DNA major groove.

2.2 Preparation of site-specifically derivatized DNA fragment

Figure 1A summarizes the chemical reactions involved in preparation of the site-specifically derivatized DNA fragment (*Protocol 1*). An oligodeoxyribonucleotide is synthesized having a phosphorothioate (a phosphate with a single-atom oxygen→sulfur substitution) at the position of interest and having two bases 5′ and 17 bases 3′ to the phosphorothioate. The resulting oligodeoxyribonucleotide is reacted with azidophenacyl bromide, a sulfur-specific bifunctional phenyl-azide photo-activatible crosslinking agent (24), under conditions that result in highly efficient, highly specific, incorporation of azidophenacyl at phosphorothioate.

Protocol 1

Preparation of site-specifically derivatized DNA fragment, chemical reactions

Equipment and reagents

- Azidophenacyl bromide (Sigma)
- Tetraethylthiuram disulfide/acetonitrile (PE Biosystems)
- dA-CPG, dC-CPG, dG-CPG, T-CPG (1 μmol, 500 Å) (PE Biosystems)
- dA, dC, dG, T β-cyanoethylphosphoramidites (PE Biosystems)
- Reagent kit for oligodeoxyribonucleotide synthesis (0.02 M iodine) (PE Biosystems)
- Dichloromethane (anhydrous) (PE Biosystems)
- Acetonitrile (anhydrous) (PE Biosystems)
- Acetonitrile (HPLC grade) (Fisher Scientific)
- Oligonucleotide purification cartridge (OPC) (PE Biosystems)
- 2 mM triethylammonium acetate, pH 7.0 (PE Biosystems)
- TE (10 mM Tris-HCl, pH 7.6, 1 mM EDTA)
- 0.5× TBE (45 mM Tris-borate, pH 8.3, 1 mM EDTA)
- 1 M potassium phosphate, pH 7.0
- 3 M sodium acetate, pH 5.2
- ABI392 DNA/RNA synthesizer (PE Biosystems)
- Varian 5000 HPLC system (Varian)
- L-3000 diode-array HPLC UV detector (Hitachi)
- LiChrospher 100 RP-18 reversed-phase HPLC column (5 μm) (Merck)
- Speedvac evaporator (Savant)

A. Synthesis of phosphorothioate oligodeoxyribonucleotide

1 Perform 17 standard cycles of solid-phase β-cyanoethylphosphoramidite oligodeoxyribonucleotide synthesis to prepare CPG-linked precursor containing residues

Protocol 1 continued

3–19 of the desired oligodeoxyribonucleotide. Use the following settings: cycle, 1.0 μM CE; DMT, on; end procedure, manual.

2. Perform one modified cycle of solid-phase β-cyanoethylphosphoramidite oligodeoxyribonucleotide synthesis to add residue 2 and phosphorothioate linkage. Replace iodine/water/pyridine/tetrahydrofuran solution (bottle 15) with tetraethylthiuram disulfide/acetonitrile solution, and use the following synthesizer settings: cycle, 1.0 μM sulfur; DMT, on; end procedure, manual.
3. Perform one standard cycle of solid-phase β-cyanoethylphosphoramidite oligodeoxyribonucleotide synthesis to add residue 1. Replace tetraethylthiuram disulfide/acetonitrile solution (bottle 15) with iodine/water/pyridine/tetrahydrofuran solution. Place collecting vial on the DNA synthesizer, and use the following settings: cycle, 1.0 μM CE; DMT, on; end procedure, CE.
4. Remove collecting vial, screw cap tightly, and deblock by incubating 8 h at 55 °C. Transfer sample to 6 ml polypropylene round-bottom tube, place tube in Speedvac and spin for 20 min at room temperature with Speedvac lid ajar and with no vacuum (allowing evaporation of ammonia). Close Speedvac lid, apply vacuum, and lyophilize.
5. Detritylate and purify ≈ 0.1 μmol on OPC according to supplier's protocol.
6. Lyophilize in Speedvac.
7. Resuspend in 100 μl TE, Remove 2 μl aliquot, dilute with 748 μl TE, and determine concentration from UV absorbance at 260 nm (molar extinction coefficient = 175 000 AU/M·cm).
8. To confirm purity of oligodeoxyribonucleotide, mix aliquot containing 1.5 nmol oligodeoxyribonucleotide with equal volume of formamide. Apply to 12% polyacrylamide (30 : 0.8 acrylamide : bisacrylamide), 8 M urea, 0.5× TBE slab gel (100 × 70 × 0.075 cm). As marker, load 10 μl formamide containing 0.3% xylene cyanol and 0.3% bromophenol blue in the adjacent lane. Electrophorese 30 min at 15 V/cm. Disassemble gel, place onto autoradiography intensifying screen, and view in dark using 254 nm germicidal lamp. Oligodeoxyribonucleotide should appear as dark shadow against green background and should migrate slightly more slowly than bromophenol blue. If purity is ≥ 95%, proceed to next step.
9. Divide remainder of oligodeoxyribonucleotide into 50-nmol aliquots, transfer to 1.5 ml siliconized polypropylene microcentrifuge tubes, lyophilize in Speedvac, and store at −20 °C (stable for at least 2 years).

B. Derivatization of phosphorothioate oligodeoxyribonucleotide with azidophenacyl bromide (all steps carried out under subdued lighting[a])

1. Dissolve 10 mg (42 μmol) azidophenacyl bromide in 1 ml of chloroform. Transfer 100 μl aliquots (4.2 μmol) to 1.5 ml siliconized polypropylene microcentrifuge tubes, and lyophilize in Speedvac. Wrap tubes with aluminium foil and store desiccated at 4 °C (stable for at least a year).

Protocol 1 continued

2 Resuspend 50-nmol aliquot of phosphorothioate oligodeoxyribonucleotide (*Protocol 1A*, step 9) in 50 μl water, and 42 μmol aliquot of azidophenacyl bromide (step 1) in 220 μl methanol.

3 Mix 50 μl (50 nmol) phosphorothioate oligodeoxyribonucleotide solution, 5 μl 1 M potassium phosphate (pH 7.0), and 55 μl (1 μmol) azidophenacyl bromide solution in a 1.5 ml siliconized polypropylene microcentrifuge tube. Incubate for 3 h at 37 °C in the dark.

4 Precipitate derivatized oligodeoxyribonucleotide by adding 11 μl 3 M sodium acetate (pH 5.2), and 275 μl ice-cold 100% ethanol. Invert tube several times and store at −80 °C for 30 min. Centrifuge 15 min at 13 000 **g** at 4 °C. Remove supernatant, wash pellet with 300 μl ice-cold 70% ethanol. Air dry for 15 min at room temperature. Store at −20 °C (stable for at least 1 year).

C. Purification of derivatized oligodeoxyribonucleotide (all steps carried out under subdued lighting[a])

1 Resuspend derivatized oligodeoxyribonucleotide in 100 μl 50 mM triethylammonium acetate (pH 7.0).

2. Analyse 5 μl aliquot by C18 reversed-phase HPLC to confirm efficiency of derivatization reaction. Use LiChrospher 100 RP-18 C18 reversed-phase HPLC column (5 μm), with solvent A (50 mM triethylammonium acetate, pH 7.0, 5% acetonitrile), solvent B (100% acetonitrile) and a flow rate of 1 ml/min. Equilibrate column with 10 column volumes of solvent A before loading sample. After loading sample, wash column with six column volumes solvent A, and elute with 50-min gradient of 0–70% solvent B in solvent A. Derivatized and underivatized oligodeoxyribonucleotides elute at ≈ 25% solvent B and ≈ 16% solvent B, respectively.[b–d]

3 If derivatization efficiency is ≥ 80%, purify remainder of sample using procedure in step 2, and collect peak fractions.[b–d]

4 Pool peak fractions, divide into 1-ml aliquots, and lyophilize in Speedvac. Store desiccated at −20 °C in the dark (stable for at least 1 year).

5 Resuspend one aliquot in 100 μl TE. Remove 5 μl, dilute with 495 μl water, and determine concentration from UV absorbance at 260 nm (molar extinction coefficient = 177 000 AU/M·cm).

6 Divide remainder of derivatized-oligodeoxyribonucleotide/TE solution from Step 5 into 20 10-pmol aliquots and one larger aliquot, lyophilize in Speedvac, and store desiccated at −20 °C in the dark (stable for at least 1 year).

[a] Fluorescent light and daylight must be excluded. Low to moderate levels of incandescent light (e.g. from single task lamp with 60 W tungsten bulb) are acceptable.

[b] If using the PerSeptive Biosystems BioCADSPRINT HPLC system and POROS 20 R2 C18 reversed-phase column, use a flow rate of 5 ml/min and 5-min gradient of 0–40% solvent B in solvent A. With this HPLC system, derivatized and underivatized oligodeoxyribonucleotides elute at ≈ 30% solvent B and ≈ 15% solvent B, respectively.

Protocol 1 continued

[c] The derivatized oligodeoxyribonucleotide tolerates exposure to the Hitachi L-3000 diode-array HPLC UV detector (and also to the BioCADSPRINT HPLC UV detector). The derivatized oligodeoxyribonucleotide can be identified unambiguously by monitoring the UV absorbance spectrum from 200–350 nm. The derivatized oligodeoxyribonucleotide exhibits an absorbance peak at 260 nm, attributable to DNA, and a shoulder at 300–310 nm, attributable to the azidophenacyl group.

[d] The derivatization procedure yields two diastereomers in an ≈ 1-to-1 ratio: one in which azidophenacyl is incorporated at the atom corresponding to the phosphate O1P, and one in which azidophenacyl is incorporated at the atom corresponding to the phosphate O2P (see refs. 10 and 12). Depending on oligodeoxyribonucleotide sequence and HPLC conditions, the two diastereomers may elute as a single peak, or as two peaks (e.g. at 24% and 25% solution B). In most cases, no effort is made to resolve the two diastereomers, and experiments are performed using the unresolved diastereomeric mixture. This permits simultaneous probing of protein–DNA interactions in the DNA minor groove (probed by the O1P-derivatized diastereomer) and the DNA major groove (probed by the O2P-derivatized diastereomer).

Figure 1B summarizes the subsequent enzymatic reactions involved in preparation of the site-specifically derivatized DNA fragment (*Protocol 2*). The oligodeoxyribonucleotide conjugate is radiophosphorylated using T4 polynucleotide kinase and [γ-^{32}P]ATP, resulting in incorporation of a ^{32}P radiolabel two bases from the azidophenacyl group. The radiolabelled oligodeoxyribonucleotide conjugate is annealed to a circular single-stranded DNA template containing the sequence of interest, extended using T4 DNA polymerase, and ligated using T4 DNA ligase. The resulting derivatized circular double-stranded DNA is digested using a pair of restriction enzymes, yielding the derivatized DNA fragment.

In work to date, these methods have been used to generate derivatized DNA fragments 100–300 bp long (6–8, unpublished data). In principle, these methods could be used to generate derivatized DNA fragments up to several thousand base pairs long.

Protocol 2

Preparation of site-specifically derivatized DNA fragment, enzymatic reactions

Equipment and reagents

- Derivatized oligodeoxyribonucleotide (*Protocol 1*)
- M13mp18-AdMLP or M13mp19-AdMLP ssDNA[a] (or analogous ssDNA)
- T4 polynucleotide kinase (10 units/μl) (cat. no. 2312, Ambion)
- T4 DNA polymerase (3 units/μl) (cat. no. 203S, New England Biolabs)
- T4 DNA ligase (5 units/μl) (Roche Molecular Biochemicals cat. no. 799009, Roche Molecular Biochemicals)
- *Eco*RI (10 units/μl) (cat. no. 15202–021, GIBCO-BRL) (or other restriction enzyme appropriate for theDNA sequence)
- *Sph*I (10 units/μl) (cat. no. 1026534, Roche Molecular Biochemicals) (or other restriction enzyme appropriate for the DNA) sequence)

Protocol 2 continued

- [γ-^{32}P]ATP (10 mCi/ml, 6000 Ci/mmol) (NEN)
- 100 mM ATP (Pharmacia)
- 100 mM dNTPs (Roche Molecular Biochemicals)
- 17-mer M13 universal primer (Pharmacia)
- Poly(dG-dC)/poly(dG-dC) (MW_{av} = 500 kDa) (Pharmacia)
- 20× annealing buffer (300 mM Tris-HCl, pH 8.0, 500 mM KCl, 70 mM $MgCl_2$)
- 10× phosphorylation buffer (500 mM Tris-HCl, pH 7.6, 100 mM $MgCl_2$, 15 mM β-mercaptoethanol)
- Ligase dialysis buffer (20 mM Tris-HCl, pH 7.5, 60 mM KCl, 1 mM EDTA, 10 mM β-mercaptoethanol, 500 μg/ml BSA, 0.1 mM PMSF, 50% glycerol)
- 10× digestion buffer (500 mM Tris-HCl, pH 8.0, 1 M NaCl, 100 mM $MgCl_2$)[b]
- Elution buffer (10 mM Tris-HCl, pH 8.0, 200 mM NaCl, 1 mM EDTA)
- Denaturing loading buffer (0.3% bromophenol blue, 0.3% xylene cyanol, 12 mM EDTA in formamide)
- TE (10 mM Tris-HCl, pH 7.6, 1 mM EDTA)
- Non-denaturing loading buffer (0.3% bromophenol blue, 0.3% xylene cyanol, 30% glycerol in water)
- 0.5× TBE (45 mM Tris-borate, pH 8.3, 1 mM EDTA)
- 0.5 M EDTA, pH 8.0
- 10% SDS
- CHROMA SPIN+TE-10 spin column (Clontech)
- CHROMA SPIN + TE-100 spin column (Clontech)
- PicoGreen dsDNA quantification kit (cat. no. P-7589, Molecular Probes)
- Spin-X centrifuge filter (0.22 μm, cellulose acetate) (Fisher)
- Dialysis membranes (6–8 kDa molecular weight cut off) (Fisher)
- Glogos II autorad markers (Stratagene)
- Speedvac evaporator (Savant)

A. Radiophosphorylation of derivatized oligodeoxyribonucleotide (all steps carried out under subdued lighting[c])

1. Resuspend 10 pmol derivatized oligodeoxyribonucleotide in 12 μl water. Add 2 μl 10× phosphorylation buffer, 5 μl [γ-^{32}P]ATP (50 μCi) and 1 μl (10 units) T4 polynucleotide kinase. Incubate 15 min at 37 °C. Terminate reaction by heating 5 min at 65 °C.[d]
2. Add 15 μl water.
3. Desalt radiophosphorylated derivatized oligodeoxyribonucleotides into TE using CHROMA SPIN+TE-10 spin column according to supplier's protocol.
4. Proceed immediately to next step, or, if necessary, store radiophosphorylated derivatized oligodeoxyribonucleotide solution at −20 °C in the dark (stable for up to 24 h).

B. Annealing, extension, and ligation of radiophosphorylated derivatized oligodeoxyribonucleotide (all steps carried out under subdued lighting[c])

1. In a 1.5 ml siliconized polypropylene microcentrifuge tube, mix 31 μl radiophosphorylated derivatized oligodeoxyribonucleotide, 1 μl 10 μM 17-mer M13 universal primer, 1 μl 1 μM M13mp18-AdMLP or M13mp19-AdMLP ssDNA (or analogous ssDNA)[a], and 2 μl 20× annealing buffer.

Protocol 2 continued

2 Heat 5 min at 65 °C.[d] Transfer to 500 ml beaker containing 200 ml water at 65°C, and place beaker at room temperature to permit slow cooling (65 °C to 25 °C in ≈ 60 min).

3 Add 1 μl 25 mM dNTPs, 1 μl 100 mM ATP, 1 μl (3 units) T4 DNA polymerase, and 1.5 μl (7.5 units) dialysed T4 DNA ligase.[e,f] Perform parallel reaction without ligase as a 'no-ligase' control.

4 Incubate for 15 min at room temperature, followed by 35 min at 37 °C. Terminate reaction by adding 1 μl 10% SDS, followed by 10 μl water.

5 Desalt into TE using CHROMA SPIN+ TE-100 spin column according to supplier's protocol.

6 Proceed immediately to next step.

C. Digestion and purification of derivatized DNA fragment (all steps carried out under subdued lighting[c])

1 In 1.5 ml siliconized polypropylene microcentrifuge tube, mix 43.5 μl product from *Protocol 2B*, 5 μl 10× digestion buffer, 1 μl (10 units) *Eco*RI, and 1 μl (10 units) *Sph*I (or other restriction enzymes appropriate for the DNA sequence). Incubate 1 h at 37 °C. Terminate reaction by adding 1.5 μl 0.5 M EDTA (pH 8.0).

2 Perform parallel reaction using 43.5 μl of 'no-ligase' control from *Protocol 2B*.

3 Mix 3 μl aliquot of reaction from step 1 with 7 μl denaturing loading buffer, and mix 3 μl aliquot of the 'no ligase' control reaction of step 2 with 7 μl denaturing loading buffer. Heat for 5 min at 65°C, and then apply to 12% polyacrylamide (30 : 0.8 acrylamide : bisacrylamide), 8 M urea, 0.5× TBE slab gel (100 × 70 × 0.075 cm). As marker, load 5 μl denaturing loading buffer in adjacent lane. Electrophorese 30 min at 15 V/cm. Dry gel, expose to X-ray film 1 h at −80 °C, and process film. Estimate ligation efficiency by comparing products in reaction and 'no-ligase' control lanes. If ligation efficiency is ≥ 80%, proceed to next step.

4 Mix remainder of reaction from step 1 (49 μl) with 10 μl 50% glycerol. Apply to non-denaturing 8% polyacrylamide, 0.5× TBE slab gel (100 × 70 × 0.15 cm). As marker, load 5 μl non-denaturing loading buffer in adjacent lane. Electrophorese at 7 V/cm until bromophenol blue reaches bottom of the gel.

5 Remove one glass plate and cover gel with plastic wrap. Attach two autorad markers to gel. Expose to x-ray film for 1–1.5 min at room temperature and process film. Cut out portion of film corresponding to derivatized DNA fragment. Using a light-box, superimpose cut-out film on gel, using autorad markers as alignment references. Using disposable scalpel, excise portion of gel corresponding to derivatized DNA fragment.

6 Transfer excised gel slice to 2 ml siliconized polypropylene microcentrifuge tube. Add 500 μl elution buffer, wrap tube with aluminium foil, and rock gently for 12 h at 37 °C.[g]

Protocol 2 continued

7. Transfer supernatant to Spin-X centrifuge filter, centrifuge 1 min at 13 000 **g** at room temperature in fixed-angle microcentrifuge.
8. Transfer filtrate to 1.5 ml siliconized polypropylene microcentrifuge tube. Add 1 μl 1 mg/ml poly(dG-dC)/poly(dG-dC). Precipitate derivatized DNA fragment by addition of 1 ml ice-cold 100% ethanol. Invert tube several times and place at −80 °C for 30 min. Centrifuge 10 min at 13 000 **g** at 4 °C in a fixed-angle microcentrifuge. Remove supernatant, wash pellet with 1 ml ice-cold 70% ethanol, and air dry 15 min at room temperature.
9. Resuspend in 30 μl TE. Determine radioactivity by Cerenkov counting. Remove 1 μl aliquot, and determine DNA concentration using PicoGreen dsDNA quantification kit according to supplier's protocol. Calculate specific activity (expected specific activity = 2000–5000 Ci/mmol).
10. Store derivatized DNA fragment at 4 °C in the dark (stable for ≈ 1 week).

[a] M13mp18-AdMLP and M13mp19-AdMLP are described in (6). M13mp18-AdMLP and M13mp19-AdMLP ssDNA templates are prepared as in (13). For preparation of derivatized DNA fragments < 120 bp long, synthetic ssDNA can be used in place of phage ssDNA (9).

[b] This 10× digestion buffer is for *Eco*RI and *Sph*I. Use 10× digestion buffer recommended by supplier, omitting DTT (25), for other restriction enzymes.

[c] Fluorescent light and daylight must be excluded. Low to moderate levels of incandescent light (e.g. from single task lamp with 60 W tungsten bulb) are acceptable.

[d] Phenyl-azides are unstable at temperatures above 70 °C. Avoid heating above this temperature.

[e] The T4 DNA ligase stock solution contains 5 mM DTT, which efficiently reduces phenyl-azides to photo-inert phenyl amines (25), and therefore the T4 DNA ligase stock solution must be dialysed before use[f]. T4 DNA polymerase, T4 polynucleotide kinase, *Eco*RI, and *Sph*I stock solutions purchased from the suppliers specified in Equipment and reagents contain lower concentrations of DTT (1 mM), and can be used without dialysis.

[f] One-hundred microlitres of T4 DNA ligase stock solution (5 units/μl) is dialysed against 2 l ligase dialysis buffer for 4 h at 4°C. Dialysed ligase is stored in 10-μl aliquots at −20 °C, and is stable for at least 1 month.

[g] Alternatively, place excised gel slice in a 1.5 ml siliconized polypropylene microcentrifuge tube, and crush with 1 ml pipette tip. Add 300 μl elution buffer, centrifuge 5 s at 5000 **g**, vortex, and rock for 12 h at 37 °C.

2.3 Photocrosslinking

The site-specifically derivatized DNA fragment is used to prepare the multiprotein–DNA complex of interest. The resulting multiprotein–DNA complex is UV-irradiated, either directly in solution ('solution photocrosslinking'; *Protocol 3*) (6–9), or after isolation by non-denaturing PAGE, with UV-irradiation *in situ* in the gel matrix ('in-gel photocrosslinking'; *Protocol 4*) (7). In-gel photocrosslinking is more labour-intensive, but is more reliable. In-gel photocrosslinking is recommended in all cases and is strongly recommended in cases in which non-specific binding occurs, and/or in which multiple, different complexes coexist in solution.

Protocol 3
Solution photocrosslinking

Equipment and reagents

- Derivatized AdMLP DNA fragment (*Protocol 2*)
- Affinity purified human RNA polymerase II (RNAPII)[a]
- Recombinant human TATA-element binding protein (TBP)[a]
- Recombinant human transcription factor IIB[a]
- Recombinant human transcription factor IIF[a]
- ^{14}C-labelled protein molecular weight standards (GIBCO-BRL)
- DNase I (140 units/μl) (cat. no. 18047-019, GIBCO-BRL)
- S1 nuclease (930 units/μl) (cat. no. 18001-024, GIBCO-BRL,)
- S1 nuclease dilution buffer (GIBCO-BRL, provided with S1 nuclease)
- Buffer A (20 mM Tris-HCl, pH 7.9, 100 mM KCl, 0.2 mM EDTA, 1 mM DTT, 0.2 mM PMSF, 20% glycerol)
- Buffer B (20 mM Tris-HCl, pH 7.9, 100 mM KCl, 0.2 mM EDTA, 1 mM DTT, 0.2 mM PMSF, 40% glycerol)
- Buffer C (200 mM HEPES-KOH, pH 7.9, 80 mM $MgCl_2$)
- 1.5 M Tris-HCl, pH 8.8.
- 5× SDS loading buffer (0.3 M Tris-HCl, pH 6.8, 10% SDS, 25% β-mercaptoethanol, 0.1% bromophenol blue, 50% glycerol)
- SDS running buffer (25 mM Tris, 250 mM glycine, pH 8.3, 0.1% SDS)
- 10× protease inhibitor solution (2 mM PMSF, 10 μg/ml aprotinin, 5 μg/ml leupeptin, 7 μg/ml pepstatin, 1 mg/ml chymostatin)
- PMSF (Sigma)
- Aprotinin (Roche Molecular Biochemicals)
- Leupeptin (Roche Molecular Biochemicals)
- Pepstatin (Roche Molecular Biochemicals)
- Chymostatin (Roche Molecular Biochemicals)
- Poly(dG-dC)/poly(dG-dC) (MW_{av} = 500 kDa) (Pharmacia)
- Polyethylene glycol (MW_{av} = 8 kDa) (Sigma)
- 10% SDS
- Dialysis membranes (6–8 kDa molecular weight cut off) (Fisher)
- 1 ml polystyrene microcentrifuge tubes with cap (cat. no. 04-978-145, Fisher)
- 13 × 100-mm borosilicate glass culture tubes (cat. no. 60825-571, VWR)
- Rayonet RPR-100 photochemical reactor equipped with 16 RPR-3500 Å tubes and RMA400 sample holder (Southern New England Ultraviolet)
- Speedvac evaporator (Savant)

A. Formation of multiprotein–DNA complex (all steps carried out under subdued lighting[b])

1. Add the following, in order, to a 1 ml polystyrene microcentrifuge tube: 1 μl 4 nM derivatized AdMLP DNA fragment (2000–5000 Ci/mmol), 7 μl buffer A, 1 μl 0.1 μM TBP in buffer A, 1 μl 0.1 μM IIB in buffer A, 1 μl 0.1 μM IIF in buffer A, 2 μl 0.1 μM RNAPII in buffer B, 2 μl buffer C, 2.6 μl 20% polyethylene glycol, 0.5 μl 1 mg/ml poly(dG-dC)/poly(dG-dC), and 1.9 μl water[c].

Protocol 3 continued

2 Incubate 30 min at 28 °C in the dark.

3 Immediately proceed to next step.

B. UV irradiation

1 Place polystyrene tube inside 13 × 100 mm borosilicate glass culture tube.[d] Place in RMA400 sample holder of Rayonet RPR-100 photochemical reactor equipped with 16 RPR-3500 Å tubes.

2 UV irradiate for 2 min at 25 °C (11 mJ/mm^2 at 350 nm).

3 Proceed immediately to next step.

C Nuclease digestion

1 Transfer sample to siliconized polypropylene microcentrifuge tube.

2 Add 1 μl 115 mM $CaCl_2$, 0.1 μl (13 units) DNase I, and 2 μl 10× protease inhibitor solution. Incubate for 15 min at 37 °C. Terminate reaction by adding 1 μl 10% SDS and heating for 5 min at 65 °C.

3 After allowing the sample to cool to room temperature, add 1 μl 25 mM $ZnCl_2$, 1 μl 1 M acetic acid, 1 μl 10× protease inhibitor solution, and 1 μl (30 units) S1 nuclease (diluted to 30 units/μl in S1 nuclease dilution buffer). Incubate 10 min at 37 °C.

4 Immediately proceed to next step.

D. Analysis

1 Adjust pH by adding 2 μl 1.5 M Tris-HCl, pH 8.8.

2 Add 10 μl 9 M urea and 6 μl 5× SDS loading buffer. Heat 3 min at 65°C. Apply entire sample (46 μl) to 7–15% gradient polyacrylamide (30 : 0.8 acrylamide/bisacrylamide), 0.1% SDS, slab gel (27 × 15 × 0.15 cm). As a marker, load 5 μl ^{14}C-labelled protein molecular weight marker in 5 μl 5× SDS loading buffer into adjacent lane . Electrophorese in SDS running buffer at 13 V/cm until bromophenol blue reaches bottom of gel.

3 Dry gel, and autoradiograph or phosphorimage.

[a] RNAPII and transcription factors are prepared as described in (26). Before use, 1 ml RNAPII (0.1 μM) is dialysed against 2 l buffer B for 5 h at 4°C, and 1 ml of each transcription factor (0.2–10 μM) is dialysed against 2 l buffer A for 5 h at 4 °C. Dialysed RNAPII and transcription factors are stored in aliquots at −80 °C (stable for at least 1 year).

[b] Fluorescent light and daylight must be excluded. Low to moderate levels of incandescent light (e.g. from single task lamp with 60 W tungsten bulb) are acceptable.

[c] Buffer C, 20% polyethylene glycol, poly(dG-dC)/poly(dG-dC), and water can be premixed and added together.

[d] Polystyrene and borosilicate glass exclude wavelengths < 300 nm, minimizing photodamage to protein and DNA.

Protocol 4

In-gel photocrosslinking

Equipment and reagents

- Equipment and reagents for solution photocrosslinking (*Protocol 3*)
- Cystamine dihydrocloride (Sigma)
- Acryloyl chloride (Aldrich)
- 1 M DTT (freshly made)
- Non-denaturing loading buffer (0.3% bromophenol blue, 0.3% xylene cyanol, 30% glycerol)
- Digital thermometer with 0.7 mm diameter needle thermocouple probe (Cole-Parmer)
- 0.5× TBE (45 mM Tris-borate, pH 8.0, 1 mM EDTA)
- Silicone heating mat with extension cord (Cole-Parmer)
- Voltage controller (Cole-Parmer)
- Filter unit (22 μm pore size, 250 ml) (Millipore)
- 50 ml Büchner funnel with glass frit (10 μm-pore size) (Fisher Scientific)
- 500 ml separating funnel (Fisher Scientific)

A. Synthesis of N,N′-bisacryloylcystamine (BAC)[a]

1. Dissolve 4.0 g (18 mmol) cystamine dihydrochloride in 40 ml 3 M NaOH (120 mmol). Dissolve 4.3 ml (54 mmol) acryloyl chloride in 40 ml chloroform. Mix solutions in 500 ml flask. (Two phases will form: an upper, aqueous phase and a lower, organic phase). Place flask on hot plate stirrer, and stir 15 min at 50 °C.
2. Discontinue stirring. Immediately transfer reaction mixture to separating funnel, allow phases to separate, and transfer lower, organic phase to 250 ml beaker.
3. Cool on ice for 10 min. Collect precipitate by filtration in Büchner funnel.
4. Redissolve precipitate in 30 ml chloroform with heating to 50 °C. Again, cool on ice for 10 min, and collect precipitate by filtration in Büchner funnel.
5. Transfer precipitate (BAC) to 50 ml polypropylene centrifuge tube. Seal tube with Parafilm, pierce seal several times with needle, place tube in vacuum desiccator, and dry under vacuum for 16 h at room temperature. Expected yield is 1.5–1.9 g.

B. Preparation of polyacrylamide–BAC gel

1. Prepare 20% acrylamide–BAC (19 : 1) stock solution by dissolving, in order, 19 g acrylamide and 1 g BAC in 80 ml water at 60°C and adjusting volume to 100 ml with water.[b] Filter stock solution using 0.22 μm filter unit and store at room temperature in the dark (stable for at least 2 months).
2. Mix 17.5 ml 20% acrylamide–BAC (19 : 1) stock solution, 3.5 ml 10× TBE, 3.5 ml 50% glycerol and 45 ml water at 24–25 °C. Add 350 μl TEMED and 200 μl freshly prepared 10% ammonium persulfate.[c] Mix thoroughly, and immediately pour into a slab gel assembly (27 × 15 × 0.15 cm). Insert comb. Allow 10–20 min for polymerization. (The polyacrylamide–BAC gel is stable for up to 3 days at 4 °C.)

Protocol 4 continued

C. Formation of multiprotein–DNA complex (all steps carried out under subdued lighting[d])

1 Follow procedures in *Protocol 3A*, substituting 1.5 ml siliconized polypropylene microcentrifuge tubes for polystyrene microcentrifuge tubes.

2 Proceed immediately to the next step.

D. Isolation of multiprotein–DNA complex (all steps carried out under subdued lighting[d])

1 Place 5% polyacrylamide-BAC gel assembly in electrophoresis apparatus, and pour 0.5× TBE buffer in upper and lower reservoirs. Wash wells of gel carefully with 0.5× TBE to remove unpolymerized acrylamide and BAC.

2 Apply 20 μl of solution containing multiprotein-DNA complex (*Protocol 4B*) to gel.[e] As a marker, load 5 μl non-denaturing loading buffer into adjacent lane. Electrophorese 3 h at 7 V/cm. Monitor gel temperature at 5-min intervals by inserting thermocouple probe into the gel for 5 s (and removing immediately thereafter). Maintain gel temperature at 28 °C.[f]

3 Immediately proceed to next step.

E. In-gel UV irradiation of multiprotein–DNA complex

1 Remove the gel with both glass plates in place,[g] and mount vertically in a Rayonet RPR-100 photochemical reactor equipped with 16 RPR-3500 Å tubes.

2 UV irradiate for 3 min (17 mJ/mm^2 at 350 nm).

3 Immediately proceed to next step.

F. Identification, excision and solubilization of portion of gel containing multiprotein–DNA complex

1 Remove one glass plate, and cover gel with plastic wrap (leaving other glass plate in place). Attach two autorad markers. Expose to X-ray film for 2 h at 4 °C. Process film.

2 Cut out portion of film corresponding to multiprotein-DNA complex of interest. Using a light box, superimpose cut-out film on gel, using autorad markers as alignment references. Using disposable scalpel, excise portion of gel corresponding to multiprotein-DNA complex of interest. Transfer excised gel slice to 1.5 ml microcentrifuge tube.

3 Solubilize gel slice by addition of 15 μl 1 M DTT (≈ 0.5 M final), and incubating for 5 min at 37 °C.[h]

4 Immediately proceed to next step.

G. Nuclease digestion

1 Transfer 30 μl of sample to new 1.5 ml siliconized polypropylene microcentrifuge tube, and add 1.5 μl 115 mM $CaCl_2$, 3 μl 10× protease inhibitor solution, and 0.2 μl

Protocol 4 continued

(26 units) DNase I. Incubate for 30 min at 37 °C. Terminate reaction by adding 1.5 μl 10% SDS, and heating for 10 min at 65 °C.

2 Add 1.5 μl 25 mM $ZnCl_2$, 1.5 μl 1 M acetic acid, 1.5 μl 10× protease inhibitor solution, and 1.5 μl (45 units) S1 nuclease (diluted to 30 units/μl in S1 nuclease dilution buffer). Incubate for 15 min at 37 °C.

3 Immediately proceed to next step.

H. Analysis

1 Adjust pH by adding 4 μl 1.5 M Tris-HCl, pH 8.8.

2 Add 8 μl 5× SDS loading buffer. Heat 5 min at 65 °C. Apply entire sample (54 μl) to 7–15% gradient polyacrylamide (30:0.8 acrylamide/bisacrylamide), 0.1% SDS, slab gel (27 × 15 × 0.15 cm). As marker, load ^{14}C-labelled protein molecular weight marker in 5 μl 5× SDS loading buffer into adjacent lane 5 μl. Electrophorese in SDS running buffer at 13 V/cm until bromophenol blue reaches bottom of gel.

3 Dry gel, and autoradiograph or phosphorimage.

[a] BAC is a disulfide-containing analogue of bisacrylamide (27–30). Polyacrylamide–BAC gels can be solubilized by addition of reducing agents (27–30). The synthesis of BAC in this chapter is adapted from (27).

[b] BAC is substituted for bisacrylamide on a mole-equivalent, not mass-equivalent, basis (25–28). The solubility of BAC in water is increased by adding acrylamide before adding BAC, and by performing additions at 60 °C.

[c] TEMED and ammonium persulfate concentrations are critical variables in preparation of polyacrylamide–BAC gels (28–30). (Use of non-optimal TEMED and ammonium persulfate concentrations in preparation of polyacrylamide–BAC results, subsequently, in difficulties in solubilizing gels.) The following table provides general guidelines for TEMED and ammonium persulfate concentrations as a function of polyacrylamide–BAC gel percentage (adapted from ref. 30):

Gel (%)	3.5	5.0	7.5	10.0	12.5
TEMED (%)	0.2	0.5	2	3	5
Persulfate (%)	0.04	0.01	0.04	0.008	0.008

[d] Fluorescent light and daylight must be excluded. Low to moderate levels of incandescent light (e.g. single task lamp with 60 W tungsten bulb) are acceptable.

[e] Do not add loading buffer to the reaction mixture. The reaction mixture is sufficiently dense for loading (owing to the presence of polyethylene glycol).

[f] A heating apparatus can be constructed at minimal cost from a 10 × 25 cm silicone heating mat, a voltage controller and four large binder clips, with the heating mat clipped directly to outer glass plate of the slab gel assembly.

[g] UV irradiation is performed with both glass plates in place. The glass plates exclude wavelengths < 300 nm, minimizing photodamage to protein and DNA. It is important to verify that the plates exhibit absorbances of < 0.15 AU at 350 nm and < 1.5 AU at 320 nm (e.g. by sacrificing a glass plate and placing a piece in the cuvette holder of a UV/vis spectrophotometer). Glass plates purchased from Aladin have performed satisfactorily.

[h] 2–4 M β-mercaptoethanol can be substituted for 1 M DTT.

3 Representative data

In published work, we have used systematic site-specific protein-DNA photocrosslinking to define positions of polypeptides relative to DNA within a human transcription complex containing RNAPII, TBP, transcription factor IIB, transcription factor IIF and promoter DNA (*Figure 2*) (7).

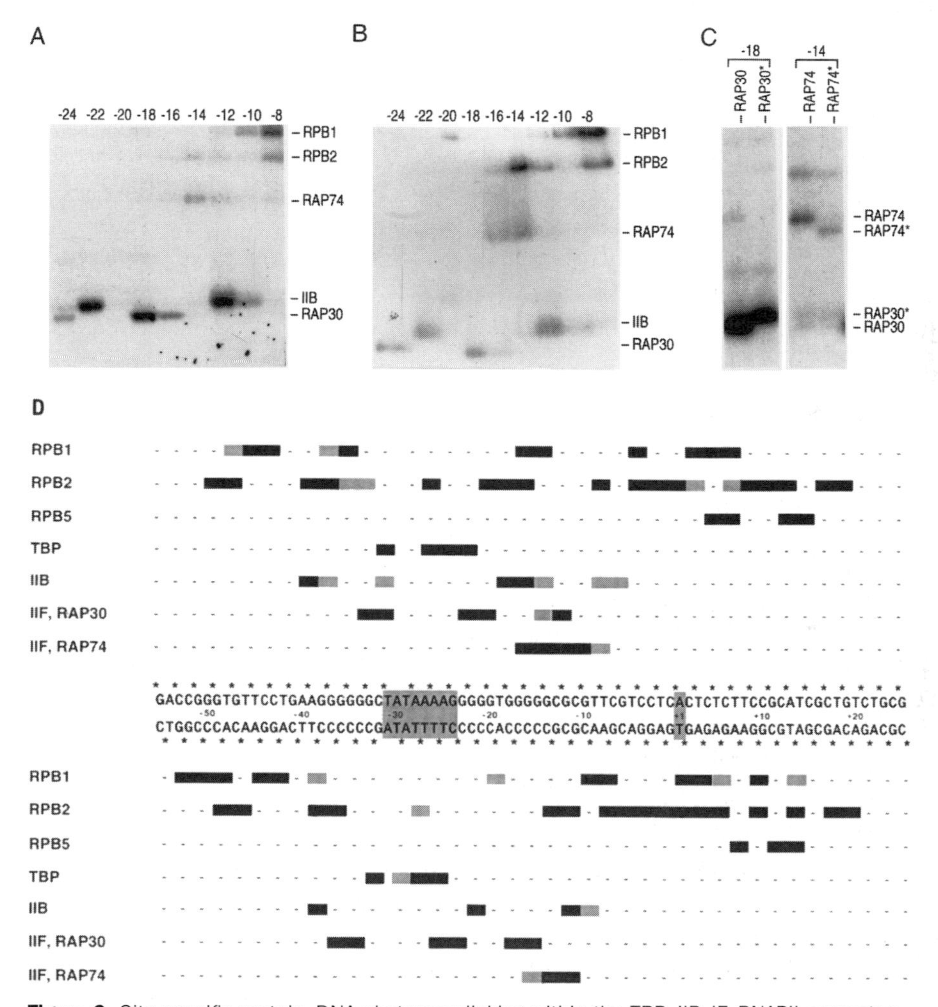

Figure 2. Site-specific protein–DNA photocrosslinking within the TBP–IIB–IF–RNAPII–promoter complex (7). (*A*) Representative data from solution photocrosslinking experiments (positions –24 to –8 of DNA template strand). (B) Representative data from in-gel photocrosslinking experiments (positions -24 to –8 of DNA template strand). (C) Representative data from control experiments using IIF-RAP30* and IIF-RAP74* (positions –18 and –14 of DNA template strand; see *Section 3*). (D) Summary of results (results for non-template strand above sequence; results for template strand beneath sequence). Phosphates analysed are indicated by asterisks. Sites exhibiting reproducible crosslinking are indicated by solid bars; sites exhibiting less reproducible crosslinking are indicated by shaded bars (scoring based on data from two to five independent standard photocrosslinking experiments). The TATA-element and transcription start site are indicated by shading.

We constructed 80 site-specifically derivatized DNA fragments, each containing a phenyl-azide photo-activatible crosslinking agent incorporated at a single, defined phosphate of the adenovirus major late promoter (positions −55 to +25; *Figure 2D*). For each of the 80 DNA fragments, we then formed the RNAPII-TBP-IIB-IIF-promoter complex, UV irradiated the RNAPII-TBP-IIB-IIF-promoter complex—in solution and, in parallel, in gel—and determined the polypeptide or polypeptides at which crosslinking occurred. The identities of crosslinked polypeptides were determined from positions of radiolabelled bands in SDS-PAGE (Figure 2*A,B*), and identities of crosslinked polypeptides were confirmed by performing parallel experiments using polypeptide derivatives with different molecular masses: i.e., TBPc (6), IIBc (6), IIF with hexahistidine-tagged RAP30 (IIF-RAP30*; see ref. 31), and IIF with hexahistidine-tagged RAP74(1–409) (IIF-RAP74*; see ref. 32) (*Figure 2C*).

The results establish that the largest subunit of RNAPII, the second-largest subunit of RNAPII, the fifth-largest subunit of RNAPII, TBP, IIB, the RAP30 subunit of IIF and the RAP74 subunit of IIF, contact, or are close to, promoter DNA in the RNAPII-TBP-IIB-IIF-promoter complex and define the positions of these polypeptides relative to promoter DNA (*Figure 2D*) (7). The results further establish that the interface between the largest and second-largest subunits of RNAPII forms an extended (≈ 240 Å) channel that interacts with promoter DNA both upstream and downstream of the transcription start (7).

4 Prospects

The site-specific protein–DNA photocrosslinking procedure in this chapter is generalizable to essentially any multiprotein–DNA complex. Straightforward extensions of the procedure permit mapping of crosslinks to individual polypeptide regions (Y. Kim, A. Revyakin, N. N., and R. H. E., unpublished data). Further straightforward extensions of the procedure—i.e., incorporation of rapid mixing and UV-laser irradiation—permits millisecond-scale kinetic analysis of individual protein–DNA interactions during assembly, function, and disassembly of multiprotein-DNA complexes (see S. Druzhinin, A. Revyakin, N. Naryshkin and R. H. Ebright, unpublished data; see Chapter 14).

Acknowledgements

This work was supported by Howard Hughes Medical Institute Investigatorships to D. R. and R. H. E. by National Institutes of Health grant GM37120 to D.R., and by National Institutes of Health grants GM53665 and GM41376 to R. H. E.

References

1. Record, M. T., Reznikoff, W., Craig, M., McQuade, K., and Schlax, P. (1996). In *Escherichia coli and Salmonella* (ed. Neidhart, F.), Vol. 1, p. 792. ASM Press, Washington, D.C.
2. Orphanides, G., Lagrange, T., and Reinberg, D. (1996). *Genes Dev.* **10,** 2657.
3. Ebright, R. (1998). *Cold Spring Harbor Symp. Quant. Biol.* **63**, in press.

4. Kelman, Z. and O'Donnell, M. (1995). *Annu. Rev. Biochem.* **64,** 171.
5. Baker, T. and Bell, S. (1998). *Cell* **92,** 295.
6. Lagrange, T., Kim, T. K., Orphanides, G., Ebright, Y., Ebright, R., and Reinberg, D. (1996). *Proc. Natl. Acad. Sci. USA* **93,** 10620.
7. Kim, T.-K., Lagrange, T., Wang, Y.-H., Griffith, J., Reinberg, D., and Ebright, R. (1997). *Proc. Natl. Acad. Sci. USA*, **94,** 12268.
8. Lagrange, T., Kapanidis, A., Tang, H., Reinberg, D., and Ebright, R. (1998). *Genes Dev.*, **12,** 34.
9. Wang, Y. and Stumph, W. (1998). *Mol. Cell. Biol.* **18,** 1570.
10. Fidanza, J., Ozaki, H., and McLaughlin, L. (1992). *J. Am. Chem. Soc.* **114,** 5509.
11. Yang, S.-W. and Nash, H. (1994). *Proc. Natl. Acad. Sci. USA*, **91,** 12183.
12. Mayer, A. and Barany, F. (1995). *Gene* **153,** 1.
13. Sambrook, J., Fritsch, E., and Maniatis, T. (1989). *Molecular cloning. A laboratory manual*, 2nd edn. Cold Spring Harbor Laboratory Press, Cold Spring Harbor.
14. Bartholomew, B., Kassavetis, G., Braun, B., and Geiduschek, E. P. (1990). *EMBO J.* **9,** 2197.
15. Bartholomew, B., Kassavetis, G., and Geiduschek, E. P. (1991). *Mol. Cell. Biol.* **11,** 5181.
16. Braun, B., Bartholomew, B., Kassavetis, G., and Geiduschek, E. P. (1992). *J. Mol. Biol.* **228,** 1063.
17. Kassavetis, G., Kumar, A., Ramirez, E., and Geiduschek, E. P. (1998). *Molec. Cell. Biol.* **18,** 5587.
18. Bell, S. and Stillman, B. (1992). *Nature* **357,** 128.
19. Coulombe, B., Li, J., and Greenblatt, J. (1994). *J. Biol. Chem.* **269,** 19962.
20. Gong, X., Radebaugh, C., Geiss, G., Simon, S., and Paule, M. (1995). *Mol. Cell. Biol.* **15,** 4956.
21. Pruss, D., Bartholomew, B., Persinger, J., Hayes, J., Arents, G., Moudrianakis, E. and Wolffe, A. (1996). *Science* **274,** 614.
22. Robert, F., Forget, D., Li, J., Greenblatt, J., and Coulombe, B. (1996). *J. Biol. Chem.* **271,** 8517.
23. Forget D., Robert F., Grondin G., Burton Z., Greenblatt J., and Coulombe B. (1997). *Proc. Natl. Acad. Sci. USA* **94,** 7150.
24. Hixson, S. and Hixson, S. (1975). *Biochemistry* **14,** 4251.
25. Staros, J., Bayley, H., Standring, D., and Knowles, J. (1978). *Biochem. Biophys. Res. Commun.* **80,** 568.
26. Maldonado, E., Drapkin, R., and Reinberg, D. (1996). *Methods Enzymol.* **274,** 72.
27. Hansen, J. N. (1976). *Anal. Biochem.* **76,** 37.
28. Hansen, J. N. (1980). *Anal. Biochem.* **105,** 192.
29. Hansen, J. N. (1981). *Anal. Biochem.* **116,** 146.
30. Bio-Rad Laboratories. *Bis-acrylylcystamine (BAC) instruction manual, LIT318, rev. B.* Bio-Rad Laboratories, Hercules.
31. Tan, S., Conaway, R., and Conaway, J. (1994). *BioTechniques* **16,** 824.
32. Wang, B. and Burton, Z. (1995). *J. Biol. Chem.* **270,** 27035.

Chapter 24
DNA–protein complexes analysed by electron microscopy and cryo-microscopy

Eric Le Cam and Etienne Delain
Laboratoire de Microscopie Moléculaire et Cellulaire, LM2C, UMR 8532 CNRS, Institut Gustave-Roussy, 39 Rue Camille Desmoulins, F–94805 Villejuif Cedex, France

Eric Larquet
Station Centrale de Microscopie Electronique, Institut Pasteur, 25 Rue du Dr Roux, France

Françoise Culard
Centre de Biophysique Moléculaire CNRS, Rue Charles Sadron, F–45071 Orléans Cedex, France

Jean A. H. Cognet
Laboratoire de Physico-chimie Biomoléculaire et Cellulaire, ESA 7033 CNRS, T22–12, Université Pierre et Marie Curie, 4 Place Jussieu, F–75252 Paris Cedex 05, France

1 Introduction

Characterization of the DNA curvature and flexibility, intrinsic or induced by protein, remains controversial and depends on the method used. Such an analysis is even more difficult with respect to the conformational changes in DNA induced by non-specific binding proteins such as histone-like proteins. Electron and recent near-field microscopes now offer different potentials to visualize and analyse individual molecules. These microscopes have the capacity to determine the average behaviour of a population of molecules observed individually and can thus provide a better appreciation of variability within the series of molecules than biochemical or other biophysical methods.

For the last 40 years, electron microscopy (EM) has been used to study the structure and the interactions of various biological macromolecules; analysis of DNA–protein interactions provides original information, complementary to that obtained by conventional biochemical and biophysical methods (1). Mapping of DNA–protein complexes, polymerization of the protein along the DNA and local variations of DNA conformation either intrinsic or induced, can be analysed.

The spreading or the adsorption of the molecules is an important step because the conformation of the spread DNA molecules can be modified (1). This adsorption of macromolecules on a support (carbon film or mica) is eliminated in cryo-electron microscopy (Cryo-EM) (2).

Preliminary studies must be initiated with traditional EM to control the quality of the preparation, DNA and protein alone, and to define the conditions of the DNA–protein complex formation. The main advantage of traditional electron microscopy is it makes it possible to check various conditions within a few hours. Moreover, we have shown that the molecules adsorbed onto a carbon film conserve their main conformational properties (4, 5), even if three-dimensional information is partly lost. When all the conditions are well defined and the analysis has been done by traditional EM, studies must be completed by cryo-EM to avoid the problem of the adsorption and staining and to obtain 3D information.

Any type of TEM can image molecules, but the different types of microscope available provide various levels of contrast and resolution. They have two main basic areas of application: imaging and analysis. The images are formed on a fluorescent screen, a photographic film or a TV camera and result from global electron irradiation of the specimen and complex interactions with its components. For cryo-observations, the TEM contains a tilting specimen cryo-holder, which allows stereoscopic views.

The protein used in the experiments described in this paper is MC1. It is a histone-like protein isolated from archaebacteria of the *Methanosarcina* family, that exhibits a strong affinity for DNA (6). Observation by TEM of the binding complexes of MC1 to any linear or circular DNA shows sharp kinks (5, 7, 8). TEM and cryo-microscopy have been used to analyse the DNA conformational changes induced by MC1.

2 Electron microscopy of DNA–protein complexes spread on carbon film support

2.1 Methods used to observe nucleic acids and nucleoprotein complexes

2.1.1 Introduction

The basic techniques for imaging isolated nucleic acid molecules or their complexes with proteins are rather similar, although specific procedures may be used for the latter. Physical procedures have been developed to avoid or to limit chemical contrasting. Although electron microscopy permits original observations of the structure of nucleic acids, it should always be used in conjunction with biochemical and other biophysical methods.

2.1.2 Spreading methods

The choice of the method is related to the biological material under study, to its reproducibility, the rapidity of the process, and the level of resolution obtained. Spreading methods can be divided into two types: those that need a film, usually made of cytochrome *c*, in which the molecules are embedded; and those that use the direct adsorption of the molecules onto a support. Contrasting procedures are often needed to observe such tiny objects, which can either be stained, by the absorption of heavy metal ions on them, or decorated with a vacuum-deposited heavy metal shadow (platinum, tungsten, etc.).

2.1.2.1 Spreading with a protein film

These spreading methods (*Figure 1*) are mainly designed for DNA visualization, but can be adapted to DNA–protein complexes.

Cytochrome *c* is a basic protein which complexes nucleic acids and, when spread on water, produces a film by partial denaturation. The simplest of these methods, called the 'diffusion method' consists of the preparation of a buffer containing mainly DNA or DNA–protein complexes and cytochrome *c*, which allow the DNA-cytochrome *c* film to form at the air–water interface (*Figure 1A*). Such a method allows 'micro-spreading', which is very suitable for very small amounts of DNA or for very diluted solutions.

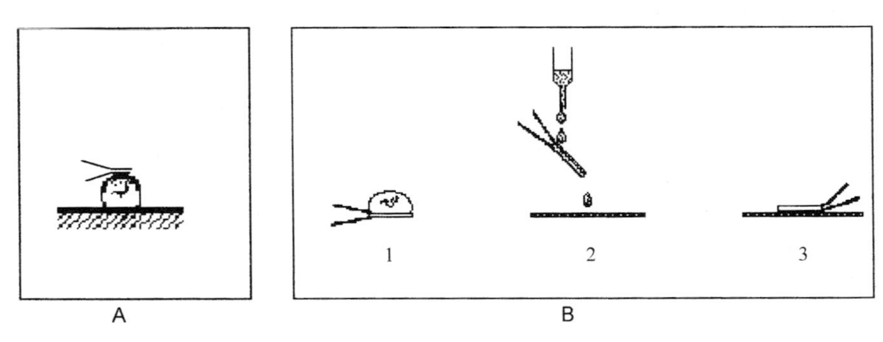

Figure 1 (A) Microversion of the diffusion method showing about a 50 μl droplet of DNA–cytochrome deposited on Parafilm. The surface of the droplet with the DNA–cytochrome film is touched with a carbon-coated grid, held with a forceps, then plunged into absolute ethanol before shadowing or positive staining with uranyl acetate. (B) Spontaneous adsorption of DNA on grids activated by glow discharge. The 5 μl droplet is left for 1 min on the support (1) then the excess of solution is washed away with uranyl acetate (2). The grid is carefully dried by its edge and finally deposited on a filter paper (3).

2.1.2.2 Adsorption methods

DNA and DNA–protein complexes can be easily adsorbed directly on carbon films or mica supports. Numerous assays have been developed using 'sticking' substances as mediators to bind the molecules to the carbon substrate, such as polylysine or divalent cations such as Mg^{2+}, used mainly for adsorption onto mica. These procedures have been described, forgotten and resurrected to prepare DNA molecules for observation with near-field observation with atomic force (AFM) and scanning tunnelling (STM) microscopes .

Direct adsorption onto activated carbon film is the method of choice to spread DNA and DNA–protein complexes and to analyse such complexes in optimal conditions. This very simple, reproducible method consists of treatment of the carbon-coated grids with a glow-discharge in the presence of pentylamine (9). This leads to the 'spontaneous' attachment of the negatively charged DNA molecules to the positively charges amine groups deposited on the carbon. The success of the method is based mainly on the use of very thin carbon films (which is necessary to obtain a good contrast in dark-field), on the quality of the treatment of the carbon and on the positive staining of the specimens.

Protocol 1

Preparation of the carbon support

Equipment and reagents

- Balzers MED 010 or other vacuum evaporator
- Copper grids (600 mesh)
- Mica

Method

1. To make carbon films of a few nm thick, evaporate a carbon thread onto mica in the vacuum evaporator at a distance of about 100 mm from the carbon source at 10^{-4} Pa. Control the thickness of the carbon deposit with a piece of paper or a porcelain chip. It varies according to the square of the carbon source-mica distance.
2. Plunge the carbon-coated mica in a water trough containing the copper grids to float off the carbon deposit. A good test for estimating the thickness of carbon films is the degree of difficulty in detecting them by observation in reflected light.
3. Draw up slowly the water with a vacuum pump to cover with the carbon film the grids previously set at the bottom of the trough.

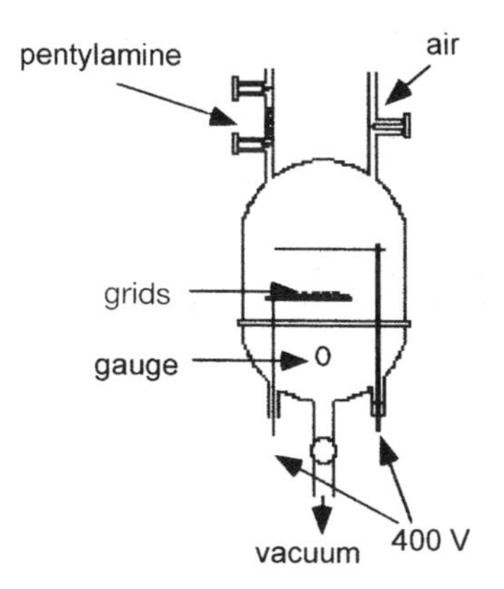

Figure 2 The bell jar for glow-discharge activation of the grids has two glass taps for pentylamine or air. Vacuum is obtained from the bottom with a rotary pump through a large tap for vacuum control according to the gauge. Pentylamine is introduced and the glow discharge operated as described in the text. Grids are deposited on the lower electrode on a Parafilm-coated glass slide.

2.1.3 Activation of the carbon-coated grids

Ionization is obtained in a laboratory-made apparatus (*Figure 2*).

It comprises a detachable 0.5-l bell jar, on which two Teflon-glass taps are welded allowing access to pentylamine or air. The base of the bell, fixed to the support, has four welded glass passages for two vertical conductors, a gauge allowing vacuum measurement and a connection for the pumping system, through a large Teflon-glass tap, which ensures the control of the pumping speed. The extremities of the two electrodes have two stainless steel plates about 8 cm in diameter and about 3 cm apart. Measurements in the range of 1–50 kPa is performed with the vacuum gauge. The mechanical pump should be able to sustain pentylamine vapour. An adjustable transformer delivers an alternative tension of 2–400 V to the steel plates.

Protocol 2

Activation of the carbon-coated grids

1. Grids are deposited on a glass slide coated with Parafilm and placed on the lower steel plate.
2. At 1 kPa vacuum, a small amount of pentylamine is introduced, up to about 2 kPa.
3. Ionization is performed over 20–30 s by switching on the tension (400 V). A faint violet glow discharge light should be visible around the plates.
4. Grids prepared in this manner keep their DNA binding efficiency for about 1 day.

2.1.4 Deposit of the sample on the grids: adsorption and contrasting methods

The contrast of very thin deposits is faint in bright-field but they can be observed in dark-field, which offers very good imaging except when the platinum deposits are too thick, resulting in a granulous background. The contrast and the quality of the images are highly dependent on the fineness of both the carbon support and the object. Thus, unshadowed specimens provide the best-resolved images.

Protocol 3

Spreading and staining

Equipment and reagents

- Uranyl acetate (2% in water)

Method

1. Spreading: 5 μl of the DNA solution at a concentration of 0.5 μg/ml is deposited on the grid, left for about 1 min, and rinsed with 3 drops of a 2% solution of aqueous uranyl acetate, then gently dried on a filter paper (*Figure 1B*).

Protocol 3 continued

2 Observation: The preparation can be observed directly by TEM operated in dark-field, as the contrast is very high. Metal shadowing (rotary or directional) is also possible with platinum, or pellets of platinum–carbon, evaporated by heating a tungsten basket, or by electron-gun deposition with platinum–iridium or tungsten–tantalum, which provides thinner deposits.

2.1.5 Methods of observation

Since TEM makes it possible to visualize all the elements present in a reaction mixture, the main prerequisite for EM observations is the use of adequately purified material, and spreading and contrasting procedures which do not interfere, as far as possible, with the interactions under study. Annular dark-field is mainly used to observe molecules which are not decorated with an evaporated metal coating, since they lack the contrast needed to be easily visible in traditional bright-field illumination. This mode of observation can be obtained by tilting the electron beam via a specific coil located in the condenser, thereby providing a tilted illumination of the specimen, with the direct electron beam being arrested by the objective aperture, and the image being formed by the electrons scattered at wide angles by the heavy atoms present in the specimen. Another means of acquiring a dark-field image consists of illumination of the specimen with a conical electron beam obtained through an annular condenser aperture. Once again, contrast is provided by impeding the direct electrons, and only allowing scattered electrons to pass through the objective aperture, the size of which should be chosen so that direct electrons are totally stopped. In both dark-field methods, thin film gold apertures are preferred for inducing a limited astigmatism.

2.2 Characterization and method for the analysis of the binding of protein MC1 to DNA. Measurement of DNA curvature and flexibility

2.2.1. Linear DNA fragments

Observation by TEM of the binding complexes of MC1 to any linear DNA shows sharp kinks randomly located along the fragments (*Figure 3*).

MC1 is too small (11 kDa) to be visualized by TEM and only the conformational change can be correlated with the binding of the protein. It is difficult to correlate unambiguously the observed bends to the binding of MC1 in a linear DNA fragment which does not exhibit a specific or preferential binding site. Thus, we have analysed the bending of DNA induced by MC1 with a 176 bp DNA fragment containing a preferential binding site located at its centre (10). Such a small DNA fragment allows one to restrict the angle measurements within a narrow region of the fragment. This has proved to be essential to perform quantitative measurements. Optimal conditions for visualization of the complexes re-

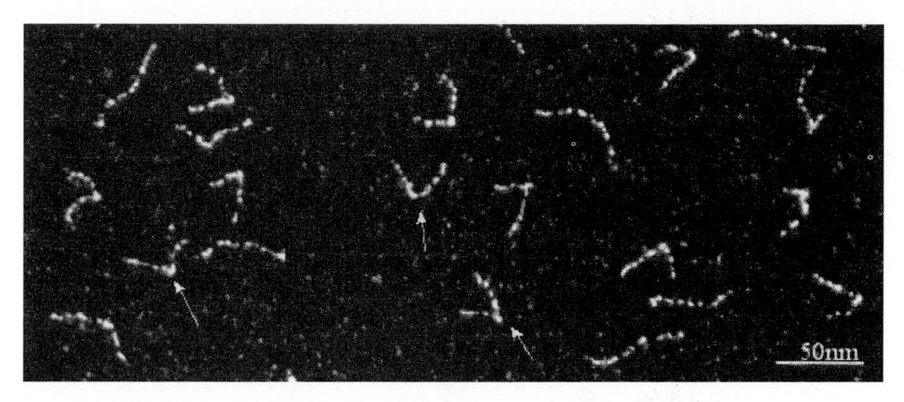

Figure 3 Visualisation of MC1–DNA complexes by annular dark field electron microscopy. MC1 (11 kDa) induces a kink (arrows) which can be correlated with the binding of the protein. The 176 bp linear DNA fragment contains a preferential binding site for MC1 located at its centre. The MC1 : DNA ratio is 20 and the final DNA fragment concentration is 8.6 nM.

quires their purification by gel filtration to remove unbound protein and reduce non-specific binding.

We have recorded the contour length of the 176 bp DNA fragment and measurements were performed on a large number of molecules (> 500). From the digitized contours, the method consists of measuring the angles at the centre of all duplexes observed by TEM. The angle population distributions are very well fitted by a worm-like chain model with isotropic flexible junctions and an equilibrium between the two conformations of DNA with bound or unbound MC1. The angle population distributions obtained with this model are a sum of two folded Gaussian distributions, which allowed us to measure the fraction of DNA with bound MC1 as a function of MC1 concentrations (*Figure 4*).

This interaction was quantitatively characterized, by dark-field TEM and PAGE, with 176 bp DNA duplexes that contained a preferential binding site for MC1, located at the centre of the DNA. An equilibrium dissociation constant of $K_d = 100$ nM and a kink angle equilibrium value of $\theta_0 = 116°$ were obtained (5).

To perform these analyses, a more general method was defined based on statistical polymer chain analysis. This method allows one to quantify the average angle value of intrinsic or induced bent as well as the flexibility associated to a given locus in a DNA sequence imaged by TEM. Consider that a DNA fragment is composed of two branches, A_0A_m and A_0B_n, with m and n base pairs, respectively, where the standard deviations of the angle formed by two consecutive base pairs are uniform over each branch, $\sigma_{\sigma A}$ and $\sigma_{\theta B}$, respectively (*Figure 5*).

We show that the average flexibility of the angle $A_mA_0B_n$ is given by the standard deviation:

$$\sigma^2_{A_mA_0B_n} = \frac{(m-1)(2m-1)}{6m}\sigma^2_{\theta_A} + \frac{(n-1)(2n-1)}{6n}\sigma^2_{\theta_B} + \sigma^2_{\theta_0}$$

where σ is the standard deviation of the angle at locus A_0.

The validity of this equation is established when the length of the DNA

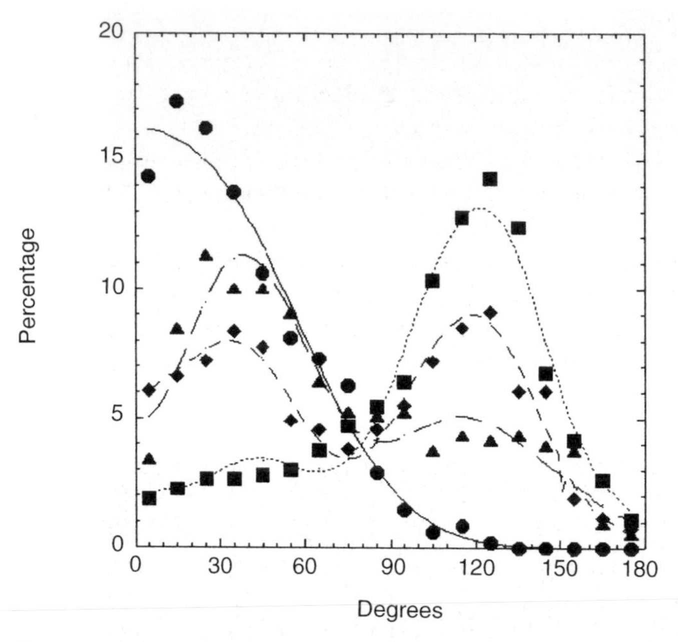

Figure 4 Distribution of kink angle values, in 10° intervals, measured on the linear 176-bp DNA fragments at variable ratios, $r = [MC1]/[DNA] = 0$ (●), 5 (▼), 10 (◆), and 50 (■). The scattered plots with their fitted curved of the percentage of the observed population are represented. These distributions were collected on the EM grids after purification by chromatography. The curve fits are a sum of two folded Gaussian distributions.

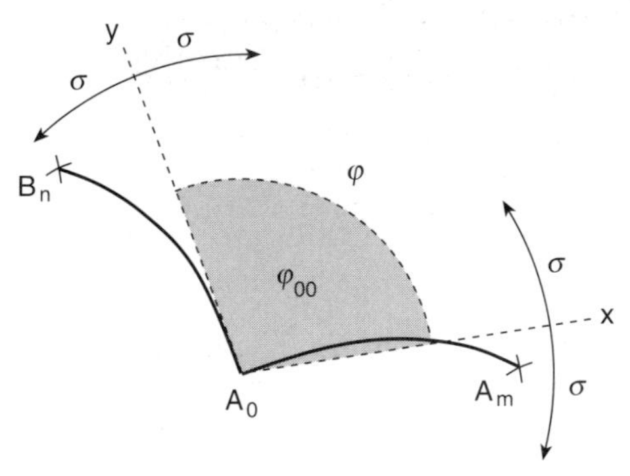

Figure 5 Measurement of the angle value , formed at the centre A_0 a 176 bp.DNA fragment. ϕ_{00} is the average angle value and σ result from the flexibility of the structure located at the centre.

fragments is of the order of, or less than, the persistence length, as determined by computer simulations. It is verified experimentally by a detailed analysis of the digitised contours of homogeneous linear 139-bp DNA fragments observed by TEM (11). This method is useful to quantify directly from microscopy, such as EM or SFM, the true bending angle, either intrinsic or induced by a ligand, and its associated flexibility at a given locus in any small DNA fragment.

2.2.2 Circular DNA fragment

The determination of the bending values does not give any information about the directionality of the curvature and more precisely about the torsional stress associated to the bending. To give information about this torsional stress associated with the bending induced by MC1, we have analysed the conformational change with relaxed DNA minicircles by traditional TEM and then by cryo-TEM (7, 8).

With relaxed 207-bp DNA minicircles, gel retardation assays and TEM have shown that conformational changes resulting from the binding of 1 and 2 MC1 molecules, in presence of low ionic strengths induce two types of complexes, C1 and C2, respectively. C1 complexes appear with a single strong kink whereas C2 complexes are tapered in shape with two diametrically opposed kinks (*Figure 6*).

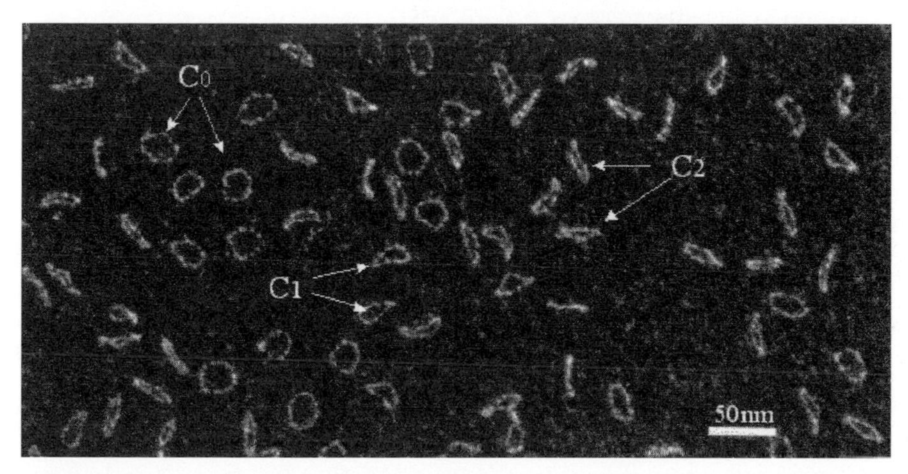

Figure 6 Relaxed DNA minicircles (203 bp) in the presence of MC1 at a protein : DNA minicircle ratio r : 7 in 80 mM NaCl. The most frequent configuration correspond to the complex C2 with two kinks diametrically opposed. C0 = minicircles without protein, C1 = minicircles with one protein.

These C2 complexes are remarkably stable with time (12). A local torsional stress seemed to be present around the MC1 binding site but could not be characterized. Only cryo-TEM can resolve the three-dimensional (3D) path of the minicircles.

3 Cryo-electron microscopy

3.1 Methods used to observe nucleic acids and nucleoprotein complexes

3.1.1 Introduction

Cryo-microscopy of macromolecules such as nucleic acids consists of the observation of a solution of the molecules, without staining, which has been rapidly frozen (usually in liquid ethane, cooled in liquid nitrogen), to obtain a thin film of vitrified water, which, as a result of very rapid cooling maintains an amorphous (vitreous) structure. With cold stages and low-dose imaging, molecules can be seen floating in their original solution. Cryo-microscopy is the method of choice for the optimal preservation of biological macromolecules. Biological objects embedded in frozen water display images with a very low contrast, which need to be taken with an underfocus to benefit from phase contrast. The geometrical and topological parameters of DNA molecules, either naked or associated with ligands such as proteins or drugs, can be analysed in a liquid environment and in various experimental conditions.

3.1.2 Frozen hydrated suspension technique

An EM grid coated with a perforated carbon film is inserted into tweezers mounted in an automatic blotter and plunger. A 4 μl drop of DNA solution (≈ 50–100 μg/ml) in the buffer of interest is applied to the grid. After blotting the excess of solution (≈ 1–2 s) with a filter paper, the plunger immediately immerses the specimen into a container of liquid ethane cooled close to its freezing point using liquid nitrogen as the secondary cooling bath. A humid chamber is used (13) to avoid evaporation phenomena after the blotting and during the plunging of the specimen in the cryogen. The vitrified specimens is transferred into a cryo-specimen holder via a cryo-transfer system and introduced into the cryo-TEM (14, 15).

3.1.3 EM observations

3.1.3.1 Contamination

For observation of the DNA specimen in the microscope, the cryo-holder is needed; this provides cooling of the specimen and keeps it below the devitrification point (−120 °C). Another important piece of equipment for the cryo-EM is an anticontaminator or cryo-blades, cooled to the lowest possible temperature (nominally −180 °C). It is, in principle, a cold trap for the residual water molecules present in EM vacuum, which would otherwise condense on the specimen. Principally two major types of contamination can be observed: ice crystals and vitreous water layer formation. The first originates from small ice crystals present in the cryogen or liquid nitrogen, or from the transfer of the specimen into the microscope. During the transfer, the specimen is, for a short time, subjected to humid room atmosphere, which results in the condensation of humidity in the form of small hexagonal ice crystals on the specimen. Contamination with vitreous water is caused by slow condensation of residual water vapour in the vacuum of the microscope. It is a slow but continuous process, which increases effective thickness of the water layer and causes the deterioration in the image quality.

3.1.3.2 Surface tension and water layer thickness

The adsorption of DNA on a supporting film is avoided in cryo-TEM, but the blotting process before the vitrification can introduce surface tension at air–water interface; the related forces can strongly influence the distribution and the orientation of the molecules. Not only the surface effect is involved, but also the influence of the thin layer thickness, which is not uniform. The properties of the perforated film can affect the variation of the layer thickness. In the case of a hydrophilic film, the water is attracted to the surface of the supporting film and consequently the water layer is the thinnest in the centre of the hole. Hydrophobic supporting film makes the layer thicker in the centre and thinner at the edge of the holes. This non-uniformity in the layer thickness is at least partly responsible for uneven distribution of samples within the layer.

3.1.3.3 Influence of temperature

The time of vitrification is generally estimated to be 10^{-4} s. However, some of the typical relaxation times of DNA molecules coincide with or are within the time-scale attributed to the vitrification process and molecules have time to react to the change of temperature before their complete immobilization due to the growing viscosity of supercooled media.

3.1.3.4 Image formation and DNA contrast

The contrast obtained with an unstained and frozen hydrated specimen is inherently very low and dependent on the molecular density. Moreover, the contribution of inelastically scattered electrons in a vitreous water layer does not allow recording of an image of appropriate quality and requires strong underfocusing. Thus, the low-dose mode is necessary and correct underfocusing has to be used to enhance the contrast. The contrast of DNA not only depends only on underfocus but also on the thickness of the water layer in which it is embedded. Therefore, to obtain a good contrast, it is necessary to observe DNA molecules contained in holes, with the thinnest layer of vitreous water.

3.2 3D-reconstruction of DNA molecules

3.2.1 Image acquisition

From two pictures of an object seen under different angles, it is possible (with certain limitation) to make a 3D reconstruction of any object. The basic procedure for DNA 3D reconstruction can be found in Dustin *et al.* (1991) (15). When stereo pair of micrographs are digitized, a system of coordinates is chosen so that the x–y plane corresponds to the plane of the first micrograph and y-axis is parallel to the tilt axis. The micrographs are mutually aligned along the Oy direction based on a reference point, which is a well-defined feature, clearly visible on both micrographs. This reference point is taken as the origin of the system of coordinates.

3.2.2 Calculation

The projection of the object results in the set of x–y values for each point in the first image and in x'–y' in the second. Since the y-axis is parallel to the tilt axis, $y = y'$ for each point. Then a simple trigonometric calculation gives the z coordinate. The determination of the path of the DNA molecule is obtained by following the molecule with the pointer. Successive points on the first image

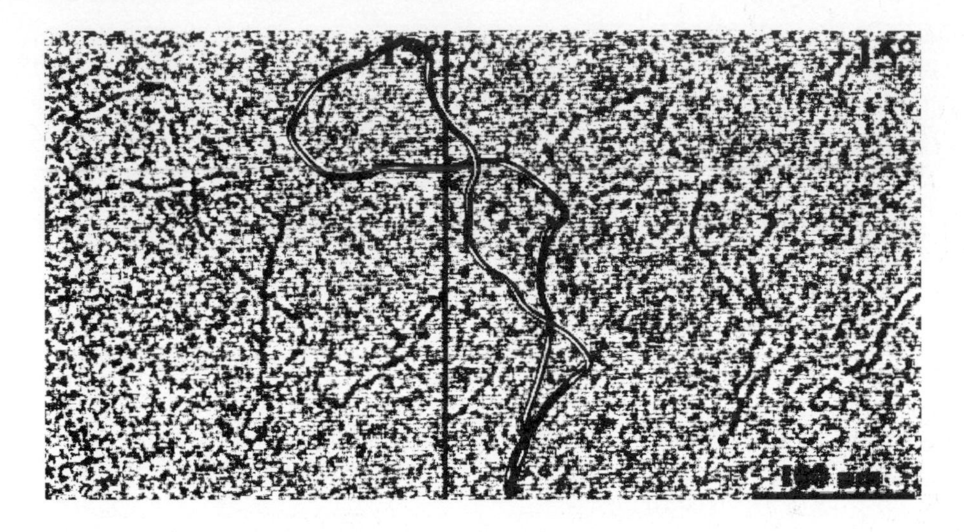

Figure 7 A 3D reconstruction of supercoiled plasmidic DNA (pUC18) from two stereo pair of micrographs obtained under different angles (–15°, +15°).

defines their coordinates (x, y). This also defines the y' coordinate for the corresponding points on the second image. The movement of the pointer is then restricted to the x direction at $y = y'$. Bicubic spline interpolation based on the chosen points results in a smooth path overlapping each displayed projection of the molecule. The reconstructed molecules are visualized as flexible cylinders, obtained by attributing a 'user defined' thickness representing the DNA diameter.

For a demonstration of 3D-reconstruction, supercoiled plasmid (pUC18) was used (*Figure 7*). This object has several advantages: it has a high contrast, thus the problems with beam damage are reduced; it can preserve sufficient contrast for the second exposure; and it probably has higher mechanical rigidity that could make it more resistant to all possible forces trying to flatten it or change its shape. Stereo-pairs of images were taken by a Slow Scan CCD Camera connected directly to the microscope, with a tilt angle 15° at minimum-dose exposure with a direct magnification of 80 000. Reconstruction of this plasmid shows that DNA preserves its 3D path at least at the level of superhelicity. More precisely, comparison of the average superhelical diameter in two mutually perpendicular projections (one at the plane xy, the second xz) does not exhibit significant differences. This suggests that the distortion of DNA does not take place at this level. However, when extension of the molecule in the z dimension was checked, its value did not exceed 40 nm. This suggests that there is a limitation in the z dimension that could be caused by a specific positioning of the DNA molecules within the vitreous ice layer or by its natural thickness. The latter does not fit much with often-referred to thickness of the water layer ($\approx$ 100 nm). Specific positioning of the DNA molecules within the layer can be caused by the effect of water–air interface expelling DNA from the surface and trying to keep it close to the centre of the film.

3.2.3 3D path of the MC1–DNA minicircles

The three-dimensional path of the minicircles with or without MC1 was calculated from cryo-microscopy stereoscopic pictures which showed that the naked minicircles were in fact flat, whereas those complexed with MC1 were folded at the level of the kinks, a torsional constraint which yielded a boat-like shape (*Figure 8*). Furthermore, these results confirmed that the binding of MC1 did not shorten the minicircles (circumference 70 nm).

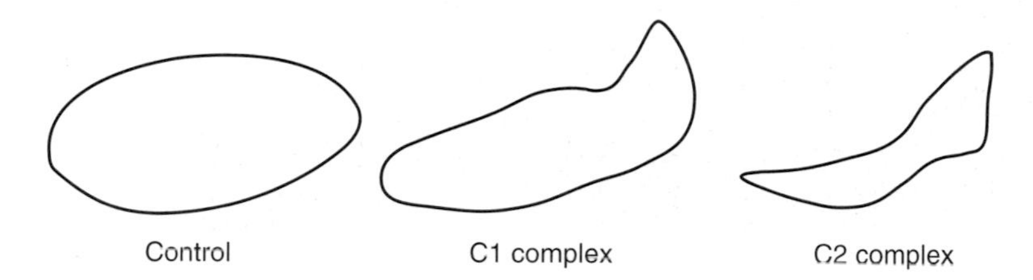

Figure 8 The 3D path of the minicircles with or without MC1 obtained from the analysis of 30 cryo-microscopy stereoscopic pictures in each class. The naked minicircles (203 bp) were in fact flat (Control). C1 complex is folded at the level of the kink. A torsional constraint at the level of the two diametrically opposed kinks yielded a boat-like shape of the C2 complex.

Topological variations of the minicircles induced by the binding of MC1 can be characterized by the writhe value, Wr. Thirty molecules in each class were analysed and distributions of Wr values relative to the two types of complexes, compared with the control were analysed:

Relaxed DNA minicircles control: $Wr = -0.009 \pm 0.002$

C1 complex: $Wr = -0.032 \pm 0.003$

C2 complex: $Wr = -0.096 \pm 0.003$

Variation of Wr can be related to the torsional stress induced by MC1. Cryo-electron microscopy can quantify the torsional stress associated with the bending in the plane determined by traditional TEM. Such analysis can be transposed to any type of protein–DNA complex characterized by local changes caused by bending upon protein binding, even if the protein itself cannot be disclosed directly owing to its small size.

Acknowledgements

Cryo-electron microscopy has been performed in the laboratory of Jacques Dubochet (Laboratoire d'Analyse Ultrastructurale, bâtiment de biologie, Université de Lausanne, CH-1015 Lausanne, Switzerland). We thank Jacques Dubochet and his collaborators. C. Pakleza is acknowledged for his contribution to angle analysis of DNA.

References

1. Le Cam, E. and Delain, E. (1995). In *Visualization of nucleic acids* (ed. Morel, G.), p. 333. CRC Press. Boca Raton.
2. Kellenberger, E. (1987). In *Cryotechniques in biological electron microscopy, R.A.a.Z.* (ed. Steinbrecht, K.), p. 35. Springer-Verlag, Berlin.
3. Joanicot, M. and Révet, B. (1987). *Biopolymers* **26**, 315.
4. Le Cam, E., Fack, F., Ménissier-de Murcia, J., Cognet, J. A. H., Barbin, A., Sarantoglou, V., Révet, B., Delain, E., and de Murcia, G. (1994). *J. Mol. Biol.* **235**, 1062.
5. Le Cam, E., Culard, F., Larquet, E., Delain, E., and Cognet, J. A. H. (1999). *J. Mol. Biol.* **285**, 1011.
6. Culard, F., Laine, B., Sautière, P., and Maurizot, J.-C. (1993). *FEBS Lett.* **315**, 335.
7. Larquet, E., Le Cam, E., Fourcade, A., Culard, F., Fuller, A., and Delain, E. (1996). *C.R. Acad. Sci. Paris III* **319**, 461.
8. Toulmé, F., Le Cam, E., Teyssier, C., Delain, E., Sautière, P., Maurizot, J.-C., and Culard, F. (1995). *J. Biol. Chem.* **270**, 6286
9. Dubochet, J., Ducommun, M., Zollinger, M., and Kellenberger, E. (1971). *J. Ultrastruct. Res.* **35**, 147.
10. Teyssier, C., Laine, B., Gervais, A., Maurizot, J.-C., and Culard, F. (1994). *Biochem. J.* **303**, 567.
11. Cognet, J. A. H., Pakleza, C., Cherny, D., Delain, E., and Le Cam, E. (1999). *J. Mol. Biol.* **285**, 997.
12. Teyssier, C., Toulmé, F., Touzel, J. P., Gervais, A., Maurizot, J.-C., and Culard, F. (1996). *Biochemistry* **35**, 7954.
13. Bednar, J., Furrer, P., Stasiak, A, Dubochet, J., Egelman, E. H., and Bates, A. D. (1994). *J. Mol. Biol.* **235**, 825.
14. Adrian, M., Dubochet, J., Lepault, J., and McDowall, A. W. (1984). *Nature* **308**, 32.
15. Dustin, I., Furrer, P., Stasiak, A., Dubochet, J., Langowski, J., and Egelman, E. H. (1991). *J. Struct. Biol.* **107**, 15.

Chapter 25

Characterization of T7 RNA polymerase protein–DNA interactions during the initiation and elongation phases.

Dmitry Temiakov, Pamela E. Karasavas and William T. McAllister

Department of Microbiology and Immunology, Morse Institute for Molecular Genetics, State University of New York, Health Science Center, Brooklyn, New York 11203-2098, USA

1 Introduction

Although the DNA-dependent RNA polymerase (RNAP) encoded by bacteriophage T7 consists of a single subunit, this enzyme is able to carry out all of the steps in the transcription cycle as the more complex multisubunit RNAPs. Its relatively simple structure, together with the availability of structural data, makes the phage RNAP an attractive model for studies of transcription at the molecular level (for review, see refs. 1 and 2). This chapter outlines techniques that have been developed to halt T7 RNAP at defined positions as it moves away from the promoter during the early stages of transcription initiation, as well as during elongation, and methods to characterize interactions between the polymerase and the DNA and/or the nascent RNA product within these halted complexes.

As with other RNAPs, transcription by T7 RNAP proceeds in two phases. During the initial stage the enzyme forms an unstable initiation complex that undergoes a process of abortive initiation in which short oligoribonucleotides 2–12 nt long are continuously synthesized and released. After the synthesis of 8–12 nt, the initiation complex isomerizes to a highly processive elongation complex that is capable of polymerization at a rate of 200 nt/s until it encounters a pause or termination signal or reaches the end of the template. The transition from an unstable initiation complex to a stable elongation complex is accompanied the by release of upstream promoter contacts and a change in the footprint of the enzyme on the DNA. However, the point at which this transition occurs varies depending upon the promoter sequence, as does the efficiency of promoter clearance.

Little is known about the nature of the elongation complex or the steps that proceed its formation. A number of methods to halt T7 transcription complexes have been described previously. One approach has been to carry out transcription under conditions of limiting substrate. For example, the T7 late promoter $\phi 10$ initiates transcription with the sequence +1 GGGAGACCACAACGGU... In the presence of GTP alone, the active site may be extended to +3 at this promoter. However, due to slippage of the nascent transcript on the template strand the enzyme synthesizes and releases a ladder of oligo-G products 2–14 nt long under these conditions (3). In the presence of GTP and ATP the active site is extended to +6, and in the presence of GTP, ATP, and CTP transcription is extended to +15. Complexes halted at these positions have been probed by footprinting with agents such as DNaseI and methydium propyl Fe(II)-EDTA (4, 5). Prior to isomerization (which at this promoter occurs between +6 and +10) the polymerase maintains contacts with the upstream region of the promoter while the footprint is extended in the downstream direction. After isomerization, the enzyme releases the upstream contacts and assumes a more compact footprint on the DNA template. Similar studies have been carried out with other promoters, but owing to variations in the transcribed sequence it has been difficult to compare results among these experiments.

Another approach towards halting the progress of the elongation complex is to use blocking agents placed along the DNA template. One type of experiment used bound proteins such as *lac* repressor or a non-cleaving derivative of *Eco*RI to block transcription (6–8). However, once T7 RNAP has formed a stable elongation complex its progress seems to be impeded little by the presence of bound proteins, except under reduced substrate conditions. Sastry and Hearst (9) explored the use of psoralen adducts that were crosslinked to the template strand, or to both strands of the DNA, and found that the progress of the RNAP was arrested at the site of the crosslink. This approach allowed the characterization of a halted complex, but because the block in the progress of the RNAP was not reversible the significance of these findings with regard to normal transcription is not clear.

To circumvent these problems, we developed plasmid templates that allow T7 RNAP to be halted at various stages during initiation and elongation in a preserved sequence context (10). By this approach, we hoped to avoid the problems arising from variable sequences and/or perturbations in the structure of the template. These complexes have proven to be suitable for analysis by a variety of techniques including UV-laser crosslinking, DNaseI footprinting, probing with potassium permanganate, etc. (11).

In a further development, we constructed templates that allow photoreactive analogues to be placed into the nascent transcript at defined positions upstream from the 3′ end of the RNA in a halted complex. To map the sites in the protein at which RNA–protein crosslinks are formed, we characterized the action of various proteolytic agents on T7 RNAP and defined their sites of cleavage. Although these techniques have been applied to the characterization of RNA–potein crosslinks, the methods are well suited to characterization of DNA–protein crosslinks.

The integrity of the amino terminus of T7 RNAP is not required for catalytic activity and a number of *N*-terminal fusion proteins have been characterized. Taking advantage of this, we previously fused a histidine leader sequence to T7 RNAP to facilitate its purification (12). Here, we describe a modified protocol for the large-scale purification of high quality enzyme. This His-tagged RNAP has also proven useful in mapping crosslinked sites in the RNAP. By cleaving the crosslinked protein and purifying the *N*-terminal fragments by chromatography on a metal-ligand column it is possible to identify *N*-terminal crosslinks, or in the case of incomplete digestion, to obtain a nested set of *N*-terminal partial digestion products. Both of these approaches are described in this chapter.

Finally, we have found that functional elongation complexes formed with histidine tagged-T7 RNAP may be immobilized on Ni^{2+}-agarose beads. Removing substrates by washing the bound complexes then allows the RNAP to be 'walked' various distances downstream from the promoter in any sequence context, and permits the incorporation of novel substrates into the RNA product at specific position.

2 Purification of T7 RNA polymerase

We have described the construction of histidine-tagged T7 RNAPs and methods for rapid purification of these enzymes on a small scale (12). For crosslinking studies, high concentrations of enzyme with greater purity are required. Here, we describe a modified protocol that includes steps to remove nucleic acid contaminants and results in enzyme preparations of high purity and concentration.

Protocol 1

Purification of His-tagged polymerase

Equipment and reagents

- *Escherichia coli* BL21/pDL21
- Ni^{2+}-NTA agarose (Qiagen)
- Polymin P (Sigma, 10%, v/v in water); pH 7.9
- Lysis buffer: 40 mM Tris-HCl, pH 7.9, 0.3 M KCl, 10 mM EDTA, 1 mM PMSF (Sigma)
- Binding buffer: 40 mM Tris-HCl, pH 7.4, 20 mM imidazole, 150 mM KCl, 0.05% Tween-20, 5% glycerol, 5 mM 2-mercaptoethanol.
- Wash buffer: binding buffer containing 60 mM imidazole
- Elution buffer: binding buffer containing 200 mM imidazole.
- Storage buffer: PBS, pH 7.4 (Gibco BRL) containing 5% glycerol and 10 mM dithiothreitol
- Ammonium sulfate: saturated solution in water at room temperature
- PD-10 desalting column (Pharmacia)
- Empty disposable PD-10 column (Pharmacia)
- Centricon-50 centrifuge concentrator (Millipore)
- Sonicator (Sonic Dismembrator, Fisher Scientific)

Method

1 Prepare, induce and harvest a 300 ml culture of BL21/pDL21 cells as described in (12).

Protocol 1 continued

2 Resuspend the cell pellet in 25 ml of cold lysis buffer. Perform all subsequent steps at 0–4 °C.

3 Sonicate the sample for 30 min using alternate cycles of 40 s power, 40 s idle.

4 Centrifuge the sample at 15 000 **g** for 15 min and collect the supernatant.

5 Add KCl to 0.6 M and 1/40 volume 10% of polymin P and mix for 15 min.

6 Centrifuge the sample at 15 000 **g** for 15 min and collect the supernatant.

7 Slowly add an equal volume of ammonium sulfate solution and mix for 30 min.

8 Centrifuge the sample at 15 000 **g** for 15 min and collect the pellet. You may stop at this stage and store the precipitate frozen at −70 °C.

9 Take up the pellet in 10 ml binding buffer. Centrifuge at 15 000 **g** for 10 min to remove any insoluble material.

10 Place 4 ml of Ni^{2+}-agarose bead suspension into an empty PD-10 column. Wash the beads with distilled water (15 ml) and then with binding buffer (15 ml).

11 Load the protein solution onto the column. Wash the column with 15 ml binding buffer, or until the flow-through has an OD_{280} of less then 0.05.

12 Wash the column with 6–8 ml wash buffer.

13 Apply elution buffer to the column and collect 2 ml fractions.

14 Load aliquots of each fraction onto a 10% PAGE gel.

15 After electrophoresis, select fractions of desired purity and desalt them by chromatography on a PD-10 column equilibrated with 25 ml of storage buffer.

16 Concentrate the polymerase by centrifugation in an Amicon-50 concentrator at 5000 **g**. Invert the unit after each 15-min centrifugation cycle to prevent precipitation of the protein on the membrane.

17 Determine the concentration of the protein using a molar extinction coefficient of $\varepsilon_{280} = 1.4 \times 10^5$/M (13).

18 Add an equal volume of 100% glycerol, and DTT to a final concentration of 10 mM. Store the enzyme at −20 °C.

For this application, as well as for the walking and immobilization experiments described below, we use the plasmid pDL21, which encodes an RNA polymerase with a His12-leader sequence (12). This provides greater affinity for the Ni^{2+}-agarose column than the His6-tagged enzymes used previously. For optimal yields, it is important to streak out single colonies of this culture prior to isolation to ensure that a high percentage of cells carry the desired plasmid. Following induction and harvesting, the cells are lysed by more aggressive sonication than was previously described to optimize release of the protein. An ammonium sulfate precipitation step to remove undesired protein contaminants and a polymin P precipitation step to remove nucleic acids were also introduced. The sample is loaded onto a metal-ligand column as previously described, but because of the longer histidine-tag a higher concentration of imidazole is used to wash the sample, which further eliminates impurities. Imidazole is removed by passing the sample over a PD-10 filtration column and the sample is concentrated by centrifugation in an Amicon concentrator. A typical yield from 300 ml of culture is 8–10 mg of highly pure T7 RNAP at a concentration of 4 mg/ml.

3 Formation of halted complexes in a conserved sequence context

We have developed a series of plasmid templates that allow T7 RNAP to be halted at defined distances downstream from the start site of the promoter in a conserved sequence context (10). These plasmids (pPK3–7) have an initially transcribed sequence of $(GGGA)_n$GACT.... so that incubation of polymerase in the presence of GTP and ATP as the sole substrates allows transcription complexes to be halted at +6, +10, +14, +18, and +22 (see *Table 1*). The stability and properties of these complexes have been described elsewhere, and reveal changes in the complexes as they progress from an unstable initiation complex to a stable elongation complex (10, 11).

A second generation of templates was developed that permits the introduction of a unique UMP-residue into the transcript at defined positions (*Table 1*). Note that the reactive U-residue may be positioned at a variable distance upstream from the 3′ end of the RNA in a transcript of variable length (templates pPK10, pPK12 and pPK13) or of constant length (templates DT01 to DT03).

Protocol 2

Formation of halted complexes

Equipment and reagents

- 10× HT buffer (300 mM HEPES, pH 7.8, 2.5 mM EDTA, 10 mM DTT, 150 mM $Mg(OAc)_2$, 0.5% Tween-20
- RNAP purified as described in *Protocol 1*, 20 ng/μl
- Ultrapure rNTPs (Pharmacia)
- 2× stop solution: 7 M urea, 0.01 M EDTA, 0.01% bromophenol blue, 0.01% xylene cyanol FF)

Method

1. Premix 1 μg linearized plasmid template, 1 μl 10× HT buffer, 1 μl 5 mM GTP, 1 μl 1 mM ATP, 0.5 μl [α-^{32}P]ATP (800 Ci/mmol) and 4.5 μl water.
2. Incubate at 37°C for 2 min and then add 20 ng RNAP (1 μl) to a final volume of 10 μl.
3. Incubate at 37 °C for 1–5 min.
4. Use reaction for crosslinking or terminate by the addition of an equal volume of 2× stop solution and perform analysis by PAGE.

4 Walking of T7 RNAP on a DNA template

The method described above allows T7 RNAP to be halted at defined distances downstream from the promoter in a conserved sequence context. However, for a variety of purposes it is useful to halt the polymerase at specific positions within a non-conserved sequence. Studies with *E. coli* RNA polymerase have benefited greatly from the ability to 'walk' this polymerase to specific positions on a template by immobilizing the enzyme on affinity beads, and then adding and removing limited mixtures of substrates by rapid washing steps (14). Here, we describe a method that allows T7 RNAP to be walked to specific positions on a template.

To form an elongation complex, His12-T7 RNAP is first incubated with a suitable template and with a mixture of substrates that allows the formation of a stable elongation complex. This usually requires that the enzyme extend to at least +15 on the template. As an example of this, we provide a protocol that allows walking of the polymerase on the plasmid pPK10, which starts transcription with the sequence + 1 GGGAGAGGGAGGGAUCC...

T7 RNAP requires a high concentration of GTP to initiate, and hence formation of the starting complex is carried out in the presence of 0.3 mM GTP and 0.1 mM ATP (which, on this template, permits extension to +14). We have found that formation of the complexes prior to the addition of Ni^{2+}-agarose beads is necessary to provide optimal retention of transcriptionally competent complexes. After immobilization, unincorporated substrates are removed by brief cycles of centrifugation and removal of the supernatant. Subsequent steps of elongation require lower concentrations of substrate and may be carried out using 2–10 μM rNTP (which facilitates more complete removal of unincorporated substrates). The complexes formed in this manner are rather stable, and we have walked T7 RNAP successfully through at least five cycles with little loss of transcription competence between cycles (see *Figure 1*).

Protocol 3

Walking of T7 RNAP

Equipment and reagents

- Ni^{2+}-NTA agarose (Qiagen)
- Histidine-tagged T7 RNA polymerase (DL21; see *Protocol 1*)
- Transcription buffer: 20 mM Tris-HCl, pH 7.9, 8 mM $MgCl_2$, 5 mM 2-mercaptoethanol, 0.1 mM EDTA
- Ultrapure rNTPs (Pharmacia)
- DNA template (PCR product from pPK10 or other appropriate plasmid)
- Table top minicentrifuge (National Labnet Co.)

Method

1 Place 10 μl Ni^{2+}-agarose beads suspension into an Eppendorf tube. Wash twice with distilled water (1 ml) and twice with transcription buffer (1 ml) at room temperature.

Protocol 3 continued

2 Incubate 2 μg of polymerase with an equimolar amount of PCR template (PK10), 0.3 mM GTP, and 0.1 mM ATP in 20 μl transcription buffer at 37 °C for 1 min to allow extension to +14. Perform all subsequent steps at room temperature.

3 Transfer reaction from step 2 into the tube containing the Ni^{2+}-agarose beads and incubate for 5 min with gentle agitation.

4 Resuspend beads in 1 ml of transcription buffer, centrifuge at maximum speed for 10 s and remove supernatant. Repeat four times.

5 Take up beads in 10 μl transcription buffer and extend complex to +15 by adding [$^{32}\alpha$-P]-UTP to final concentration of 2 μM. Incubate for 2 min.

7 Wash as described in step 4.

8 Take up beads in 10 μl transcription buffer and extend complex to +21 by adding CTP, GTP and UTP to final concentrations of 10 μM. Incubate for 2 min.

9 Wash as described in step 4.

10 Repeat steps 7 and 8 with other substrates, as needed, to continue walking the RNAP.

11 Analyse the RNA products made after each step gel electrophoresis.

In our experiments, we have used short DNA templates ≈ 100–150 bp long. To obtain appropriate concentrations of template, we have found it convenient to amplify a segment of a plasmid by standard PCR protocols. For example, using the plasmid pPK10 (see *Table 1*) we amplify a segment of 130 bp in which the

Table 1 Templates to allow the formation of halted elongation complexes

Template[a]	Sequence of transcript from +1	Transcript length[b] GTP, ATP	GTP, ATP, Uth	U at:[c]
pPK3	GGGAGAC	6	6	
pPK4	GGGAGGGAGAC	10	10	
pPK5	GGGAGGGAGGGAGAC	14	14	
pPK6	GGGAGGGAGGGAGGGAGAC	18	18	
pPK7	GGGAGGGAGGGAGGGAGGGAGAC	22	22	
pPK10	GGGAGAGGGAGGGAUC	14	15	–1
pPK12	GGGAGAQGGAGGGAUGAC	14	17	–3
pPKI3	GGGAGAGGGAGGGAUGGAGC	14	19	–5
pPKI4	GGGAGAGGAGQGAUGGGAGAC	13	20	–7
DTO1	GGGAGAGGAGGGAUAGGGAGC	13	20	–7
DTO2	GGGAGAGGAGGUAGGGAGGGC	11	20	–9
DTO3	GGGAGAGGAUAQQGAGGGAGC	9	20	–11

[a] pPK3–pK14 are plasmid templates that allow initiation as indicated; DT01–DT03 are PCR products amplified from pPKJ4 using a mismatched primer complementary to the template strand of the plasmid. The sequence of the transcribed region up to the first incorporation of CMP is given.

[b] The distance downstream from the start site (at +1) to which transcription is extended in the presence of the indicated substrates.

[c] The position of the unique UMP residue relative to the 3′ end (at -1) of the nascent RNA in the baited complex.

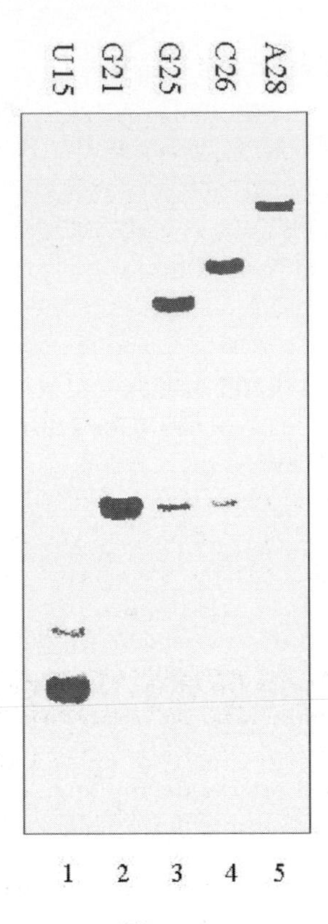

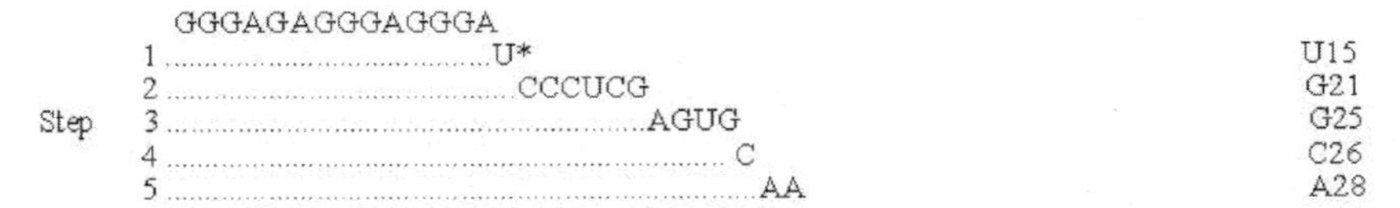

Figure 1 Walking of T7 RNAP. A 130-bp fragment of pPK10 that includes the promoter region and initiates with the sequence shown was amplified by polymerase chain reaction. This template was then incubated with T7 RNAP in the presence of GTP and ATP, causing the polymerase to halt at +14. The halted complexes were immobilized on Ni^{2+}-agarose beads, washed, and extended to +15 by the addition of ^{32}P-UTP (step 1). The complexes were washed again and extended to +21 by the addition of CTP, UTP, and GTP (step 2). Additional walking steps were performed with ATP, GTP, and UTP (step 3); with CTP (step 4); or with ATP (step 5). The products from each step were analysed by electrophoresis on 20% polyacrylamide gel.

promoter is located approximately equidistant from either end. The protocol given below uses 2 μg of polymerase in a 10 μl reaction for each step to be analysed. Thus, if five separate steps are to be analysed, the reaction is scaled up five times (to 50 μl).

5 Purification of crosslinked protein

Methods to crosslink protein to nucleic acids have been widely used to characterize the sites of interaction between these two components. However, after crosslinking, the reactions often contain undesirable byproducts that may confound the interpretation of subsequent analysis (for example, RNA–DNA hybrids). Here we describe two methods for purification of the desired protein–nucleic acid crosslinked species. In the first, the reaction products are resolved by electrophoresis in protein gels in the presence of SDS. The crosslinked fragment is then excised from the gel and eluted by macerating the gel slice, followed by extensive incubation in buffer containing SDS. This method provides good yield and high purity, but the enzyme is denatured.

A second method involves the use of histidine-tagged polymerase and Ni^{2+}-agarose beads to retain the protein and any crosslink that has been made. This method is advantageous because it takes place under native conditions, permitting the use of proteases that require native substrates.

Protocol 4
Purification of crosslinked RNAP by gel electrophoresis

Equipment and reagents

- Laemmli gel electrophoresis system
- Centrifuge tube filters (Spin-X, Costar)
- Microcon-10 microconcentrators (Amicon)
- Mini shaker (Labquake Shaker, Fisher Scientific)

Method

1. Load sample containing crosslinked RNAP and Laemmli loading buffer onto gel of appropriate percentage.
2. After electrophoresis, locate desired band by autoradiography and excise it from the gel.
3. Place the gel piece ($\approx 0.75 \times 1 \times 5$ mm) into 15 ml tube and wash it twice with 10 ml of distilled water, incubating for 10 min each time on the shaker.
4. Remove the gel piece from the tube and place it in the crease of a folded overhead transparency film (5×5 cm).
5. Crush the gel slice by squeezing.
6. Cut out the portion of transparency film which contains the gel mass and place it into an Eppendorf tube, leaving a small part of the strip on the outside of the mouth of the tube.
7. Close the cap and slowly pull the strip through the closed tube, leaving the gel mass in the tube.
8. Extract the RNAP from the gel by incubating with 0.4 ml 0.2% SDS for 1 h at 37 °C with constant mixing on the shaker.

Protocol 4 continued

9 Using a centrifuge tube filter, separate the gel pieces from liquid.

10 Concentrate the liquid phase containing the RNAP to a volume of 5–10 μl using a microconcentrator at 10 000 **g**.

Protocol 5

Purification of crosslinked T7 RNAP using Ni^{2+}-agarose beads

Equipment and reagents

- Ni^{2+}-NTA Agarose, (Qiagen)
- Table top minicentrifuge (National Labnet Co.)
- Wash buffer: 50 mM Tris-HCl, pH 8.0; 0.8 M NaCl, 5 mM 2-mercaptoethanol
- Elution buffer: 50 mM Tris pH 8.0, 150 mM NaCl, 400 mM imidazole, 5 mM 2-mercaptoethanol
- Mini shaker (Labquake shaker, Fisher Scientific)

Method

1 Transfer 10 μl Ni^{2+}-agarose beads suspension into an Eppendorf tube.

2 Add 1 ml of distilled water and centrifuge 10 s Remove supernatant and repeat washing procedure with water (1 ml) and then twice with wash buffer (1 ml).

3 Adjust crosslinking reaction (which involves His12-RNAP, see *Protocol 1*) to 0.8 M NaCl, and transfer it into the tube containing washed beads.

4 Incubate mixture with gentle agitation at 4 °C for 3 h.

5 Spin down and remove supernatant. Wash beads four times with wash buffer (1 ml) on shaker for 20 min each time.

6 Remove wash buffer and add 30 μl of elution buffer to the beads.

7 Incubate for 2 h at room temperature with gentle agitation.

8 Centrifuge, and save supernatant containing RNAP.

6 Mapping of crosslink sites of nucleic acids to T7 RNA polymerase by protease digestion

Information may be obtained from protein–nucleic acid crosslinked complexes in a number of ways. First, it is of use to know the site of crosslinking in the nucleic acid. For example, where does the polymerase contact the DNA relative to the start site for transcription? Elsewhere, we describe characterization of crosslinked complexes halted at a variety of positions through the use of DNA polymerase, which extends a DNA primer up to the site at which the crosslink occurs (11)

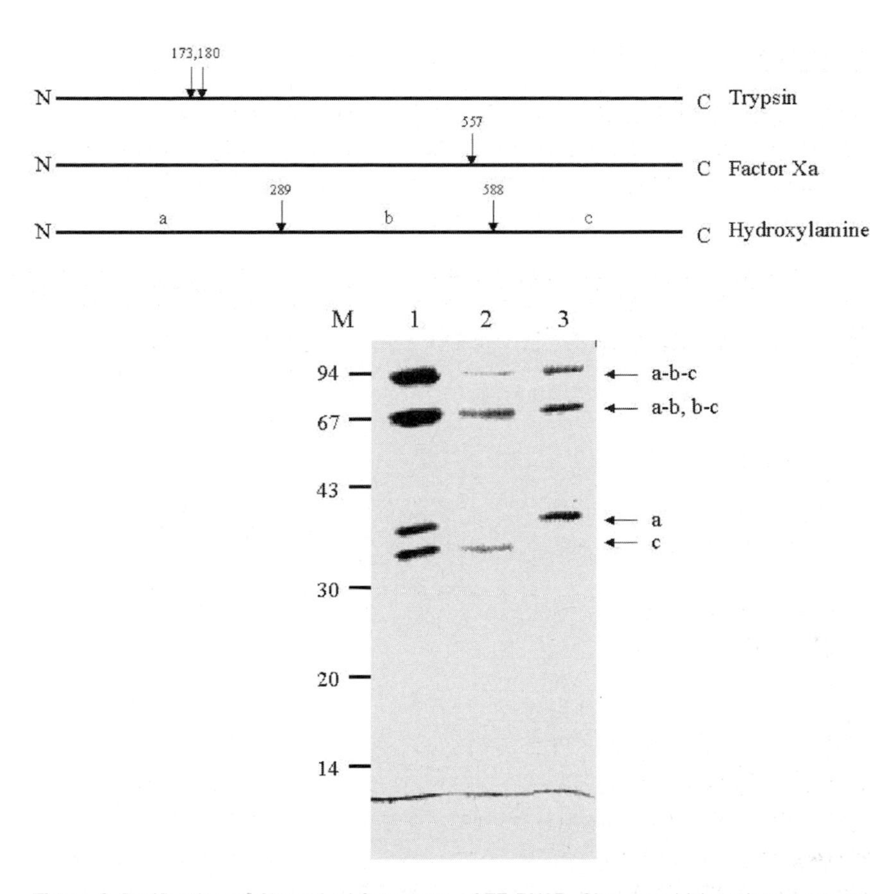

Figure 2 Purification of *N*-terminal fragments of T7 RNAP. Sites at which various proteases cleave T7 RNAP are shown (top). A sample of T7 RNAP was loaded onto a Ni^{2+}-agarose column and digested with hydroxylamine as described in *Protocol 7*. The column was washed with buffer A and the flow-through was collected. The column was then washed with buffer B and subsequently eluted with elution buffer. The total reaction products (lane 1), flow-through products (lane 2) and elution products (lane 3) were analysed by gel electrophoresis. Note that the *N*-terminal fragment (a) is retained by the column whereas the *C*-terminal fragment (c) is found in the flow-through. The two partial digestion products (a–b and b–c) are not resolved by this gel; note that fragment a–b is expected to be retained by the column, while fragment b–c is not. The positions of protein molecular weight markers are shown in lane M.

However, it is also useful to determine the site in the protein to which the crosslink is made. For this, one needs to have methods to resolve and identify fragments of the protein that are generated by digestion with various proteolytic agents. Using templates similar to those described above, we have crosslinked transcripts that carry photoreactive UMP analogues at various positions and have developed methods to resolve and characterize the cleavage fragments (D. T., P. K. and W. T. M., in preparation). While these methods were originally developed to characterize RNA–protein crosslinks, they are equally applicable to the analysis of DNA crosslinks.

It has previously been shown that trypsin and the outer membrane (ompT) protease of *E. coli* cleave T7 RNAP in the *N*-terminal region. For trypsin, two nearby cleavage sites at aa173 and aa180 were identified (15, 16); for the ompT protease the primary site of cleavage appears to be at aa178 (17) (see *Figure 2*). We have found that Factor Xa cleaves T7 RNAP once, and have mapped the site of cleavage to Arg557 (D. T, P. K. and W. T. M., unpublished observations). Hydroxylamine is predicted to cleave T7 RNAP after residues 289 and 588; we have confirmed this and have identified the resulting fragments (see *Figure 2*). Conditions for cleavage of T7 RNAP with each of these reagents are given in *Protocols* 6 and 7.

Protocol 6

Enzymatic cleavage of T7 RNAP

Equipment and reagents

- Trypsin (Sigma)
- Factor Xa (ProZyme, 200 U/mg)
- Cleavage buffer: PBS (Gibco BRL), pH 7.4
- Stop buffer: 4× Laemmli sample buffer for SDS-PAGE
- Laemmli gel electrophoresis system

Method

1. Purify crosslinked RNAP by chromatography on Ni^{2+} agarose beads (see *Protocol 5*). The sample should contain about 2 μg RNAP.
2. Dilute the sample 10 times with cleavage buffer.
3. Add trypsin to a final ratio of polymerase–trypsin = 1000:1 (w/w) or Factor Xa to a final polymerase–Xa ratio of 20 : 1 (w/w).
4. Incubate 5–30 min at room temperature.
5. Analyse products by gel electrophoresis.

7 Purification of *N*-terminal fragments of T7 RNAP

Since the histidine-tag is attached to the *N*-terminus of T7 RNAP, digestion products that include the *N*-terminus are retained on a metal-ligand column. This feature may be used to determine whether a nucleic acid has been crosslinked to the *N*-terminal fragment, or, in the case of partial digestion products, to purify a nested set of *N*-terminal fragments (*Figure 2*). It is known that various cleavage products of T7 RNAP remain tightly associated with one another under native conditions, and thus purification of the *N*-terminal fragments must be carried out under denaturing conditions. Below, we describe a method for purification of the *N*-terminal fragments generated by hydroxylamine, and a method to carry out cleavage of protein that has been immobilized on Ni^{2+}-agarose beads.

Protocol 7

Cleavage of T7 RNAP by hydroxylamine and purification of resulting *N*-terminal fragment

Equipment and reagents

- Ni^{2+}-NTA agarose (Qiagen)
- Histidine-tagged T7 RNA polymerase (DL21; see *Protocol 1*)
- Cleavage buffer: 4.5 M LiOH, 2 M hydroxylamine; 6 M urea, 15 mM Tris-HCl, pH 9.3
- Wash buffer A: 50 Tris-HCl, pH 8.0; 7 M guanidine-HCl
- Wash buffer B: 50 mM Tris-HCl, pH 8.0, 6 M urea, 0.5 M NaCl, 5 mM 2-mercaptoethanol
- Elution buffer: 50 mM Tris pH 8.0; 150 mM NaCl; 400 mM imidazole, 5 mM 2-mercaptoethanol
- Table top minicentrifuge (National Labnet Co.)
- Mini shaker (Labquake Shaker, Fisher Scientific)

Method

1. Place 10 μl Ni^{2+}-agarose bead suspension in an Eppendorf tube. Wash twice with distilled water (1 ml) and with 2× transcription buffer (1 ml).
2. Immobilize crosslinked RNAP on Ni^{2+}-agarose beads as described in *Protocol 5*.
3. Spin down beads and remove supernatant.
4. Add to the immobilized RNAP 40 μl of cleavage buffer. Incubate for 3 h at 45 °C.
5. Resuspend beads in 1 ml of wash buffer A, centrifuge for 10 s and remove supernatant. Repeat three times.
6. Resuspend beads in 1 ml of wash buffer B, centrifuge 10 s and remove supernatant. Repeat two times.
7. Add 30 μl of elution buffer and incubate at 37 °C for 30 min with gentle agitation.
8. Analyse the eluate by gel electrophoresis.

References

1. McAllister, W. T. (1997). In *Nuclear acids and molecular* biology (ed. Eckstein, F. and Lilley, D.), Vol II, p. 14. Springer, Berlin.
2. Sousa, R. (1997). In *Nuclea*r acids and molecular biology (ed. Eckstein, F. and Lilley, D.) Vol II, p. 1, Springer, Berlin.
3. Martin, C. T., Muller, D. K., and Coleman, J. E. (1988). *Biochemistry* **27**, 3966.
4. Basu, S. and Maitra, U. (1986). *J. Mol. Biol.* **190**, 425.
5. Ikeda, R. A. and Richardson, C. C. (1986). *Proc. Natl. Acad. Sci. USA* **83**, 3614.
6. Giordano, T. J., Deuschle, U., Bujard, H., and McAllister, W T (1989). *Gene* **84**, 209–219.
7. Lopez, P., Guillerez, J., Sousa, R., and Dreyfus, M. (1998). *J. Mol. Biol.* **267**:861
8. Pavco, P. A. and Steege, D. A. (1991). *Nuc. Acids Res.* **19**, 4639
9. Sastry, S. S. and Hearst, J. E. (1991). *J. Mol. Biol.* **221**, 1091

10. Karasavas, P. E. (1999) Ph.D Thesis, State University of New York, Brooklyn, N.Y. USA.
11. Place, C., Oddos, J., Buc, H., McAllister, W. T., and Buckle, M. (1999) *Biochemistry* **38**, 4948.
12. He, B., Rong, M., Lyakhov, D. L., Gartenstein, H., Diaz, G. A., Castagna, R. C., McAllister, W. T., and Durbin, R. K. (1997). *Prot. Express. Purif.* **9**, 142
13. King, G. C., Martin, C. T., Pham, T. T., and Coleman, J. E. (1986). *Biochemistry* **25**, 36.
14. Kashlev, M., Nudler, E., Severinov, K., Borukhov, S., Komissarova, N., and Goldfarb, A. (1996). In *Methods in enzymology* (ed. Wo, R., Grossman, L., and Moldave, K.), Vol. 274, p. 326. Academic Press, London.
15. Muller, D. K., Martin, C. T., and Coleman, J. E. (1988). *Biochemistry* **27**, 5763.
16. Ikeda, R. A. and Richardson, C. C. (1987). *J. Biol. Chem.* **262**, 3790.
17. Grodberg, J. and Dunn, J. J. (1988). *J. Bacteriol.* **170**, 1245.

List of suppliers

Aladin Enterprises, Inc., 1255 23rd Avenue, San Francisco, CA 94122, USA

Aldrich Chemical Company, Inc., 1001 West Saint Paul Avenue, Milwaukee WI 53233, USA

Amersham Pharmacia BioTech
Pharmacia Biotech (Biochrom) Ltd., Unit 22, Cambridge Science Park, Milton Road, Cambridge CB4 0FJ, UK
Tel: 01223 423723 Fax: 01223 420164;
Web site: www.biochrom.co.uk
Pharmacia and Upjohn Ltd., Davy Avenue, Knowlhill, Milton Keynes, Buckinghamshire MK5 8PH, UK
Tel: 01908 661101 Fax: 01908 690091
Web site: www.eu.pnu.com

Anderman and Co. Ltd., 145 London Road, Kingston-upon-Thames, Surrey KT2 6NH, UK
Tel: 0181 541 0035
Fax: 0181 541 0623

BDH
Poole, Dorset BH15 1TP

Beckman Coulter
Beckman Coulter Inc., 4300 N Harbor Boulevard, PO Box 3100, Fullerton, CA 92834-3100, USA
Tel: 001 714 871 4848 Fax: 001 714 773 8283; Web site: www.beckman.com
Beckman Coulter (UK) Ltd., Oakley Court, Kingsmead Business Park, London Road, High Wycombe, Buckinghamshire HP11 1JU, UK
Tel: 01494 441181; Fax: 01494 447558; Web site: www.beckman.com

Becton Dickinson
Becton Dickinson and Co., 21 Between Towns Road, Cowley, Oxford OX4 3LY, UK
Tel: 01865 748844; Fax: 01865 781627;
Web site: www.bd.com
Becton Dickinson and Co., 1 Becton Drive, Franklin Lakes, NJ 07417-1883, USA.
Tel: 001 201 847 6800;
Web site: www.bd.com

Becton Dickinson Labware, 2 Bridgewater Lane, Lincoln Park, NJ 07035, USA

Biacore AB, Rapsgaten 7 SE 75450 Uppsala Sweden.

Bio 101 Inc.
Bio 101 Inc., c/o Anachem Ltd., Anachem House, 20 Charles Street, Luton, Bedfordshire LU2 0EB, UK
Tel: 01582 456666; Fax: 01582 391768;
Web site: www.anachem.co.uk
Bio 101 Inc., PO Box 2284, La Jolla, CA 92038-2284, USA Tel: 001 760 598 7299; Fax: 001 760 598 0116; Web site: www.bio101.com

Bioprobe Systems, 26bis, rue Kleber F-93100 Montreuil sur bois, France

Bio-Rad Laboratories Ltd
Bio-Rad Laboratories Ltd, Bio-Rad House, Maylands Avenue, Hemel Hempstead, Hertfordshire HP2 7TD, UK
Tel: 0181 328 2000; Fax: 0181 328 2550; Web site: www.bio-rad.com
Bio-Rad Laboratories Ltd, Division Headquarters, 1000 Alfred Noble Drive, Hercules, CA 94547, USA
Tel: 001 510 724 7000; Fax: 001 510 741 5817; Web site: www.bio-rad.com

Biosoft, 22 Hills Road, Cambridge C82 1JP, UK
Tel: 01223 312873

Boehringer Mannheim
Boehringer Ingelheim Bioproducts, Biowhittaker, Biowhittaker Home, 1 Ashmill Way, Wokingham RG4l 2PC, UK

Boehringer Mannheim Corporation, 9115 Hogue Road, PO Box 50414, Indianapolis, IN 46250-0414 USA.

Cambridge Bioscience, Cambridge Bioscience: 24–25 Signet Court, Newmarket Road, Cambridge CB5 8LA, UK

Clontech Laboratories, Inc., 1020 East Meadow Circle, Palo Alto CA 94303, USA
Web site: www.clontech.com

Cole-Parmer Instrument Co., 625 East Bunker Court, Vernon Hills IL 6006I-1844, USA

Corning Costar, The Valley Centre, Gordon Road, High Wycombe HP13 6EQ, UK

CP Instrument Co. Ltd., PO Box 22, Bishop Stortford, Hertfordshire CM23 3DX, UK
Tel: 01279 757711 Fax: 01279 755785
Web site: www.cpinstrument.co.uk

CSC
Colorimetry Sciences Corp., 515 East 1860 South, PO Box 799, Provo, Utah 84603-0799, USA

Digital Instruments Inc
Veeco Metrology Group, 112 Robin Hill Road, Santa Barbara, CA 93117
Tel: 800-873-9750
Fax: 805-967-7717 (general)
E-mail: Info@di.com http:www.di.com

Dupont
Dupont (UK) Ltd., Industrial Products Division, Wedgwood Way, Stevenage, Hertfordshire SG1 4QN, UK
Tel: 01438 734000; Fax: 01438 734382; Web site: www.dupont.com
Dupont Co. (Biotechnology Systems Division), PO Box 80024, Wilmington, DE 19880-002, USA
Tel: 001 302 774 1000;
Fax: 001 302 774 7321;
Web site: www.dupont.com

Dynal
Dynal UK Ltd, 10 Thursby Road, Croft Business Park, Bromborough, Wirral, Merseyside L62 3PW, UK
Dynal, Inc., 5 Delaware Drive, Lake Success, NY 11042, USA

Eastman Chemical Co., 100 North Eastman Road, PO Box 511, Kingsport, TN 37662-5075, USA
Tel: 001 423 229 2000;
Web site: www.eastman.com

Fisher Scientific
Fisher Scientific UK Ltd., Bishop Meadow Road, Loughborough, Leicestershire LE11 5RG, UK
Tel: 01509 231166; Fax: 01509 231893; Web site: www.fisher.co.uk
Fisher Scientific, Fisher Research, 2761 Walnut Avenue, Tustin, CA 92780, USA
Tel: 001 714 669 4600; Fax: 001 714 669 1613; Web site: www.fishersci.com

Flowgen, Lynn Lane, Shenstone, Lichfield, Staffordshire WS 14 OEE, UK
Web site: flowgen.co.uk

Fluka
Fluka, PO Box 2060, Milwaukee, WI 53201, USA
Tel: 001 414 273 5013; Fax: 001 414 2734979; Web site: www.sigma-aldrich.com
Fluka Chemical Co. Ltd., PO Box 260, CH-9471, Buchs, Switzerland
Tel: 0041 81 745 2828; Fax: 0041 81 756 5449; Web site: www.sigma-aldrich.com

FMC Bioproducts, 191 Thomaston Street, Rockland, ME 04841, USA; UK distributor is Flowgen

Gibco BRL Life Technologies, 3 Fountain Drive, Inchinnan Business Park, Paisley PA4 9RF, UK

Hitachi Instruments, 3100 North First Street, San Jose CA 95134, USA

Hybaid
Hybaid Ltd, Action Court, Ashford Road, Ashford, Middlesex TW15 1XB, UK
Tel: 01784 425000; Fax: 01784 248085; Web site: www.hybaid.com
Hybaid US, 8 East Forge Parkway, Franklin, MA 02038, USA
Tel: 001 508 541 6918; Fax: 001 508 541 3041; Web site: www.hybaid.com

HyClone Laboratories, 1725 South HyClone Road, Logan, UT 84321, USA
Tel: 001 435 753 4584; Fax: 001 435 753 4589; Web site: www.hyclone.com

IBI (International Biotechnologies Inc)
25 Science Park, New Haven, CT 06511, USA

Invitrogen
Invitrogen, De Schelp 12, NL-9351 NV Leek, The Netherlands
Invitrogen BV, PO Box 2312, 9704 CH Groningen, The Netherlands
Tel: 00800 5345 5345 Fax: 00800 7890 7890
Web site: www.invitrogen.com
Invitrogen Corp., 1600 Faraday Avenue, Carlsbad, CA 92008, USA
Tel: 001 760 603 7200; Fax: 001 760 603 7201; Web site: www.invitrogen.com

Jobin-Yvon Ltd, 2–4 Wigton Gardens, Stanmore, Middlesex HA7 1BG

Kintek Corporation, 760 Spruce Road, Clarence, PA 16829, USA

Life Technologies
Life Technologies Ltd., PO Box 35, Free Fountain Drive, Inchinnan Business Park, Paisley PA4 9RF, UK
Tel: 0800 269210 Fax: 0800 838380
Web site: www.lifetech.com
Life Technologies Inc., 9800 Medical Center Drive, Rockville, MD 20850, USA.
Tel: 001 301 610 8000;
Web site: www.lifetech.com

Logix Consulting Inc, 11408 Audelig Ste 4944 Dallas TX 75243, USA
Tel: 001 214520 7310
Web site: www.lgx.com

Mathworks, 24 Prime Park Way, Natick, MA 01760-1500, USA

Matlab (Matlab is a registered trademark of Mathworks (q.v.))

Merck Sharp & Dohme
Merck Sharp & Dohme Research Laboratories, Neuroscience Research Centre, Terlings Park, Harlow, Essex CM20 2QR, UK
Web site: www.msd-nrc.co.uk
Merck & Co. Inc. Whitehouse Station NJ, USA Web site: www.merck.com
MSD Sharp and Dohme GmbH, Lindenplatz 1, D-85540, Haar, Germany
Web site: www.msd-deutschland.com

Microcal
Microcal Inc., One Roundhouse Plaza, Northampton MA 01060-2327, USA

Micromath Research
PO Box 71550-0550 Salt Lake City, Utah 84171-0550, USA
Tel: 00 1801 483 2949
Web site: www.micromath.com

Millipore
Millipore (UK) Ltd, The Boulevard, Blackmoor Lane, Watford, Hertfordshire WD1 8YW, UK
Tel: 01923 816375; Fax: 01923 818297;
Web site: www.millipore.com/local/UK.htm
Millipore Corp., 80 Ashby Road, Bedford, MA 01730, USA Tel: 001 800 645 5476;
Fax: 001 800 645 5439;
Web site: www.millipore.com

Molecular Dynamics
9828 E Arques Avenue, Sunnyrale, CA 94086-4520, USA
Web site: www.moleculardynamics.com
5 Beech House, Chiltern Court, Asheridge Road, Chesham, Bucks HP5 2PX

Molecular Probes, Inc., 4849 Pitchford Avenue, Eugene, OR 97402-9144, USA

Nanoscope (Nanoscope is a registered trademark of Digital Instruments Inc. (q.v.))

Nanosensors GmbH
IMO-Building, Im Amtmann 6, D-35578 Wetzlar-Blankenfeld, Germany
Fax: (+49) 6441-978841
Tel: (+49) 6441-978840
e-mail: nanosensors@compuserve.com
www.nanosensors.com

National Labnet Co
PO Box 841, Woodbridge, NJ 07095, USA

NEN Life Science Products, 549 Albany Street, Boston MA 02118 USA

New England Biolabs, 32 Tozer Road, Beverley, MA 01915-5510, USA
Tel: 001 978 927 5054
Web site: www.neb.com

Nikon
Nikon Corp., Fuji Building, 2–3, 3-chome, Marunouchi, Chiyoda-ku, Tokyo 100, Japan
Tel: 00813 3214 5311; Fax: 00813 3201 5856;
Web site: www.nikon.co.jp/main/index_e.htm
Nikon Inc., 1300 Walt Whitman Road, Melville, NY 11747-3064, USA
Tel: 001 516 547 4200; Fax: 001 516 547 0299; Web site: www.nikonusa.com

Novagen, 601 Science Drive, Madison, WI 53711, USA; UK distributor is **Cambridge BioScience** (q.v.)

Nycomed
Nycomed Amersham plc, Amersham Place, Little Chalfont, Buckinghamshire HP7 9NA, UK
Tel: 01494 544000; Fax: 01494 542266;
Web site: www.amersham.co.uk
Nycomed Amersham, 101 Carnegie Center, Princeton, NJ 08540, USA
Tel: 001 609 514 6000; Web site: www.amersham.co.uk

PE Biosystems, 850 Lincoln Center Drive, Foster City, CA 94404, USA
Tel: 001 650 638 5800
Web sites: www.pebio.com, www.pedirect.co.uk

Perbio Science (UK) Ltd.,
Pierce Chemical Company, 3747 Merdian Rd., PO Box 117, Rockford IL 61105, USA
Pierce & Warriner (UK) Ltd., 44 Upper Northgate St., Chester, CH1 4EF, UK

Perkin Elmer Ltd., Post Office Lane, Beaconsfield, Buckinghamshire HP9 1QA, UK.
Tel: 01494 676161; Web site: www.perkin-elmer.com

Pharmacia (please see Amersham Pharmacia Bio Tech)

Pharmingen, Inc., 10975 Torreyana Road, San Diego, CA 92121, USA; UK distributor is **Cambridge Bioscience** (q.v.)

Promega
Promega UK Ltd., Delta House, Chilworth Research Centre, Southampton SO16 7NS, UK Tel: 0800 378994; Fax: 0800 181037; Web site: www.promega.com
Promega Corp., 2800 Woods Hollow Road, Madison, WI 53711-5399, USA
Tel: 001 608 274 4330; Fax: 001 608 277 2516; Web site: www.promega.com

Qiagen
Qiagen UK Ltd., Boundary Court, Gatwick Road, Crawley, West Sussex RH10 2AX, UK
Tel: 01293 422911; Fax: 01293 422922; Web site: www.qiagen.com
Qiagen Inc., 28159 Avenue Stanford, Valencia, CA 91355, USA
Tel: 001 800 426 8157; Fax: 001 800 718 2056; Web site: www.qiagen.com

Quantum Biotechnologies Inc., 1801 de Maisonneure Blvd Montréal, Quebec H3H 1J9 Canada
Prozyme 1933 Davis St., Smite 207, San Leandro CA 94577-1258, USA
Web site: www.prozyme.com

Roche Diagnostics
Roche Diagnostics Ltd., Bell Lane, Lewes, East Sussex BN7 1LG, UK
Tel: 01273 484644; Fax: 01273 480266; Web site: www.roche.com
Roche Diagnostics Corp., 9115 Hague Road, PO Box 50457, Indianapolis, IN 46256, USA
Tel: 001 317 845 2358; Fax: 001 317 576 2126; Web site: www.roche.com
Roche Diagnostics GmbH, Sandhoferstrasse 116, D–68305 Mannheim, Germany
Tel: 0049 621 759 4747; Fax: 0049 621 759 4002; Web site: www.roche.com

Roche Molecular Biochemicals
see Roche Diagnostics

Sartorius
Sartorius Ltd, Longmead Business Centre, Blenheim Rd., Epsom, Surrey KT19 9QN
Web site: ww.sartorius.co.uk

Savant Instruments, Inc., 100 Cohn Drive, Holbrook NY 11741, USA

Schleicher and Schuell Inc., Keene, NH 03431A, USA
Tel: 001 603 357 2398

Setaram, 7 rue d l'Oratoire. F-69300 Caluaire, France

Shandon Scientific Ltd, 93–96 Chadwick Road, Astmoor, Runcorn, Cheshire WA7 1PR, UK
Tel: 01928 566611;
Web site: www.shandon.com

Sigma-Aldrich
Sigma-Aldrich Co. Ltd., The Old Brickyard, New Road, Gillingham, Dorset XP8 4XT, UK
Tel: 01747 822211; Fax: 01747 823779; Web site: www.sigma-aldrich.com
Sigma-Aldrich Co. Ltd., Fancy Road, Poole, Dorset BH12 4QH, UK
Tel: 01202 722114; Fax: 01202 715460; Web site: www.sigma-aldrich.com
Sigma Chemical Co., PO Box 14508, St Louis, MO 63178, USA
Tel: 001 314 771 5765; Fax: 001 314 771 5757; Web site: www.sigma-aldrich.com

Southern New England Ultraviolet Co., 550-29 East Main Street, Bradford, CT 06405, USA

Spectra-Physics, 1335 Terra Bella Ave, Mountain View, CA 94043, USA
Web site: www.spectrphysics.com

SPSS Science Inc., 233 S. Wacker Dr. Chicago, IL 60606-6307, USA
Tel: 001 800 543 5815
Web site: www.spss.com/software/science

Stratagene
Stratagene Europe, Gebouw California, Hogehilweg 15, NL-1101 CB Amsterdam Zuidoost, The Netherlands
Tel: 00800 9100 9100;
Web site: www.stratagene.com
Stratagene Inc., 11011 North Torrey Pines Road, La Jolla, CA 92037, USA
Tel: 001 858 535 5400; Web site: www.stratagene.com

Synergy Software, 2457 Perkiomen Ave., Reading. PA 19606-2059, USA

Techne
Techne, Duxford, Cambridge CB2 4PZ, UK
Techne, Inc., 3700 Brunswick Pike, Princeton, NJ 08540-6192, USA

Thermometric
Thermometric Ltd., 10 Dalby Court, Godbrook Business Park, Northwich, Cheshire CW9 7TN, UK
Thermometric AB, Spjutvägen, 5A S-175 61 Järfälla, Sweden

UltraViolet Products Ltd., The Science Park, Milton Road, Cambridge C84 4FH, UK
Tel: 0044 1223 42022
UVP Inc 5100 Walnut Grove Avenue, PO Box 1501, San Gabriel CA 91778, USA
Tel: 001 800 452 6788

United States Biochemical, PO Box 22400, Cleveland, OH 44122, USA
Tel: 001 216 464 9277

Varian, Inc., 3050 Hansen Way, Palo Alto, CA 94304, USA

VWR, 1310 Goshen Parkway, West Chester, PA 19380, USA

Index